Unmatched

The rapid and extraordinary progress of supercomputing over the past half-century is a powerful demonstration of our relentless drive to understand and shape the world around us. In this book, David Barkai offers a unique and compelling account of this remarkable technological journey, drawing from his own rich experiences working at the forefront of high-performance computing (HPC).

This book is a journey delineated as five decade-long 'epochs' defined by the systems' architectural themes: vector processors, multi-processors, microprocessors, clusters, and accelerators and cloud computing. The final part examines key issues of HPC and discusses where it might be headed.

A central goal of this book is to show how computing power has been applied, and, more importantly, how it has impacted and benefitted society. To this end, the use of HPC is illustrated in a range of industries and applications, from weather and climate modeling to engineering and life sciences. As such, this book appeals to both students and general readers with an interest in HPC, as well as industry professionals looking to revolutionize their practice.

David Barkai started his HPC career shortly after receiving a PhD in theoretical physics in the early '70s and was active in the field for over 40 years. A central theme of his work was the relationships between applications and architecture, with numerous publications over the years. David's employment at several HPC companies during their heydeys – Control Data, Floating Point Systems, Cray Research, Supercomputing Systems Inc., Intel, and SGI – as well as a stint at NASA, offered him a front-row view of the evolving HPC scene over the last few decades.

Chapman & Hall/CRC
Computational Science Series

Series Editor: Horst Simon and Doug Kothe

Data-Intensive Science
Terence Critchlow, Kerstin Kleese van Dam

Grid Computing
Techniques and Applications
Barry Wilkinson

Scientific Computing with Multicore and Accelerators
Jakub Kurzak, David A. Bader, Jack Dongarra

Introduction to the Simulation of Dynamics Using Simulink
Michael A. Gray

Introduction to Scheduling
Yves Robert, Frederic Vivien

Introduction to Modeling and Simulation with MATLAB® and Python
Steven I. Gordon, Brian Guilfoos

Fundamentals of Multicore Software Development
Victor Pankratius, Ali-Reza Adl-Tabatabai, Walter Tichy

Programming for Hybrid Multi/Manycore MPP Systems
John Levesque, Aaron Vose

Exascale Scientific Applications
Scalability and Performance Portability
Tjerk P. Straatsma, Katerina B. Antypas, Timothy J. Williams

GPU Parallel Program Development Using CUDA
Tolga Soyata

Parallel Programming with Co-arrays
Robert W. Numrich

Contemporary High Performance Computing
From Petascale toward Exascale, Volume 3
Jeffrey S. Vetter

Unmatched
50 Years of Supercomputing
David Barkai

For more information about this series please visit:
https://www.crcpress.com/Chapman--HallCRC-Computational-Science/book-series/CHCOM-PUTSCI

Unmatched
50 Years of Supercomputing

David Barkai

CRC Press
Taylor & Francis Group
Boca Raton London New York

CRC Press is an imprint of the
Taylor & Francis Group, an **informa** business

A CHAPMAN & HALL BOOK

First edition published 2024
by CRC Press
6000 Broken Sound Parkway NW, Suite 300, Boca Raton, FL 33487-2742

and by CRC Press
4 Park Square, Milton Park, Abingdon, Oxon, OX14 4RN

CRC Press is an imprint of Taylor & Francis Group, LLC

ISBN: 978-0-367-48113-1 (hbk)
ISBN: 978-0-367-47961-9 (pbk)
ISBN: 978-1-003-03805-4 (ebk)

DOI: 10.1201/9781003038054

Publisher's note: This book has been prepared from camera-ready copy provided by the author.

To Nomi

Who stood by me, tolerated the many hours of my absence while working on this book, and offered her loving support during these long few years

Contents

PART I   The Epoch of Big Iron

PART II   The Epoch of Multiprocessors

PART III The Epoch of Microprocessors

PART IV The Epoch of Clusters

PART V The Epoch of Accelerators and Cloud

PART VI Wrap-Up and Outlook

List of Figures

Foreword

The rapid and extraordinary progress of supercomputing over the past half-century is a testament to human ingenuity and a powerful demonstration of our relentless drive to understand and shape the world around us. In *Unmatched: 50 Years of Supercomputing*, David Barkai offers a unique and compelling account of this remarkable technological journey, drawing from his own rich experiences working at the forefront of high-performance computing (HPC).

David Barkai's career has spanned five decades, during which he has had the rare opportunity to be part of some of the most significant developments in the field of supercomputing. His personal and professional insights, combined with his deep knowledge and passion for the subject matter, make this book an invaluable resource for anyone interested in the evolution of HPC and its impact on our lives.

David Barkai outlines the incredible rate at which supercomputing has advanced, outpacing even the exponential growth predicted by Moore's Law. He highlights the complexity of HPC systems, from advances in processor architecture to the use of accelerators and hierarchical memory systems. Throughout the book, David emphasizes the critical role supercomputing has played in advancing our understanding of various scientific disciplines, as well as its direct impact on our daily lives through applications in weather prediction, engineering design, and life sciences.

This book is divided into five epochs, each focusing on a specific era in the history of supercomputing. The first part delves into the early days of big iron mainframes and vector processors, followed by an exploration of the rise of multiprocessors and their impact on various industries. The third part highlights the epoch of microprocessors, which brought massive parallelism and enabled HPC to reach new heights.

The fourth part discusses the advent of clusters and the standardization of HPC, while the fifth part examines the integration of accelerators and cloud computing. Lastly, the sixth part provides a wrap-up and outlook on the future of supercomputing, as well as the role of programming languages like Fortran in this ever-evolving landscape.

The use of the term "epoch" is particularly fitting for this book, as it captures the essence of the significant and transformative eras through which supercomputing has evolved over the past 50 years. Each epoch in the book represents a distinct period marked by substantial advancements in technology, architecture, and application of high-performance computing. These epochs, much like their geological counterparts, serve as milestones in the history of supercomputing and help delineate the progression of this powerful tool. Just like "epochs" in the Earth's history, the end of a supercomputing epoch is accompanied by a massive extinction of old technologies. By organizing the book into epochs, David Barkai effectively conveys the magnitude and impact of the changes that have occurred, while also underscoring the continuous nature of innovation and adaptation in the field of supercomputing. The use of "epoch" imparts a sense of grandiosity and importance to the story, emphasizing that the journey of supercomputing is, in itself, a monumental tale of human achievement and progress.

Looking to the future, David identifies several transformative trends in the HPC landscape. Expanding beyond numerical simulations, HPC now encompasses data analysts, scientists, and AI researchers, driven by data and AI applications. The following technology trends will shape the future of HPC:

1. AI-targeted chips and lower precision arithmetic: Newly designed chips cater to machine learning computations using lower precision arithmetic.

2. Quantum computing: QC casts a large shadow over the HPC world, potentially serving as accelerators and necessitating new approaches to computational problems.

3. General-purpose processors: ARM and RISC-V architectures are anticipated to challenge x86 architecture's dominance in HPC.

4. Chiplets: A modular approach using chiplets provides greater flexibility, customization, and optimization in semiconductor manufacturing.

5. Innovative new processors: NextSilicon, a startup, has developed an architecture featuring adaptive hardware and telemetry implementation, eradicating instruction streams and dynamically optimizing algorithms. As concurrency in HPC applications continues to grow, these technology trends will influence HPC's future.

David's personal journey grants us not only a deep appreciation for the technical challenges and accomplishments of HPC practitioners but also a profound understanding of how supercomputing has enriched our lives. The book celebrates those who have dedicated their careers to pushing HPC boundaries, and their stories inspire and remind us of the power of human creativity.

Unmatched: 50 Years of Supercomputing is a captivating and insightful examination of supercomputing's past, present, and future, as seen through the eyes of a true pioneer in the field. This book is essential reading for anyone interested in the enthralling world of HPC and its potential to revolutionize our lives in the coming decades.

Horst Simon   Abu Dhabi, March 27, 2023

Preface

No other application of technology has advanced at the rate high-end scientific computing, or supercomputing, has done steadily for over 50 years now.

This is best illustrated by comparing the advances of supercomputers' performance to that predicted by Moore's Law ([1]). To remind you: Moore's Law is a prediction, based on past observations, dated back to 1965, and revised in 1975. In its revised form, the "law" predicts that the number of transistors that can be put in a given area of an integrated circuit doubles every two years. Yes, the original meaning of Moore's Law was about the density of transistors made out of silicon. However, people interpreted it, not without reason, also to mean that we can expect the performance of the processor(s) inside our computing devices to double every couple of years, approximately. Moore's prediction was realized, pretty closely, throughout the last 50+ years. The density of silicon components in processors and memory chips increased exponentially at a steady rate. Take a period of 12 years. The doubling would occur six times during that period. This means that from the beginning of any 12-year period till its end we would have observed an amazing increase of roughly 64 times in terms of density and performance. I have a reason for choosing a 12-year period, as will be clear momentarily. Of course, we will return to the accumulated impact over the last 50 years (an easy exercise for the reader).

Now, let's consider supercomputers: We're going to look at the rate of advance in terms of the total system, which is comprised of many chip-level parts. And in order to compare this rate to that of Moore's Law, we will look at the system's peak performance – how many calculations per second it can perform. Here are some data points and milestones: In the early '70s, there was the CDC 7600 rated at 36 Million-floating-point-operations-per-second (MFLOPS, or megaflops). In the '80s, we had several systems (to be presented in this book) that peaked at the range of a little under 1 Billion-FLOPS (GFLOPS, or gigaflops) to 2.5 GFLOPS. Then, in 1996, we saw the first system to reach Tera-FLOPS – 10^{12} operations per second (teraflops). In 2008, another system achieved a peak of 10^{15} FLOPS, or Peta-FLOPS (petaflops). It took longer, 14 years, for the first (official) exasflops (10^{18} FLOPS, or exaflops) to be launched in 2022. In the book, I discuss how these advances were made possible and describe how they were utilized. Here, I'm simply making the point that, since the '80s, supercomputers' peak computational speeds increased a thousand-fold every 12 years![1]

Every 12 years or so, supercomputers' power advanced at a 15 times higher rate than the technology rate of progress according to Moore's Law. That's my justification for the claim made in the opening sentence. To be clear, the peak performance figures above are those for the single top system of its time. Is this rate of advance true for the collective of High-Performance Computing (HPC) systems? Well, we have historical data of the top 500 HPC systems in the world, as measured by a linear-algebra test, since 1993 ([2]). Turns out, the graph showing system #1 in the world is *parallel* to that of the cumulative performance of all 500 top systems. This provides a pretty solid indication that all of HPC benefitted from an approximately 1,000-fold increase in potential performance about every 12 years. In fact, working through the arithmetic of compound annual growth rate (CAGR), this means that over the last 40 years HPC has outpaced technology advances a la Moore's Law by over 9,500 times! – an incredible achievement by the HPC community and the industry, considering that technology itself (per Moore's Law) moved us ahead by a factor of over 1,000,000 during the same period.

What is the 'secret sauce', or what are the ingredients that enable the capabilities of HPC systems to grow so much faster than even the exponential rate at the processor chip level? In a word: Complexity. Advances in processor architecture; ever more processors working in tandem; use of accelerators; hierarchical memory systems; high performance interconnect; and, commensurate integrated software components as a foundation for increasingly sophisticated applications.

The extent to which the community of supercomputing users took advantage of this amazing rate of increase in capability, due to technology, is the subject of this book.

As is often the case, the stories we tell represent a continuation to work done before. Several themes in this book both run parallel and move forward past works by others. Notable examples are a '90s book by the founder-director of NCSA (National Center for Supercomputing Applications) Larry Smarr and William Kaufmann, titled "*Supercomputing and the Transformation of Science*"[3], and several chapters in a book commissioned by the National Research Council in 2005, titled "*Getting Up to Speed: The Future of Supercomputing*" [4].

On a personal note: Like many other practitioners of HPC, I arrived here from the world of theoretical physics. That was in the early '70s. Coming up on 50 years of involvement in this field, and slouching into retirement, it seems natural to look back at the journey my career has taken me through. I was fortunate, some would say unlucky, to have worked at several companies and organizations that built and used supercomputing.[2] During much of my 15 years at Intel (1996–2011), I worked closely with most all of the major players in the Industry. In the more distant past,

---

[1]The step from 1 MFLOPS systems (for example, CDC 3600, 1963) to a 1 GFLOPS system (e.g., Cray X-MP, 1982) took almost 20 years.

[2]I worked for, in that order: Control Data Corporation, Floating Point Systems, Cray Research,

I worked at supercomputing companies that don't exist anymore – at least not in their original form. They will be remembered in the pages of this book. Coming into the HPC community as an applications user, but working closely with engineers and computer architects, has given me a great appreciation of the interdependence between the hardware and how the application is presented to the computer.

It is this passage of time, and the opportunities I had to observe and participate, that give me the platform from which to tell a story. Living much of my professional life at the intersection of the HPC user community and the system houses that build the computers allows me to examine HPC from both perspectives. More importantly, along the way I have met many wonderful, smart, and creative people. Their recollections and observations bring to life the saga of scientific supercomputing.

The story I wish to tell is a tribute to the people who used supercomputing to solve real-life important problems and to gain deeper understandings of basic science. It is as much a story of the evolution of 'applications' (the codes that run on the computers) as it is about the progress in technology and the transformations of computer systems architecture that made it possible. The many people I have had the fortune to know, and who created the HPC community, come from academia, research labs, and the computer industry.

The chronological bookends of this story are separated by 50 years. And what a journey it was! At one end is the user laboring with punched cards, handing over the box of early Fortran cards to a computer operator, returning hours or days later to pick up reams of printed results (in the event all went well, for which there was no guarantee). The computer was a monolithic "big iron" system with one processor, the computational power of which was about one thousandth of our present mobile phones. At the other end of the story we usually find a multidisciplinary team working online and remotely, using interactive languages and modern programming languages, on problems that run on hundred of thousands of processors (we now call "cores").

My goal for this book is to cover the, sometimes twisted, journey that brought us to where we are today in HPC. In the telling, we will encounter both evolution and revolutions in the computer architectures themselves; how the interface and interaction with the systems changed over time; and, mostly, the people who developed the applications and pushed the boundaries of what can be done with any period's HPC systems. A most relevant part of the story is the transition of HPC from that of numerical simulations and modeling to encompassing data analytics and a start at what we ambitiously refer to as *artificial intelligence* but is more correctly termed *machine learning*.

The technical challenges and achievements of HPC practitioners were, and are, impressive. But, what good did they do? Well, in the following pages, when describing how supercomputers are applied, the recurring theme will be how it all

---

Supercomputer Systems Inc., NASA Ames Research Center, Intel, Appro, Cray, SGI, and about six weeks at HPE

benefits society and humankind. Yes, scientific computing is instrumental in gaining deeper understanding in several science disciplines, and the book will touch on a few topics in physics and chemistry. But the emphasis is on computer use that affects our daily lives. The areas chosen are: Predicting weather and climate, Engineering design, and the variety of activities that fall under the umbrella of Life Sciences.

I wish to emphasize that the book's narrative is *not* that of an objective historian, but influenced and biased by my personal encounters and experiences. My apologies for the significant events and perspectives that are missing. They are not less important than what is included. Only that I thread my story without them.

Let the journey begin.

Acknowledgments

This book has been, to a large extent, a collaborative project. Many in the HPC community supported me and contributed insights and information. They have my thanks and gratitude.

Two friends who trusted and encouraged me from the start and along the way: Horst Simon and Shahin Khan.

The book was inspired by the past work of people I call the *giants of HPC*. Standing head and shoulders above is Seymour Cray. Around him are the innovators, mentioned in this book, Steve Chen, Neil Lincoln, Jim Thornton, Burton Smith, John Cocke, Steve Wallach, David Patterson, Justin Rattner, Al Gara, and Jack Dongarra. Each, in their own way and style, influenced high-performance computing in a very unique and fundamental way.

In the process of writing the book I have interviewed and communicated with many HPC practitioners. Most are old friends and colleagues. Some are welcome new acquaintances. All were very generous in sharing their knowledge and experiences with me.

Much of the book deals with the history of supercomputing. For filling in the backstories, clarifying architectural transitions, programming models' evolutions, and their rationale, I am indebted to Shahin Khan, Horst Simon, John Shalf, Ken Miura, Tim Mattson, Thomas Sterling, John Gustafson, Mark Seager, Bill Gropp, Rick Stevens, Irene Qualters, and Elad Raz.

In the area of weather and climate modeling, I benefitted from long conversations with experts in the field. The following people patiently provided explanations and data: John Michalakes, Bill Skamarock, Roy Rasmussen, Milan Curcic, Deborah Salmond, Tom Rosmond, and Peter Bauer.

For the application of HPC in engineering, I turned to subject matter experts from academia and the automotive and aerospace industries: Bob Lucas, James Ong, Sharan Kalwani, Al Erisman, and Tom Grandine. They are credited with demonstrating the role of HPC in the design of objects we take for granted.

In writing about the relationship between life sciences and HPC, I relied on several scientists: Kjiersten Fagnan, Rick Stevens, Katherine Yelick, Lenny Oliker, and Yuan-Ping Pang. They brought insights to the progression of applying computations in this field, its scope, and its future prospects.

I thank my past colleague, the physicist Claudio Rebbi, for showing me the state of QCD today.

When it came to programming languages, I solicited information and advice from some of Fortran's and MPI's current thought leaders: Milan Curcic, Bob Numrich, Ondřej Čertjk, John Reid, Tom Clune, Jon Steidel, Steve Lionel, Damian Rouson, and Bill Gropp. Their candor about sometimes-regrets and self-doubt was refreshing. I am grateful for the long conversations we had.

My CRC Press editor, Randi Slack, showed much patience with the slow progress of creating this book. For this, for her guidance, and for the technical and artistic support of the CRC team, I am grateful.

Finally, I wish to recognize the many people in our HPC community I crossed paths with over the years. You have all left a mark on my perspectives, even if not mentioned by name here. My work was more pleasant because friends and family members who heard about this book project, some of whom know nothing about HPC, encouraged me to keep going.

Thank you all.

Short Introduction to Scientific Computing

The Third Pillar of Science and How It Works

BEFORE delving into the world of high performance computing (HPC), it may be useful to visit, or revisit, several concepts and practices that are foundational to this domain of computing.

The terms *supercomputers* and *supercomputing* are used almost interchangeably with the term *HPC*. Almost, but not quite. In early chapters *supercomputers* is in more common use, because these most powerful computers of the time were the platforms on which scientific computing was done. *HPC* is a more appropriate designation now since scientific computing is often done on systems that are not among the top few hundreds systems in the world, such as workstations in the '80s and '90s and a single server today. Nevertheless, such processing is best done on platforms that have high performance features. These may include high end floating-point arithmetic performance, high memory bandwidth, high-performance interconnect, and large storage – all scaled to the size of the system.

The Third Pillar

Scientific computing, or Computational Science, can justifiably be thought of as the *third pillar* of science. The other two being *theory* and *experiment*. First, a few remarks about the interaction between Experiment and Theory.

Scientists design experiments to observe what happens under controlled set of conditions. The outcome may be what they would intuitively expect. Many times it isn't. In either case, science is about coming up with a theory that explains the observed outcome. Scientific theory is fundamentally different from the everyday, or non-scientific, use of the word. It is not meant as 'a guess' or 'speculation'. Scientific theory is an explanation based on principles or 'laws' that apply beyond the specific experiment. It needs to explain other, related, phenomena as well. That is, and this is most important, the theory has *predictive power*. It enables the calculation (prediction) of the outcome to an experiment that has not been performed

DOI: 10.1201/9781003038054-0

yet. And it has to be *repeatedly* correct. Which begs the question: What if an experiment's outcome contradicts the prediction of the theory? – This can happen when an experiment is designed in an environment or subject to conditions that were not tested before. When this happens, it means the theory was incomplete or even flawed. Science's task is, then, to come up with a more complete or a brand new theory. What is even more exciting is when the theory is applied to conditions that have not been tested experimentally – perhaps because these conditions were too difficult to find or create. Later, sometimes many years later, a way to test this unobserved prediction is found. And the theory earns another validation. A wonderful example of this is the prediction of General Relativity of the existence of Black Holes. A prediction that was confirmed in 1971, 55 years after being predicted by Einstein in 1916. And for the first actual image of such an object we had to wait until 2019.

History presented us with an example of how experiment brought about an overturn of a theory: Galileo Galilei, in late 16th century, climbed to the top of the leaning tower of Pisa and dropped two objects of different weight. He observed that the objects hit the ground at the same time.[3] Until that time the prevailing theory was the one conjectured by Aristotle almost 2,000 years earlier. He theorized that a heavier object falls faster than the lighter one. It took almost a century after Galilei's experiment before Isaac Newton expressed gravitation mathematically, and as a universal law. Newton's Second Law, as it is known, applies to any two objects in our universe (not just to objects falling to Earth). And, of course, Newton's laws were superseded in 1915 by Einstein's General Relativity.

So goes the journey of scientific discovery, into which computational simulation was added in the mid 20th century as the Third Pillar of science. Before there were computers, scientists still had to calculate in order to derive predictions from the theory so it can be compared to experimental results. What computers enabled is not just to calculate much faster, but to compute problems that were far more complex and of much larger scale, making them beyond what was possible for humans to work out. This is, though very useful in advancing science and making it practical to everyday life, a quantitative progress. Think of calculating tomorrow's weather from the weather model equations. Well, if it takes more than 24 hours to compute tomorrow's weather, it is of no good.

But by no means this is all that the third pillar of science affords us. It allows us to 'see' the very small and the very large, the very slow and the very fast. Once a theory, expressed mathematically, is validated, we can 'watch' the simulated behavior of sub-atomic interactions, as well as the formation of galaxies and black holes. We can 'watch' in slow motion steps in a process that takes a fraction of a nanosecond. We can watch processes that go for hundreds of years, such as climate change. Computer simulations allow us to design experiments that cannot be done in the lab. For example, measure what would happen if the Earth warmed at this

---

[3]Or so the story goes. Some say it was a 'thought experiment.' In any event, the outcome turned out to be correct.

pace or another. Or, what would happen if a celestial object's trajectory tilts a little. And if parameters are chosen right, and a simulated event happens, it will validate or contradict a current theory.

Computer modeling allows us to perform "experiments" digitally. Think of crash analysis of a car visualized on a computer screen, instead of physically running a car into a wall. More than that, computer modeling allows us to very quickly repeat the experiment with modified input parameters, so we can figure out the correct shape or material for an engineering product, for example.

In the area of medical care, computational chemistry is applied to study protein binding which tells us the degree to which a proposed medication attaches to protein in our blood. This determines the efficacy of the drug. The simulations allow to quickly examine hundreds of variations. They eliminate in advance many human clinical tests that would have failed. This will be a recurring theme throughout this book.

In summary, computational science enables us to make new uses of scientific knowledge. It allows us to examine the very small and very large, the very fast and the very slow. And it opens up opportunities to 'experiments' we cannot do in real life.

Numerical Simulation

For the uninitiated, or a little 'rusty', this chapter provides an overview of what is involved in *scientific computing*. We cannot say that all scientific, or numerical, computation requires supercomputers, but it was certainly true that until recently supercomputers were used almost exclusively for what falls under the umbrella of simulation and modeling of scientific phenomena. The notable, mostly classified, exception to this statement is a class of applications used for intelligence purposes, such as code breaking, electronic surveillance, etc. The latter, which will not be discussed in this book for obvious reasons, nevertheless influenced design aspects of supercomputers at various times. HPC today includes types of processing beyond number crunching, but this overview is about computational simulation of a physical process.

Physical phenomena are modeled using a set of mathematical equations. The mathematical formulation has to be adapted so it can be performed on a computer. That is, written in a way that can be coded as a computer program. When executed it simulates a real world phenomenon. What we mean by that is that the output of the program is a set of numbers that represent values of certain physical quantities or attributes over space and over time.

The following is an overview of the process for creating a computer model of a real-world environment. This is a significant computational field, but by no means the only type of scientific computing. There are statistical models for many inquiries, data and graph analysis problems, and more. For the purposes of this introduction we will stay with the concept of simulation set in the physical space and evolving in time.

The process starts with the subject-matter-expert, a scientist or an engineer, who provides the mathematical formulation. This will be, typically, a set of equations that describe the evolution of a phenomena. Think of a model for the atmosphere, or airflow over an aircraft, or a car crash. This is likely to be a set of differential equations that prescribe how physical entities change in space and over time.

Let us consider a simple example to illustrate the process that turns the mathematical formulation into a form that can be programmed for execution on a digital computer. It should be emphasized that this is the process for constructing what is referred to as the *solver* part of the program. And a drastically simplified case at that. A complete scientific application is much more complex. A few words on that later on in this chapter.

The example illustrated here is the simplest form two-dimensional wave equation. Of course, our world has three spatial dimensions, but here we are looking at the simpler case that involves fewer terms so we can focus on the process (there is an analytic solution to this form of the equation, but the goal here is to show how to solve it numerically):

$$\frac{\partial^2 u}{\partial t^2} = c^2 \left(\frac{\partial^2 u}{\partial x^2} + \frac{\partial^2 u}{\partial y^2} \right)$$

It says that the amplitude of the wave function u changes in time, in every point inside the region defined by end values of the $x - y$ plane, such that its second derivative with respect to time is proportional to the sum of the second derivative in the x and y directions of the plane. The proportionality constant is the square of the wave's propagation velocity c (not necessarily the speed of light). To *picture* the math, think of the first derivative as a measure of how the function changes when moving on the plane or in time. The second derivative amounts to taking the derivative of the function resulting from the first derivative. What it measures is the curvature of the graph we get from the original function's value when tracking the x, y, or t directions.

Again, a typical simulation problem is described by a set of equations that are more complex than the wave equation. They would cover aspects such as motion, pressure, thermal exchange, convection, and more. Here we follow a single simple equation so we can focus on the process that starts with a mathematical representation and ends with a procedure that can be submitted to a computer for execution.

The math is expressed as a continuum. There are infinite number of values over any interval. This is required for the correct formulations of derivates (and the consistency of mathematical theory).

However, computers don't have an infinite memory or storage. More important for the process, they don't have an infinite accuracy. Another way of saying this is that computers work on a set of discrete values. Therefore, the space over which the problem is defined has to be drawn as a collection of points. In our example it would be a grid, a two dimensional mesh. The natural way to set it up is to draw the points with equal distances between them. It greatly simplifies the programming. In real-life applications it sometimes becomes useful, or necessary, to have regions with different density of points for reasons of numerical stability that are beyond the scope of this overview. In other problems the sparsity of points is a function of where more physics occurs – think of simulating a galaxy, for example. The grid can take a shape different than a rectangular or a cube. Defining the *geometry* of the problem to be solved is a major step in constructing the code.

The figure below shows a small region of a two-dimensional plane on which the wave equation can be solved.

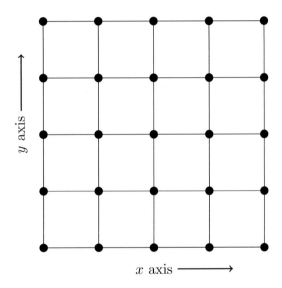

Discretization over a Plane

The program assigns or computes values for nodes, or points, on the grid. This is what we will assume here. However, the location of the values does not always have to be at the intersection of the lines. For numerical reasons, they can be defined to be midway on the line connecting two points, or at the center of the cell. Often the problem involved several physical attributes simulated over the same grid, where values for different attributes use different location schemes on the grid.

The wave equation written above has to be transformed to a form that expresses values on grid points. Getting there is the process of *discretization*. For the wave equation we need to transform the continuous second order derivative. We are asking *what is the second order rate of change of the wave function when we make small incremental spatial moves in the x, y, or in time, t.* We will use Δx, Δy, and Δt, to indicate small displacements in the x, y, and t variables, respectively.

First, we write an expression that is an approximation to a second derivative, and that can be calculated. This can be done as a Taylor series power expansion over the displacement value. Consider the definition of a (first) derivative (or differential):

$$\frac{\partial u(x, y, t)}{\partial x} = \lim_{\Delta x \to 0} \frac{u(x + \Delta x, y, t) - u(x, y, t)}{\Delta x}$$

For the purpose of computations Δx is not infinitesimal. It is the relatively small, but finite, interval between the grid points. We will skip some math manipulations, but assert that the right-hand-side represents the accurate first derivative to within an error that is proportional to Δx^2. The smaller Δx is the smaller the error. Think of the distance units scaled so that the dimensions of the plane are 1 unit by 1 unit, and with many more grid points than drawn above. Then $\Delta x << 1$ and the error is relatively small. The reader can easily construct the equivalent expressions for y and t, using Δy and Δt. The expression below states that the right-hand-side is an approximation of the derivative of the function u with respect to x.

$$\frac{\partial u(x, y, t)}{\partial x} \approx \frac{u(x + \Delta x, y, t) - u(x, y, t)}{\Delta x}$$

Now we use the expression of the first derivative to get an approximation for the second derivative with the same level of accuracy. In the following we make use of a similar expression to the above where we displace x in the other direction by substituting $-\Delta x$ for an expression that contain $u(x, y, t) - u(x - \Delta x, y, t)$. The same process that was applied to u is now applied to its (approximate) derivative:

$$\frac{\partial^2 u(x, y, t)}{\partial^2 x} \approx \frac{\frac{u(x+\Delta x, y, t) - u(x, y, t)}{\Delta x} - \frac{u(x, y, t) - u(x - \Delta x, y, t)}{\Delta x}}{\Delta x}$$
$$= \frac{u(x + \Delta x, y, t) - 2u(x, y, t) + u(x - \Delta x, y, t)}{\Delta x^2}$$

Just a little more math. The finite-differences form of the second derivative of x can be replicated for y and t. The wave equation is now represented as:

$$\frac{u(x, y, t + \Delta t) - 2u(x, y, t) + u(x, y, t - \Delta t)}{\Delta t^2} \approx$$
$$c^2 \left(\frac{u(x + \Delta x, y, t) - 2u(x, y, t) + u(x - \Delta x, y, t)}{\Delta x^2} \right.$$
$$\left. + \frac{u(x, y + \Delta y, t) - 2u(x, y, t) + u(x, y - \Delta y, t)}{\Delta y^2} \right)$$

The form above allows us to switch to a notation that will be suitable for mapping to the grid with discretized values. This requires the following transformations:

Assume $\Delta x = \Delta y = h$ for the distance between grid points. Use the discrete value i to index the x direction and j for the y direction. Then, $x + \Delta x$ becomes $i + 1$, $y - \Delta y$ is assigned the index $j - 1$, etc.

Instead of $u(x, y)$ for the value of u at the spatial coordinates x and y, we write $u_{i,j}$ for its value at the (i, j) node on the grid. The illustration below shows all the grid points, or nodes, that participate in the calculation involving an inner node (i, j).

Note that this is not the only mapping possible. Here we used what is termed *central finite differences* scheme, with a second-order error size. There are schemes, with the same order size of error, that use just *forward* or just *backward* differences. In fact, such schemes are necessary for the nodes on the boundaries of the grid. And it is possible to construct difference schemes that employ higher order expansion of terms for a smaller error size. The nature of the problem, the relative scales of values, and the desired or necessary accuracy dictate these details.

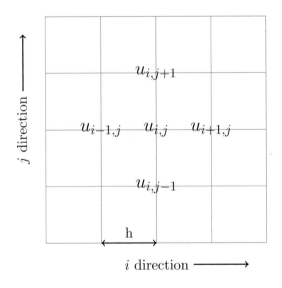

Mapping program indices onto the grid

There is one more matter of notation to deal with. The simulation is done over time, as evident from the differential equation – the wave equation we started with. The computation starts with initial values over the spatial grid. This corresponds to $t = 0$ in the simulation process. When computing we set a time step, what we marked as Δt. To complete the discretization process we set the variable n to denote the time step. The value of the wave function u at a grid point (i, j) and at time step n is $u_{i,j}^n$.

In what follows the approximated relationship in the expression above is replaced with an equality. The $\approx$ becomes $=$. We can now write down the discretized form of the wave equation using the simple finite-differences scheme we chose:

$$\frac{u_{i,j}^{n+1} - 2u_{i,j}^n + u_{i,j}^{n-1}}{\Delta t^2} =$$

$$\frac{c^2}{h^2} \left[\left(u_{i+1,j}^n - 2u_{i,j}^n + u_{i-1,j}^n \right) \right.$$
$$\left. + \left(u_{i,j+1}^n - 2u_{i,j}^n + u_{i,j-1}^n \right) \right]$$

The expression above involves the 4 immediate neighbors of an internal grid point (i, j), as drawn in the figure above. It also contains 3 time steps, marked as $n - 1, n, n + 1$.

The computation simulates the progression in time of the wave. The start values are given. These are all the u values for every (i, j) point. This initial state corresponds to $n = 0$ of the time direction. The computation produces subsequent time steps. Note that there is one term that belongs to the $n + 1$ time step. It can be kept at the left-hand-side, to be calculated by the known values, from previous time steps, at the right-hand-side:

$$u_{i,j}^{n+1} = 2u_{i,j}^n - u_{i,j}^{n-1}$$
$$+ \left(\frac{c\Delta t}{h} \right)^2 \left(u_{i+1,j}^n + u_{i-1,j}^n + u_{i,j+1}^n + u_{i,j-1}^n - 4u_{i,j}^n \right)$$

This works for time $n = 2$ onward, and for all the internal, non-boundary, grid points, since we need time steps n and $n - 1$ in order to compute the values for the next time step, marked $n + 1$. A special procedure is needed for the first time step, and there are finite differences schemes that are tailored for that (but not discussed here). In addition, the expression above cannot be used for the boundaries of the grid, since each point requires its neighbors' values in all directions. There are ways to handle these grid points that are equivalent to the methods available for the first time step. However, often the edges of the grid are kept fixed. That is, their initial values stay constant – typically, zero. This corresponds, for example, to a membrane that is kept in place and allowed to vibrate. In this case, all the computed (i, j) points are internal; they have near neighbors in all directions.

The last expression above can be transformed to code quite easily. Here is a walk-through in words, rather than code, to both explain the algorithmic procedure and to be independent of the programming language.

The solution at each step of the simulation is the value of the wave function at each grid point. In the code these can be presented in a two-dimensional array. Call it U. It is a square or a rectangle with the size of the x dimension equal the number of i points (along a fixed j), and the y direction spans the number of j points (along a fixed i). A 'new' U is computed every time step. For real application, an array such as U can be quite large, and if a copy of the array will be kept in memory for every step, it will overflow very quickly. Also, past copies are not needed anymore for the simulation. The formula above uses U^{n-1} and U^n to compute U^{n+1}. For the next time step we drop U^{n-1} (it may be saved in a storage file). U^n becomes U^{n-1} and U^{n+1} becomes U^n, and are used to compute the next time step. And so the process proceeds with three copies of U. The programmer decides to write the array out to storage every time step or every fixed number of steps, indicating what time step it corresponds to. The values in these files can be visualized to show the progression of the wave.

The computation kernel is embedded in nested loops over the indices i and j, or expressed with the language's array notation. The right-hand-side of the formula above is computed to give the (i, j) element of the $n+1$ time step. Start with twice the latest value minus the value from the previous time step. Then the 4 nearest neighbors of (i, j) are added up, and 4 times the latest (i, j) value subtracted from the sum. This latest result is multiplied by the constant $(\frac{c\Delta t}{h})^2$. It is computed once, and used as a simple multiplier throughout the computation. The result is added to subtraction done at the beginning, and this concludes the calculation of the latest value for element (i, j).

The pseudocode below sketches the computation of the main kernel. The arrays used are P, C, and N, to indicate *previous*, *current*, and *next* time steps instantiations of the U array.

The example described above should not be confused with a real-life scientific application. It is a very simple example of a one-equation solver core of an application. Applications of interest involve a set of equations for a number of physical quantities that describe a complex natural phenomenon. These problems deal with multiple sets of data with possibly different grids for different variables that require clever numerics to map the results and produce smooth and accurate simulation. There are many problems that are not amenable to regular grids; think of simulating the airflow over an airplane, for example. Not only is the shape irregular, but also the spacing between points is not a constant. Creating the grids and the datasets for each variable, and managing the data between steps in the simulation process often takes a greater effort than programming the solver. Especially when the users can resort to a rich set of general numerical routines simply by calling on external libraries. More on such issues in the introduction of weather models in Chapter 4.

It should also be clearly stated that not every numerical simulation problem can be expressed by partial differential equations, though many can.

A note about precision: The accuracy of numerical computations is clearly an important consideration. One factor is the limited precision inherent in the finite number of bits that hold the values. And then there is the additional approximation introduced by discretizing continuous equations. Numerical analysis and error analysis coexist and ever present in numerical HPC. The intersection of performance, algorithms, accuracy, and numerical stability is a matter of both science and "marketing" (by computer vendors who are engaged in performance bragging rights).

Pseudocode for the 2D Wave equation

```
 : Initialization
Read L , h , c , dt , amp, T, k
 : L is the no. of grid points in each direction
 : T is the no. of timesteps/iterations
 : Save results every k iterations
 : Set P to all zeros; but for the center grid point
for i = 1, ... ,L and j = 1, ... ,L do
    P_{i,j} = 0
end for
P_{N/2,N/2} = amp   : Pebble thrown to center of square pool
Const = (c * dt/h)^2
 : Skipping the first time step
 : (modified diff scheme for first step N values)

 : Main computational kernel
for t = 2, ... , T do : Count timesteps
    for i = 2, ... , L-1 do   : Edges are held fixed at 0
        for j = 2, ... , L-1 do   : Interior point are computed
            N_{i,j} = 2 * C_{i,j} - P_{i,j}
            N_{i,j} = N_{i,j} + Const * (C_{i+1,j} + C_{i-1,j} + C_{i,j+1} + C_{i,j-1} - 4 * C_{i,j})
        end for
    end for
 : Save results if t is a multiple of k
    if t mod k = 0 then Write N
    end if
     : Prepare for the next timestep
    P = C   : 'Current' state becomes 'Previous' state
    C = N   : New state, 'Next', becomes 'Current' state
end for
```

There are ways to reduce numerical errors (this is not 'error' as in 'mistake', but a determination of a range around the computed result within which the analytically correct result is):

A wider word size will capture more digits of real numbers. Typically, in current computer architectures, the use of *double-precision* 64-bit words over the 32-bit *single-precision* word. There are cases where even higher precision is warranted, where, with the help of compiler or library software, 128-bit words can be used. Floating-point numbers, as defined by the IEEE-754 standard, are represented by an exponent, a fraction, and a sign bit. For 64-bit word 11-bit field is for the exponent, leaving, with 1-bit sign, 52 bits for the fraction part. This translates to 15 to 17 decimal digits that can be carried for the number. This is a pretty high precision by everyday standard, say, for measurements of temperature or wind speed. However, in simulating a phenomenon over time, there are important cases where a tiny numerical error can quickly amplify and lead to inaccurate or unstable solution. This happens when we are dealing with non-linear equations. A case relevant in weather modeling, for example.

Loss of precision due to discretization can be mitigated by developing more complex finite differences schemes (refers to as *stencil*). For example, using terms that involve more grid points than the immediate neighbors, and higher-order differences.

Employing one or both of the remedies above come at the cost of more computations, and longer time to solution. More on the subject in the chapter about performance – Chapter 31.

There Is (much) More to Scientific Computing

Let us move beyond the generalities of the Third Pillar and a simple digital simulation into the wider space of scientific computing. Gene Golub and James Ortega define scientific computing in their seminal book on numerical analysis[5]:

> "*Scientific Computing is the collection of tools, techniques, and theories required to solve on a computer mathematical models of problems in Science and Engineering.*"

This introductory chapter doesn't do justice to the depth of what is involved in scientific computing. The modeling and simulation aspect alone is supported by a vast reservoir of inquiry under the heading of Numerical Analysis, of which we saw one simple example. The field is a major branch of mathematics that started well before computers existed.

To this we should add that the discussion so far, and including the 'definition' of scientific computing above, refers to the 'classic' (that is, of the past) notion of the field. In the last decade or so, two other major domains of investigation were added to what we still call *computational science* and the world of HPC. These are *data analytics* and *artificial intelligence*, mostly in the form of *machine learning*. These areas of inquiry are now an integral part of high-performance computing, when digital processing is involved, not only because they require capabilities and features developed for scientific computing. Perhaps more significant than adding non-numeric applications and workloads to HPC systems is the fact that both data analytic and AI methods are in a fast-growing use in modern science and engineering research. More on that in a later chapter – the one about the changing face of HPC (Chapter 24).

To better comprehend the scope of scientific computing it is helpful to think in terms of a multi-dimensional taxonomy of *application domains*, *problem types*, and *solution methods*. The result is far from a one-to-one mapping. A given problem type has, typically, a multitude of approaches for solving it. A particular solution method may apply to several problem types. A problem type or a solution method may be applicable to more than one application domain.

Application domains contain more than fields of science and engineering. Several industries rely on some of the same kinds of problem types as do scientists in academia. Of course, there is much research being done in the private sector, even when we limit our attention only to where high-performance computing is involved.

A high level view of the main areas of academic research that apply to HPC can be arranged into four disciplines:

Physics	Life Sciences	Earth Sciences	Engineering
Astrophysics	Comp Chemistry	Weather Modeling	Mechanical
Condensed Matter	Comp Biology	Climate Modeling	Aerospace
Plasma Physics	Evolutionary Biology	Oceanography	Civil
Particle Physics	Molecular Biology	Geophysics	Earthquake
	Ecology		
	Neuroscience		

In reality, we find other domains resorting to HPC-class systems in academia, even if to a lesser extent. Examples are *economic modeling* and *statistical analysis* for social studies. Going beyond simulation and modeling, we now include data sciences as a discipline of scientific computing that relies on high end computing systems. The work done by researchers propagated and applied in many ways outside the sphere of academia to the benefit of all, as will be seen below and throughout the book.

As we drill down on the contents of scientific computing, consider the types of real-world problems that are encountered in simulation and modeling of the physical world. A partial list includes these (see below for associated industries):

- Weather Forecasting, Climate Modeling
- Flow (Fluid Dynamics)
- Drug Discovery, Genomics
- Vehicle Design, Crash Analysis
- Seismic Processing, Reservoir Modeling
- Currency Trading, Risk Modeling
- Rendering, Visual Simulation, Content Streaming
- Material Analysis and Discovery
- Pattern Recognition, Code Breaking
- Structural Analysis
- Product Design – Consumer Products, Mechanical, Electronic, Renewable Energy

There are various ways to classify the mathematical sub-fields of numerical analysis. I find the comprehensive list compiled in a Wikipedia article (see [6]) useful in that it covers the classes of problems as well as solution methods. A helpful aspect of this reference is the links it provides for a deeper explanation of the items listed there. From the broad categories of problem types listed there it is worth noting the following:

- Linear Algebra – systems of linear equations, eigenvalues
- Partial Differential Equations
- Ordinary Differential Equations
- Optimization
- Nonlinear Equations
- Interpolation, Approximation
- Integration
- Graph analytics

For each of the mathematical areas listed above there is a whole host of numerical algorithms that can be programmed for digital computations. Typically, there is no "best" algorithm for a given problem type. The choice of one depends on the specific problem to be solved. The form of the equations, the size of the dataset, shape of the grid, the variability of data items – all factor in the choice of a method for solution. The subject is too voluminous to attempt even a summary of it here. For a taste of what is involved, let us pick *partial differential equations* (PDEs), since it is so central for much of the simulation and modeling enterprise.

The problem to solve is finding a function when its partial derivatives are the terms in the given equations (example of which is the wave function equation examined above). Typically, there is no known explicit solution that can be written as a closed form expression, and we need to resort to numerical representation of the function over some domain of interest. A large repertoire of methods was developed over the years for the discretization of PDEs (for a brief overview, and further reference, see [7]). Among the more popular and widely used approaches are the *finite difference method*, the *finite element method*, and the *finite volume method*. Within each of these general approaches there are many specific algorithms designed for particular forms of equations and solution domains.

In addition to the methods above there are quite a few that don't fall into one of those three. A couple are worth mentioning. The *spectral method* – where the solution is expressed as a sum of frequencies (via Fourier transform, typically), solved for the coefficients of the series, then transformed back to spatial values on the physical grid. Another example is the *Monte Carlo* method that uses weighted random sampling repeatedly as a way to arrive at solution where other methods would fail or be slower. See Chapter 9 for a use case.

Many PDE problems cannot be arrived at the final result in one pass through the computation and we resort to *iterative* methods. This cuts across the algorithm classification above. A set of PDEs over a grid can, frequently, be expressed as a system of linear equations in matrix form when using any of the main methods above. Iterative methods were developed for when a direct solution to the equations is not known or when it is not practical to perform a known procedure due to the time it would take to get at a solution. There are quite a few such algorithms. What they have in common is that after a discretization scheme defines the relationships between grid points, the procedure begins with a guess or an estimate of initial values on the grid. New values are computed according to the above relationships. Knowing the equations means the numerical error can be computed at each point and averaged over the grid. The new values are used for the next iteration of the procedure and new average error is computed. So it continues until the error falls below a pre-defined tolerance value. Or, as might happen, the convergence stalls and solution cannot be reached. In that case the researcher may resort to a higher precision (wider word) and/or a better starting state, or to a different algorithm. More often, a priori numerical analysis would have determined the appropriate algorithm for the set of equations and the physical problem it applies to. Describing the various iterative methods is beyond the scope of this introduction (an application of one such method is described, without numerical analysis, in the chapter on QCD – Chapter 9).

There is an element of *art* to computing with iterative methods. The goal is to arrive at a converged solution fast. An important first step is the choice of initial values on which to iterate. For that people developed various *preconditioners*. These are transformation of the original problem into a form that allows a faster solution. There are also methods used to accelerate the convergence during the iterations phase. A notable example of that is the *Multigrid* method. Its basic principle is that coarsening the grid (which reduces the iteration compute time) eliminates lower frequency components of the error. In this method the computation cycles through coarser and finer grids.

A few words on the two additional areas of inquiry, beyond simulation and modeling, that make up today's HPC:

High Performance Data Analytics (HPDA) refers to the use of high-performance computing for *big data analytics*. HPC systems allow performing tasks, such as pattern recognition, graph analytics, streaming analytics, support of decision making, and generally gaining insights, on very large datasets. Known processes and algorithms deliver more effective and accurate results when applied to larger datasets. The availability of vast amounts of data, made possible by the digitizing of analog data, triggered new applications and new methods of data analysis. The data processed may be the output of simulation – climate modeling output, for example. It may be financial, medical records, electronic communications, population data, and more.

The other field that is now associated with HPC systems and with science and commerce is *artificial intelligence* (AI) in the areas of *Machine Learning* (ML) and *Deep Learning* (DL). These go beyond data analytics in that the application is digitally 'trained' to find an answer or a pattern given a dataset (ML), or even identify relationships within the data elements without any guidance from the human programmer (DL).

More on applications and methods of these two topics in later chapters of this book.

The HPC Ecosystem

Scientific Computing, supplemented recently with data analytics and AI, exists within an HPC ecosystem. We can think of the HPC world as made up of three groups: The *users* are the scientists, researchers, and engineers who do the processing on the systems. The second group, no less critical, is made of the people who make running the applications possible. They are responsible for the software stack – the operating system, the compilers, and system utilities, the interconnect protocols, middleware, etc. The group also include the writers and developers of all the numerical software, data analysis procedures, machine learning algorithms, and more. And, lastly, we have the vendors and technology providers who put together the systems capable of high performance. This is the group, much of it in the private sector, that is credited with the unmatched advances made in computing these last 50 years.

While it is true that most of the initial work and innovation in scientific computing and supercomputing originated in research labs and universities, the HPC user community includes a significant participation of users from the *private sector*, or *Industry*. Notable industry sectors include:

- Automotive – structural, crash simulation, fluid dynamics, noise reduction
- Aerospace – fluid dynamics, structural, materials
- Chemical – molecular dynamics, electronic structure
- Pharmaceutical – drug discovery
- Oil & Gas – seismic processing, reservoir modeling
- Financial – fast trading (Monte Carlo), risk modeling
- Entertainment – rendering, visual simulation

Early HPC, or supercomputing, use was almost exclusively in government and academic funded institutes – national research labs, government agencies, and universities. From there, HPC use propagated to the private sector (often referred to as "Industry"). This process of adoption of new compute technologies and novel applications first by researchers in the public sector, then in Industry, continues to this day. This is true about where the largest systems are initially placed, where new components (GPUs, for example) are first tried, or where new applications and methods are first created (think of numerical algorithms, ML, etc.).

The "public-to-private" relationship is similarly true for the second group of the HPC community. The "enablers" of HPC use, who write the system software, user utilities, and libraries. are a mix of public and private sector individuals and companies. There is more publicly, open-source, HPC software available than commercial. In the past, though, computer companies relied almost entirely on propriety software.

The third group, the providers of HPC systems, all come from the private sector in the U.S. and most other countries; though they enjoy basic technologies developed by government-funded labs and researchers. Stories related to these companies are scattered throughout this book.

A concluding reminder to the reader: This chapter is meant to provide some context for the world of scientific computing and HPC for the uninitiated. It certainly isn't a replacement for a text book on the subject. The remainder of the book will not contain any math, but will touch on concepts described here.

I

The Epoch of Big Iron

Vector Processors of the '70s

CHAPTER 1

In the Old Days. . .

This was supercomputing 50 years ago

U SING supercomputers in the '70s was very different than it is today. Their speed and capacity may be laughable to Millennials. The programming model and languages antiquated. The process of creating, submitting, and getting results, of compute jobs was tedious and cumbersome. Here we take a look at what life of a scientific computer user was like back then.

But first, consider the computers of the '70s.

The scientific supercomputers were about the size of the commercial mainframes of the day, perhaps a little smaller. For a comparison to today's equipment in the datacenter, think of a row of 4–6 server racks, maybe 2–3 deep. Only that was, in the late '60s – early '70s, a single processor with its memory. The disc units for storage and the magnetic tape units occupied separate space nearby (Figure 1.1, for example). The compute power of the system was measured by tens of millions operations per second (megaflops). The amount of memory was of order of a million bytes (though we always counted in 'words' in those days). See Figure 1.2. To set the scale right, consider that our smart phones today (smartphones!, not supercomputers) measure their performance and memory in the billions of both operations per second and memory capacity. Yes, our small hand-held device has, in terms of operations rate and memory capacity, three orders of magnitude greater capability than the most powerful supercomputers of 50 years ago.

At the start of the decade of the 70s, there were several companies that developed scientific computers commercially. They were designed to perform complex numerical calculations well,[1] as opposed to what was referred to as commercial computer systems – the more powerful of which were known as mainframes. These

---

[1] IBM was so much larger than the other main players in the mainframe and scientific computing markets, that these companies were sometimes referred to as "Snow White and the Seven Dwarfs," with IBM being Snow White. The "dwarfs" were Control Data Corporation, UNIVAC, Honeywell, General Electric, RCA, Burroughs, and NCR. There was another designation – the BUNCH, where GE and RCA were dropped and the ordering rearranged.

DOI: 10.1201/9781003038054-1

Figure 1.1: CDC 7600. Source: IT History Society (ithistory.org).

'commercial' computers were good at character manipulations and accounting-type calculations. The next chapter gets into the nature of numerical simulations and modeling of physical phenomena. For now, keep in mind that good floating-point arithmetic performance is not the only important feature separating the scientific supercomputer from the mainframe. Just as important was its high-performance memory system – both in terms of latency (the time to fetch a data element), and bandwidth (how much data can be delivered in a unit of time).

Supercomputing was about to undergo dramatic changes mid-decade, but not so much in terms of the mechanics of accessing and using the computers. The supercomputer resided in a closed and chilled room, with no access from outside. It was monitored and managed by operators in the computer room using a console terminal. One had to be physically at a window or counter of the computer room in order to submit a job or collect a printout. Yes, this was years before there was the Internet, WiFi, or even PCs (the latter would appear by the end of the '70s).

The high-level programming language for scientific computing was Fortran, though it was quite common to resort to assembly language coding for better performance or for calls to system functions. Fortran then was much more rudimentary than today's Fortran, with none of the object-oriented features and data structure definitions in use today. We will return to Fortran in the final part of this book. The program to be executed was punched onto thick paper cards via a specially designed punch card machine (see Figure 1.3). Each card was coded to represent one line of code; up to 80 characters. Each character was represented by a vertical

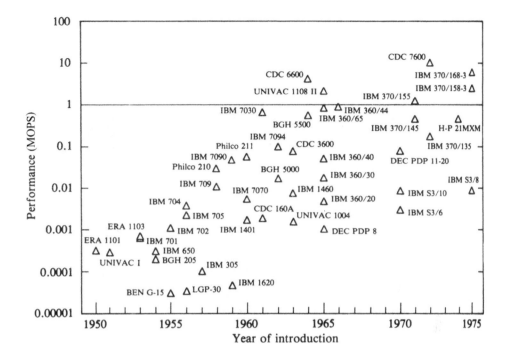

Figure 1.2: Early best-performing computers.
SOURCE: Kenneth Flamm. 1988. Creating the Computer: Government, Industry,
and High Technology. Washington, D.C.: Brookings Institution Press.

set of punched holes. Of course, any changes to any line of the Fortran program
requires new cards to be punched out.

The 'dumb' terminal used in conjunction with the card-punch device was also
used to print the source code. That was the program's 'listing.' The printer did
not have a choice of fonts. It was noisy and large, and had large wide paper with
perforation between pages and along the side margins of the paper roll. This paper
listing of the code was actually considered a form of "permanent storage" (Figure
1.4).

The user of the 70s ended up with a deck of punched cards. There were long-
sided tray boxes to place the cards in. Even so, this was a volatile situation. Placing
and replacing cards and carrying the box all too often resulted in dropping the tray-
box or a bunch of cards. It was frustrating and time consuming to place them in
order. Some machines had the ability to number the cards when they got punched.
This was probably the most valued 'feature' of a card-punch machine.

Until the days when it was common to have a terminal at the user's desk,
sometime into the 70s in most organizations, the user would go to an area adjacent
to the computer room to get the source code punched onto cards. Once the deck
was ready, it would be 'read' into a card-reader that would create an electronic
version of the source code. It would possibly include input parameters that specify

Figure 1.3: Card-Punch Machine.
Source: IBM archives.

Figure 1.4: Computer Printout.
Source: Wikipedia; https://en.wikipedia.org/wiki/Listing_(computer)

execution options and/or file names where input data resides. A front section of the deck set the load, compile, and execute ("go") sequence with its parameters – optimization level of the compiler, libraries, etc. The job would enter a batch queue. Interactive processing was very limited in those days and required access to console terminal in the computer room. Depending on the system and the organization, the job would wait execution from hours to days. The user would call or stop by to find out if the job was run. Assuming no compilation or load errors (which would kick off corrections and repeat of the process), there would be an output printout of key results. Most simulations runs would have an electronic output file beyond the summary figures printed out.

Yes, the process was time consuming and tedious. Then again, the computers were so much less capable than todays. Which meant the codes were less complex, to allow for tolerable execution time, and had to be designed to use only the limited amount of memory available in those days. The top supercomputer in 1970, the Control Data Corporation's CDC 7600, had 65K words of memory and a peak performance of just over 30 MFLOPS.

Much was about to change in the mid-70s.

CHAPTER 2

Vector Processors

The early workhorses of HPC and going beyond mainframes

THE story in this book begins in the 70s. That is not when electronic computing started. Digital computers existed since the late 40s. From the start they were designed for arithmetic calculations and data manipulations. At the beginning, they were built with vacuum tubes, and later with transistors. By the time our story begins, computers were made with integrated circuits, and a computer chip made of silicon was less than a decade away. The miniaturization factor of the basic components that took place during the period of the late '40s to the early '70s was in the millions even for that period. And the pace continues to these days.

The traditional approach to designing a computer processor for numerical computations workloads took a major turn by the mid '70s. Up until then, performing arithmetic on a computer was *sequential*. For example, to perform addition on a set of numbers with another set, the following would occur, in order: load operand 1 from memory into a register[1] on one of the computer chips (Yes, a processor in the '70s occupied multiple chips. In fact, multiple boards. More like today's chassis in a rack of servers.); load operand 2, do the addition of the two operands and place the result into another register, and finally store the result to a location in memory. The load and store operations took many more clock cycles than the addition itself. After this first result was stored, the second pair of operands would be loaded, and the sequence repeated. And so it went until the two sets of numbers were added and a set of results stored.

Then *vector processing* was invented. It was a departure from past thinking of how to process and compute, and particularly suited to numerical calculations. The vector processing design afforded considerable higher efficiency of computations. "Efficiency" here means the fraction of the theoretical peak performance that the application actually achieves. Lest the reader thinks this is only of historical interest – it is not. The concept explained on the following page has been

---

[1]A register is simply a memory location the size of a 'word,' serving as temporary storage close to the arithmetic units (the use of caches has not arrived yet).

DOI: 10.1201/9781003038054-2

resurrected in today's processors, even if the implementations are quite different. We find it both on modern server chips and as a central theme of the GPGPU.

There are several ideas behind vector processing:

- In sequential processing, also known as *scalar* processing (to contrast it with *vector*), much of the time is spent waiting on operands to load, store, and an operation to complete. This wait time, or *latency*, is wasted time and resources. It would be good to be able to have the computer do something useful during this 'dead' time.

- There are multiple functional units in the processor: Load/Store, Add/Subtract, Multiply/Divide, and Logic operations unit. There is no intrinsic reason why one unit cannot operate in service of one instruction, while another operates on a different datum on behalf of another instruction.

- As we saw in the introductory chapter, much of the simulation model involves operations over sets of numbers. A pair, or sometimes a triplet, of operands is loaded from memory, operated over, and the result stored back to memory. One such sequence of operations is followed by another. The two sequences are *independent* of each other, so there's no reason for the second not to start before the first one finishes.

- The sets of numbers, or *arrays*, operated upon are stored in consecutive locations in memory. Each of the array's elements is operated on in an identical manner. Therefore, it should be possible to design a single instruction to perform a series of identical operations on each element of the participating sets of numbers. Take adding elements of two arrays pair-wise and storing to an output array. The instruction will need as parameters the starting addresses of the input and output arrays, the lengths of the arrays (which, of course, have to be the same for all participating arrays), and the type of operation to be performed. Then, in principle, this "array operation" can be streamlined.

- To make it all work, the architecture of the processor has to be *pipelined*. Whether in the process of loading, addition or multiplication, or storing – stages that take multiple clock cycles, the operands have to move forward each clock and make room for the operands behind them. The high level functional operations have to be broken down to one-clock series of stages. The system has to provide holding places for each operand at each stage. It is similar to an assembly line.

The arrays discussed above are one-dimensional; hence, they are called *vectors*.

In the literature we find that, at times, *vector processor* and *array processor* are used interchangeably. They shouldn't be. True, from the user's point of view, both have in common an important property: one instruction applies to multiple variables and produces multiple results. This architectural feature is referred to as SIMD – Single Instruction Multiple Data. However, the very different approach to hardware implementations of vector and array processors has far reaching implications. The discussion of array processors belongs in Part II, as part of the multi-processor era. When an array processor responds to an instruction, all of its processing elements execute *simultaneously*. When a vector processor executes the same instruction, the operation is *pipelined*. After some startup time, a result pops out every cycle.[2] Perhaps the most significant difference, for the user, between the two types of architecture is that an array processor is *attached* to a host processor that controls it and is the gateway to it. A vector processor is a stand-alone system, directly connected to the outside world. The evolution of the vector processor's approach led to vector instructions on today's server chips. The array processor concept is the predecessor to the GPU and the GPGPU. The presence of an array of processing elements allows an important capability: The array can be divided so that different regions perform *different* tasks. We will return to these topics in Parts II and IV.

The ideas behind vector processing were implemented differently by the several companies and computer architects who created the early vector systems. The first ones to develop and market vector processor systems were Control Data (CDC) and Texas Instruments (TI). Both developed their designs in the late 60s and brought them to market in 1973. The architectural feature that distinguished these early systems from the better-known Cray design was the way they accessed the operands of the vector instructions. Both the Advanced Scientific Computer (ASC) from TI and the STAR-100 from CDC streamed operands directly from memory to the functional units, and back to memory. This design decision meant that the advantage of vector streaming was realized only in cases where the vector was long – made of several hundreds or thousands of elements. Due to the smaller capacity of memories in those days, this meant that the actual performance of many real applications was disappointing unless the programmer found ways to string together arrays into long one-dimensional vectors. I feel entitled to say this about the STAR-100 since I was one of its early application programmers. More on my experiences with this system and its successors at a later chapter. It is interesting to look at an image of the STAR-100 (Figure 2.1) to just get an idea of how technology advanced since the mid 70s. The two front wings in this artist rendition hold a mere 4 MB of memory, less than 1/1000th of what we carry in our hand-held devices today.

The STAR-100, and its CDC successors, were unique in another interesting aspect that was not just of interest to computer scientists and architects, but to application programmers and algorithm designers. "A Programming Language," or APL, is a name of a programming language developed in the '60s by Kenneth Iverson [8]. The language's main datatype is arrays. The feature that attracted the STAR-100 architects, Jim Thornton and Neil Lincoln, was its application-level operators.

---

[2]Some vector processors had 2 or 4 pipes that executed simultaneously. This can be seen as a hybrid vector-array, with a very small scale array component

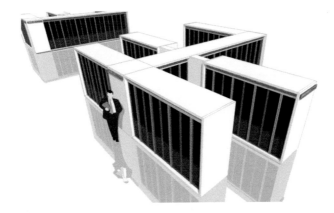

Figure 2.1: CDC STAR-100. Source:
https://en.wikipedia.org/wiki/CDC_STAR-100.

The STAR-100 contained, beyond the basic arithmetic operations, such functions as Scatter, Gather, Compress, Merge, Mask, and more. The application programmer, who used only Fortran at that time, accessed these operations through special function calls. Where appropriate, they provided a real boost to performance. The STAR-100 and the Cyber 200 line are dead now, and this anecdote is only worth mentioning because some of these operations are making a comeback. For example, the Gather/Scatter instructions are included in the repertoire of vector instructions in today's Intel processors.

These two pioneering vector designs did not quite shake the market. There were seven TI ASC systems built, and that was it. Five STAR-100 systems were built, with only three delivered to customers. However, CDC did not abandoned the product line, and went on to produce improved versions – the CDC Cyber 203, which enjoyed a short product life time of a couple of years when two systems were assembled, and the CDC Cyber 205 that was completed in 1980. The latter had the option of two or four vector pipelines. Control Data then spun out the vector product line and formed ETA Systems, which produced the Cyber 205's successor, the ETA10. By the late 80s, CDC was out of the vector processor business.

We will return to the events of the 80s in Part II.

Now, to the most dramatic event related to vector processing in the 70s: Enter the Cray-1 from Cray Research, architected and designed by the legendary Seymour Cray. Debuted in 1976, the Cray-1 was not the first vector processor in the market. It was, by general consensus, the first commercially successful vector processor. It is the best known and has been mentioned numerous times in pop culture since. It is not uncommon to say "a Cray" as a shortcut to "supercomputer" or "a very fast computing system." In the past, this was a reference to the Cray-1. As time went by this applied to later products from Cray Research, and then to those from Cray Inc. Of course, the use of "Cray" is driven by the company's name and the strength of its brand. But it is also a tribute to the genius of Seymour Cray. Cray was

Figure 2.2: Seymour Cray next to a Cray-1. Source:Computer History Museum (computerhistory.org).

foremost an engineer with a large dose of creative imagination. One of the reasons for the media popularity of the Cray-1 was its appearance. In his efforts to maximize performance, Cray focused, among other things, on minimizing the lengths of the wires and cables connecting the system's components. He achieved that by literally and figuratively thinking "outside the box." In a departure from the conventional square boxes, Cray found that a circle shape allows for shorter connections. The circle is not completely closed. This was necessary to allow maintenance of the system. Engineers had to get inside the circle to reach the boards and cables[3] (see Figure 2.2). An added nice touch was the 'bench' around the system. Well, the padding was really the added touch: The 'seat' sections housed the power supply units for the Cray-1.

Aesthetics alone did not account for the Cray-1's success (though it was a big contributor to the pop culture trending). The single most significant architectural innovation was the concept of vector registers.

---

[3]Actually, Cray started the use of non-conventional shapes for his designs with the CDC 7600, as we saw earlier.

The earlier vector processors – the CDC STAR-100 and the TI ASC – streamed vector operands directly from memory to the functional units, and stored the results directly into memory. On the face of it, this approach looks efficient; a very streamlined operation. A single fluid sequence that pipelined streaming of operands (typically, two streams) from memory to a pipelined functional unit in the CPU, and the stream of results back to consecutive (typically, but not always) locations in memory. However, the latency of getting the first operands to the CPU meant that for short vectors, executing non-vector operations from the CPU's scalar registers was often faster. It took some 40 cycles to get the vector operation started; then another 40 cycles to shut it down after the last result was processed. Operands, pre-loaded to scalar registers were accessed in one cycle. What all this means is that vectors had to be some 200–300 elements long before there was a benefit from vector processing. In reality, such long vectors rarely appear 'naturally' in applications in the '70s. It took efforts to string together long vectors. Hiding the latency – the waiting time to get the first operands to the functional unit, by doing some other useful operations, was next to impossible with the compiler technologies of that period.

Then came the Cray-1.

The idea Seymour Cray had was to expand the concept of registers in the CPU. Scalar instruction, one that performs a single operation, requires a register for each input operand and for the output result. Then a vector instruction would require a set of such scalar registers for each of the input and output vectors. That is, a *vector* register. Of course, vectors come in all lengths, and though this was still a multi-chip CPU, there were serious limitations on chip real estate – dictated mostly by distance between components. Cray settled on eight vector registers that had sixty-four 64-bit words. The compiler had to deal with the varying lengths of vectors. It would create code for chunks of the original vector that are exactly 64-elements long, plus a remainder shorter vector when the length was not divisible by 64. The user did not have to worry about managing the 'vector chunks'. But, for performance reasons, it was well advised to take advantage of matching lengths to 64-divisible numbers when possible.

This approach implied that instead of the single memory-to-memory vector instruction of the STAR-100 (and its CDC successors), on the Cray system there would be Vector Load instructions for the input vectors, followed by an arithmetic operation, and finished with a Vector Store of the result vector. Seems like an added complexity. However, this sequence of operations, combined with the existence of eight vector registers, allowed hiding the latency of kicking off a vector operation. While a previous vector operation sequence is still advancing, the loading of input vectors for the next vector sequence can proceed. This overlap means the wait time on memory access is hidden. The user sees (almost) continuous stream of results coming out. An almost non-stop operation of the arithmetic functional units.

Throughout the late '70s and into the '80s the "vector processing wars" – the battle for customers – was mainly between Control Data Corp. (CDC) and Cray Research. For fairness sake, and as an explanation of what occurred in the market, we should look at other features of both the Cray and STAR/Cyber designs. Supercomputing, at that time, was all about higher performance of numerical applications. The overarching theme was *vectorization* – feeding the vector arithmetic units. And the challenge was how to expand the space of data and compute structures that can be vectorized. Algorithmic innovation was a major driver in achieving this goal, and we'll get to it in later chapters. The computer architects, however, were able to place some hardware *tools* that were very helpful to the application programmers.

In addition to its vector registers, that helped getting good performance for short vectors, two other features are worth mentioning. The first one is referred to as "chaining." Often, one operation follows another and takes as input vector the result of the previous operation. The most common and useful example of this is a vector-multiply followed by a vector-add. It occurs, most frequently, in a vectorized version of matrix multiplication. What happens is this: Vector A and vector B get multiplied with each other. As soon as the first result element comes out of the multiply unit, it is used an input with the first element of another vector, C, for the 'add' operation. The two operations are *chained*. Storing the multiply vector result is avoided. And since the functional units (Add and Multiply) are pipelined, the processor performs *two* operation per cycle, effectively doubling its performance. This is a feature that increases performance once the vector registers are loaded.

Another very useful feature helps loading the vector registers. When dealing with multi-dimensional arrays, it is not possible that vectors in every dimension will be stored consecutively. To this end, The Cray-1 and later versions were given the ability to load and store in a *non-unit stride* fashion.[4] There was another, more controversial, feature on the Cray-1 that simplified the chip design and added to performance. Instead of including a divide unit, there was a hardware reciprocal. Rather than performing a true floating-point division, a reciprocal of the denominator, which can be computed faster, was calculated, to be then multiplied by the numerator. Theoretically equivalent, the latter was less accurate at times due to having to round the last bits twice.

To summarize: Chaining of vector operation increases the result rate. Allowing streaming data in non-unit stride manner opens up many more opportunities to vectorize and to chain operations.

The STAR-100, and its successors, also had architectural features that enabled higher performance and expanded the universe of what was vectorizable. The arithmetic units were designed so they could each act as two units when served with single-precision (32-bit) operands. It turns out there are important

---

[4]One had to watch so as not to have a stride size that is a multiple of the number of banks the memory is made of, since this would cause a serious slowdown of the load or store operations.

applications where single-precision is sufficient, notably weather models and seismic processing. In these cases, the performance, or, the result-rate, was doubled. In addition, on the Cyber-205 it was possible to do what can be called a conditional chaining. True vector chaining requires streaming of three input vectors. There were only two load pipes. However, if one of the inputs was scalar, then this form of chaining was possible. This was referred to as *linked triad*. For example: $Vector = Vector + Scalar * Vector$ is central to vectorized matrix multiplication.

As for expanding the applications space amenable to vectorization, the STAR-100 and the Cyber 205 had even more robust set of features than the Cray. Not only was it possible to operate on non-unit stride vectors, but vectors could be loaded that were composed of any combination of elements of the array. The stride did not need be constant, the list of vector members needn't be advancing in one direction – members could be fetched up and down the array, and elements could be selected more than once. This was achieved by constructing an *index list* of integers pointing to locations in the target vector. These index lists were used for the Gather and Scatter operations mentioned previously. In fact, when the desired list was one-directional and any element can be called upon just once, then a list made of zeros and ones – a bit list, would do. Such a bit string would be used in *Compress* operations to create a shorter vector of elements for a part of the computation. The reverse operation was an operation called *Expand* that would seed elements into a longer array. And there was the *Merge* instruction to create a single vector out of two separate ones. All three operations were guided by bit strings.

By the end of the '80s, it was the Cray approach that survived and won the "vector war." Though, by that time, a more formidable challenge appeared on the scene – the microprocessor.

The above overview of vector processors was not meant to be a complete architectural description of the systems mentioned. For that, there are other references more oriented toward engineers and computer scientists.[5] My goal here was to highlight some aspects of the two main U.S. competitors in the '70s. Especially those aspects that impacted the users the most. The discussion above is more than a historical footnote. The features that stood the test of time survived the demise of the proprietary big-iron vector processors. Vector instructions and vector registers show up on current commodity microprocessor servers. So does even such an esoteric operation as the *gather*. Programming techniques from the '70s and '80s have been re-learnt in the last 20 years. We will return to this theme later in the book.

Other players in the vector processor era deserving of mention are IBM with its 3090 vector processor and three Japanese companies that built vector processors. More on both in Part II. For a quick overview and brief history of vector processors, see [10].

---

[5]For example, there is concise and neat description of the Cray-1 as a 2002 writing project given at San Jose State University. See [9].

CHAPTER 3

Vectorizing Applications

Realizing Potential Performance is Challenging

V ECTOR processors get their high performance potential from the hardware features described in the previous chapter. To realize this potential the manufacturers provided *vectorizing* compilers and mathematical libraries that took advantage of vector instructions. However, for any scientific numerical application, but the very simple ones, this was not enough.

Today the subject of vectorization may seem a minor one. At least not one of the main concerns for an application developer. But in the era of vector processors this was central and unique for HPC. Vector features were used in scientific computing, but hardly ever in enterprise applications (even when IBM added vectors to its mainframes in the 3090 product line). Conferences and workshops on HPC topics in those days were consumed with the challenges of vectorizing applications. See, for example, several chapters in a 1984 conference proceedings titled "Supercomputers in Theoretical and Experimental Science" ([11]).

The importance of revisiting the topic of vectorization lies in the fact that it was a precursor to *parallelization*. Code that is correctly vectorized can also be executed in parallel. The reverse isn't true: non-vector and non-vectorizable code may still be amenable to parallelization. The methods and techniques for vectorization that were found, invented, and developed in the '70s and the '80s still apply and serve us well in today's massively distributed-processing clusters. They directly apply to intrinsic vector instructions in modern cores and GPUs.

For me personally, adapting applications to vector processors was a full-time occupation for several very fulfilling years during the period of the mid '70s to the early '90s.

Vectorization, a fine-level expression of parallelism, plays a minor role in HPC programming today relative to concurrency of execution across multiple servers and accelerators. After all, vector code in its best produces one result each clock

DOI: 10.1201/9781003038054-3

cycle, for each vector pipe. A cluster today can deliver many thousands results simultaneously each cycle. But vectorizing codes was a big deal in its day – up to about 30 years ago. The work, experimentation, and innovation that went into this activity laid significant part of the foundations upon which HPC codes are designed and developed today.

It was out of necessity that codes were modified to expose opportunities for invoking vector instructions. The Fortran compilers of that era could not be relied on to detect vector operations in most but the simplest of cases. The work done by the high-level language programmers was important not only for the modified application, but, even more importantly, as a guide for compiler writers of what transformations can be accomplished by the compiler. The "holy grail" was, and still is, *automatic vectorization.*

The following is not a tutorial, but, rather, a brief discussion of the challenges and general principles related to vectorization.

Effective use of vector-capable hardware forced the programmer to structure and *organize* both Data and Code. As was explained in the previous chapter, vector operation is one where the operands are being streamed in a pipeline fashion. In its simplest form, a vector instruction takes two variables to define a vector: a starting address, and a length (number of elements). This means that the all the vector elements have to be presented consecutively for processing. That was in-memory for systems where vector operations were memory-to-memory, such as these from Control Data and ETA, and some of the supercomputers built in Japan. Or, the vector would occupy a vector register (as in the Cray systems, for example) and the vector load and store would address consecutive locations in memory. Today's architectures use cachelines, but the principle is the same.

Most numerical computations involve multi-dimensional arrays. Only one dimension allows consecutive storage of elements. In the other dimensions (or indices, in terms of the programming language) sequential elements are separated by a fixed number of bytes or words in the physical memory. This means that the programmer better organize the code to offer the compiler a clear view of a vector operation opportunity. It was in the form of loops over indices in those early days (today the same idea would apply to array syntax), and the vectorizable index would be the one of the innermost loop in a nested loops code.

Vectorizing one-dimensional vectors or the column dimension of a dense matrix is the easy part. But doing just that would leave out most of the computation non-vectorized (also called *serial* or *scalar*). Consider matrix-multiply: It involves a series of multiplications of a row by a column. That is, multiplying consecutive pairs of elements – one from a row, the other from a column – and adding them up to construct one element of the matrix product. The first challenge is creating vectors that are rows; that is, not consecutive in memory. The programming solution was to transpose the relevant matrix, and then perform column-by-column multiply. But there was an architectural solution too: It was referred to as *non-unit stride.*

The vector instruction was supplemented with another variable, an integer, that specified how many locations apart were the elements of the matrix row. Of course, this number was determined by the length of the matrix's column. Access to memory is slow relative to computation speed, so to compensate, a range of addresses is loaded (or stored) with a single instruction call. Therefore, there is an overhead involved when using the non-unit-stride option.

The above exemplifies the challenge of choosing a strategy for vectorizing a given code. In the case of matrix multiplication, choosing between rearranging the data and using an available hardware feature. Some refer to this, and other situations, as the *art of vectorization*, but it is mostly a cost-analysis matter; albeit, not always an easy one. The considerations would include whether the operation is to be repeated many times, in which case transposing a matrix once and applying the transpose many times makes sense. The size of the arrays also comes into the calculation. In fact, for small arrays, the vectorization overhead sometimes exceed its benefit. A corollary of this is that a mathematical library routine should allow for different paths of execution depending on the size of data structures in the calling program.

Most of the computational work in simulation and modeling is done on array elements. When simulating a physical phenomenon the arrays' dimensions range from one to three (or 4-dimension for some relativistic physics applications). Exposing vector code is most straightforward when the arrays are dense – that is, most of the elements are non-zero. The data is well-structured, and vectors show themselves 'naturally' in consecutive locations or separated by a constant number of locations. The compiler can produce vector code as long as the programmer sets up the (nested) loops constructs to match the indices' conventions for the arrays. Vectorizing dense linear algebra calculations is the easy part in the process. This was the first type of code a compiler was expected to discover as vectorizable.

Unfortunately, many important real-world problems don't lend themselves to dense arrays representation. A phenomenon described by a set of partial differential equations (PDEs) is solved by discretizing the equations over some grid (as described in the introductory chapter). Each grid point is described by a mathematical expression that involves its neighbors, and, possibly, values from previous iterations. The form of the resulting expression is dependent on the numerical method chosen for the solver. Lining up all the grid points as a solution vector (or matrix or even higher-dimension array), the problem often becomes one of multiplying a *sparse* array of integers, known as the *coefficients matrix*, by a vector. The sparsity level is very high. In each column or row there might be only a handful (maybe, 3 to 5) non-zero elements. The density level of the coefficients matrix is often well under 1%. Obviously, it would be very wasteful to compute with vectors along rows or columns.

However, not all is lost. The *stencils*, the forms created by the discretization process, produce non-zero elements in the coefficient matrix along diagonals. That is, instead of working with a very sparse matrix, the programmer can extract the non-zero values and place them in a few diagonals. When the computations are modified to refer to indices of elements in those diagonal-vectors we obtain an efficiently vectorized code. We cannot expect a compiler to produce this kind of transformation in the data and the code. This is an example of where the compiler needs more than a little help from the code developer.

The density of values to be processed within a large dataset will determine if straightforward vectorization, even over zero values, is preferable to rearranging the data for more efficient processing. PDE systems tend to produce very sparse arrays, but they are structured. The non-zero elements are in fixed locations. This is not always so. For example, weather models includes processes such as cloud formation and precipitation that are applicable over parts of the area modeled. The relevant points' density is often low, and there is no special regularity of where these points are within the grid. In addition, their location changes over time as the simulation progresses. Vectorizing this kind of situation is more challenging. The idea is to still somehow group together the values that are to undergo the same computations in order to utilize vector instructions. We would want to *gather* the relevant values into consecutive locations, and thus form full-density vectors. And after performing some operations over these vectors, the results would need to be placed back, or be *scattered* into their location in the grid.

As described in the previous chapter, the Control Data architecture for vector processors, starting with the STAR-100, implemented APL operations as hardware instructions. These include the *gather* and *scatter* operations. These are vector instructions – a single instruction for processing multiple elements. Their inclusion increased the range of vectorizable code, even when their result rate was less than one element per clock (see page 31). The Cray systems of the '70s did not have the generality of collecting any set of elements into a consecutive array, but allowed for vector operations over a fixed value *non-unit stride* (where it meant that values loaded into the vector registers were skipped over during the execution).

I come back to the issue of data structure because it is applicable to most of the real-world applications. Highlighting this fact is that years after the "big iron" vector processors were replaced by microprocessors and clusters, *Gather* and *Scatter* instructions were added to the x86 server hardware instructions repertoire.

Much of the vectorization process is, then, about arranging and rearranging the data. It brings on processing overhead, of course. Therefore, the effectiveness of vectorizing segments of code is heavily dependent on the size of the datasets involved, and on how many operations are to be performed on the rearranged data.

Somewhat mitigating the vector setup overhead is the time saving of instruction decoding. Instead of decoding instructions for each pair of operands, a single vector instructions initiates multiple operations.

Quite often, in those days when vectorization was central to higher performance, both developers and customers were disappointed and frustrated by the final performance of the application. Before expanding on this, it is worth noting that the efforts to vectorize a code had the side benefit that even when eventually it was run in *scalar* mode, it did run faster than before. The code was better organized and the data better laid out.

The best way to think of vectorization is as an answer to the common challenge of the latency incurred having to wait on operands to be loaded from memory. Vector code is faster not because the arithmetic is done faster, but because the streaming hides memory latency.

Why has the concept of vector instructions been of limited success at the '70s and '80s, pushed aside in the '90s, then adopted again in present-day microchips?

The early vector processors – first a single processor system, later multi-processors (up to 8), had much smaller memories relative to today's system. That constrained the size of the problems that could be solved, which resulted in relatively short vectors. Which meant that the accelerated execution of vector operations did not make up sufficiently for the overhead in setting up the vectorization opportunities.

Vector processing has an upper limit for performance – one operation per functional unit (whereas parallel processing can, in principle, be scaled to provide ever increasing upper bound). Taking a perfectly vectorizable code, it could be sped up relative to the serial (or, scalar) execution by a factor of $O(10)$, being the number of clocks it took to fetch a unit of data from the memory. In practice, getting a 10-fold speedup for a computational kernel of the application was a good and satisfactory result.

All the above is still not the major factor in limiting performance speedup due to vectorization. The most significant factor is the portion of the code, in terms of execution time (not lines of code), that can be vectorized. The non-vectorizable part assumes a much bigger weight after vectorization. First, a simple illustration:

Take the serial execution speed to be 1, and the vector execution speed to be 10. Assume code that allows 50% of the serial execution time to be vectorized. The solution time is not the average speed of 5.5 (or $5.5x$ speedup). Far from it. What took 50% of the time originally, now takes 5% (was sped up by a factor of 10). The serial part remains at 50% of the reference time. So, the execution time after vectorization is 55% of the original. The speedup is only $1.8x$! You might recognize the reasoning above as the fact that for mixed-speed events the *geometric mean*, not the *arithmetic mean*, is the correct measure.

It is obvious that a much higher fraction of the execution time has to be vectorized in order to approach the full vector speed. On the plus side, in physical simulations much of the time is spent in a *solver* that contains relatively small number of lines of code employing nested loops, and that is iterated over many times. This means that vectorizing this small part of code will result in substantial speed improvement. In some cases, the required solver can be called from an external library, when it has been optimized for the system in use.

To illustrate a more favorable example, assume that code that amounts to 90% of the execution time is vectorized. It is then reduced to 9% of the original time. With the remaining 10% unchanged, we end up with 19% of pre-vectorization time, which is a speedup of about 5.2x. Still not close to the asymptotic 10x.

The matter of the overall speedup given that only a part of the execution can be sped up was given precise formulation by Gene Amdahl back in the late '60s. It is known as *Amdahl's Law*, and can be written as:

$$Speedup = \frac{1}{(1-f) + \frac{f}{s}}$$

where $Speedup$ = Speedup of the whole program,

f = Fraction of the execution time benefitting from

vectorization,

s = Speedup of the vectorized part of the code

We should note that the expression above can apply to parallelization too, where f refers to the fraction that can be parallelized and s is simply the number of processors applied to the job.

While Amdahl's Law appears as a precise expression, this is misleading. It belies the fact that f and s are not known exactly. One can mark the places in the code that can be potentially vectorized, and then profile the execution in serial mode, to get an approximation of f. The vector speedup, s, is even more difficult to nail down. It requires looking at the generated binary code and figure out the savings that arise from steaming. Ad hoc values are typically used, that often neglect vectorization "side effects." To get a handle on the true s value it is necessary to sample chunks of code and measure *before* and *after* vectorization. In short, getting trusted values of these variables amounts to almost do the vectorization itself. In practice, people have been using fixed values for s (often too optimistic), and ballpark figures for f.

The predictive power of Amdahl's Law is merely qualitative. In practice, people identified the time-consuming and vectorizable portions of the code, experimented with vectorization approaches, and just measured the *Speedup* variable above.

Real applications include code that cannot be vectorized – serial setup, I/O operations, testing for specific conditions, and more. We saw that even small remnants of serial code have a great negative impact on the overall performance. The memory sizes of the vector processors of that time often did not allow to increase the problem size sufficiently so that the portion of "fixed time" serial code becomes smaller, and much more of the processing time is spent in vector mode.

That said, the work done by the application engineers of the day was valuable and we reap its benefits to this day. Having to organize data and code for a better flow of the execution became commonplace in applications design. Hand coding and inventing techniques to expose independent sequences of operations provided blueprints for compiler writers. The lessons learnt at that period were applied to the parallel processing we now take for granted.

And of course, we have come a full circle with CPUs today being augmented with vector operations. Accelerators are based on performing many independent operations all at once – on vectors and arrays.

Striving for ways to extract higher performance from compute systems is at the heart of HPC. There is much more to be said about *performance*. See Chapter 31.

CHAPTER 4

Numerical Weather Prediction

The Basics of Weather Forecasting and Climate Modeling

O NE of the most direct connections between supercomputing and everyday life is the daily weather forecast we're all expecting and receiving at our finger tip. In the beginning there was the model of the atmosphere. That was sufficient for short-term, a couple of days, forecasts. In later years it was necessary to add models for the ocean (that is the whole ocean, not just surface data), land-ice and sea-ice. Otherwise credible longer-term weather forecasts and climate studies could not be done.

Weather forecasting use of supercomputers also provides us with the clearest and most specific argument for the need of ever faster computer systems. The prediction results are useless if not provided in time. Weather modeling turns out to be the perfect application to follow as we track the evolution of HPC and the accompanied scaling of applications. We will return to this topic.

Learning about weather forecasting models was my introduction to demanding HPC applications (after a more modest start in supercomputing writing a few physics programs for elementary particles studies as a graduate student). It remained a focus area for me years later in my career. More on that in Chapter 5.

The purpose of this chapter is to lay out the basics of modeling the Earth's atmosphere and surface – without equations. This is also an opportunity to go over the difference between *Weather* and *Climate*, and how it is reflected computationally. Some concepts covered here will be useful in later chapters on weather and climate.

DOI: 10.1201/9781003038054-4

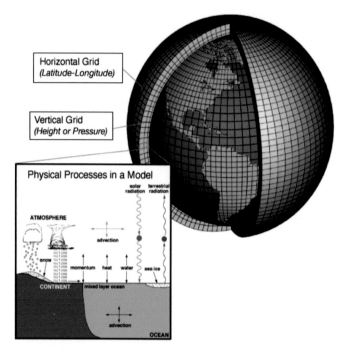

Figure 4.1: Atmospheric Model Schematic.
Source: Wikipedia.

The Grid

Before getting into the substance of the computations let us consider the space on which the simulation is done. That space is the thin layer of atmosphere we inhabit. We live on the surface of planet Earth. The part of the atmosphere affecting our weather is only 10-12 km high, the troposphere regime. That's where we find clouds and vapor and winds. The Earth's radius is over 6,000 km. Our lives are contained within a layer that is a fraction of a percent of the Earth's radius. A simulation of the weather in our planet should be made over the two-dimensional surface, and a third, vertical, dimension, that extends for a few kilometers upwards. The shape looks like the space between a sphere and another sphere enclosing it. Of course, as described in the introductory chapter, the space has to be discretized when computed upon. It would look as in Fig. 4.1.

Each dimension of the horizontal grid has many more points than the vertical dimension, as should be clear from the discussion above. The spacings between vertical layers are not even, since there is much more 'action' closer to the ground. The upper vertical points can be spaced further apart without loss of precision. It is important to note that the horizontal grid depicted here is built along latitude and longitude lines, which creates a higher density of 'cells' near the poles. This presents numerical stability issues which have to be mitigated by various numerical techniques. This was how early models were constructed. Modern models horizontal grids avoid this pitfall, and will be mentioned in later chapters on weather and climate.

What is being Computed

The flow of air is governed by the physics of fluid flow. It is captured by a set of equations known as the Navier-Stokes equations. We can think of it as, essentially, the horizontal motion of air, or wind. Its effect is called *advection* – the transfer of matter (air, moister) and energy (heat) as a result of the flow. This part of the model is referred to as the *dynamic core*. The energy that drives it is the *physics* content of the model. As depicted in Fig. 4.1, it includes the radiation from the sun and reflected radiation from the Earth's surface (land, sea, and ice). The heat generated by the solar radiation causes other physical processes that generate what we call *weather*. These include heat transfer between the surface and the atmosphere, water evaporation, cloud forming and rain, snow and ice. Computationally, the *dynamics* part applies to the whole of the well-structured grid, and can be well optimized (vectorized and parallelized). The *physics* part is less so. The processes don't apply uniformly over the grid. Think of precipitation and clouds, or radiation that occurs during daytime but not at night. Where the physical processes apply is a moving target and changes during the simulation. The physics part is computationally intensive, and it is harder to optimize it for performance.

A fine-grid model, almost always regional, allows computations of fine-scale phenomena. These can include water droplets, condensation, evaporation, ice formation and hail, convection in high resolution for sever storms and tornadoes, and such. A fine enough grid, at the level of well under 1 km grid spacing, also allows to study turbulence, an important factor for aviation, using the actual governing equations. At grid with spacing measured in multiple kilometers, turbulence is approximated through parametrization schemes.

The flow equations that govern how changes in the atmosphere propagate are non-linear. They involve exponentials. There are two impactful consequences of the non-linearity: First, there are no close-form solutions to the system of equations. They have to be solved numerically. Second, because the values computed have limited precision, any small deviation from the true value, fed into further computations (of the next time steps) *may* cause the error to grow exponentially. This situation is an example of the mathematical field called *Chaos Theory*.

This is one reason why, mostly in the past, weather forecasts were sometimes off. Inaccurate or insufficient initial values at the start of the simulation added uncertainty that could not be mitigated by careful numerics during the computations. That is because measured input data for the forecast does not exist for each grid point. Satellite data today covers the globe pretty well, but in the past the forecasters had to rely on scattered weather stations, weather balloons, and input from ships and planes. That input is extrapolated to the coordinates of the model's grid points, a process that adds further loss of precision.

Reducing such inherent uncertainties got much attention and the reliability of the forecasts was greatly improved in recent decades, which allows the extension of the forecasts period. Much work has been put into mitigating the chaotic nature of the governing equations, enabled to a large extent by the increased capabilities of HPC computer systems. We will come back to this topic (below and in Chapter 27).

The introduction chapter introduces the finite differences approach to solving differential equations, one that was used for weather models almost exclusively in the early days of modeling weather. That chapter also mentions, in passing, the *spectral method* (see page 13). Indeed, since the '90s, models of some of the major weather centers use spectral, or frequency domain, algorithms to solve the model's equations.

Types of Weather Models

There is no "standard" weather model. They differ in resolution, of course. Some are global models – simulating the whole of Earth's atmosphere; some are regional – a region may be covering a continent, sub-continent, a country, or a region within a country. Some models simulate a few days of forecast, some only a few hours, while others can forecast close to two weeks ahead.

There is a correlation between the model's purpose and its computational content and resolution. We are most familiar with the models that tell us the local weather for the next few days. Such a model predicts temperature, cloud coverage, wind, and precipitation. There are others that are concerned with air quality, or sever storms, or floods, or avalanches, etc. They tend to be of high resolution, short-time span runs, and over limited area.

As noted above, models are applied for multitude of purposes, which dictate the 'physics' content of the model. *Chemistry models*, used to study and predict air quality, would include chemical interactions – between gases, mostly. They are applied to regions, or even at the municipality level. Common applications would be the impact of air pollution due to emissions from cars and industry, or smoke from wild fires. As an extreme example, atmospheric chemistry model can be devised to simulate the spread of poisonous gas in an urban settings.

More on applications of weather models to other than weather forecasting can be found in later chapters. In particular, see Chapters 15 and 20.

As for the scope of weather models, there are two categories: Global models whose grid encompasses the globe, and regional models, also called *mesoscale models*, that cover limited area to allow for higher resolution and local customization.

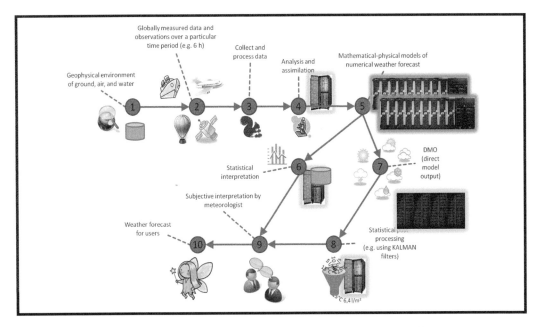

Figure 4.2: Operational Weather Forecasting Workflow.
Source: Institute of Space Systems (IRS), University of Stuttgart.

Operational Weather Centers

Most, if not all, countries in the world have their own weather agency. Some of the smaller, or poorer, countries may not have an HPC system dedicated to weather forecasting. A national weather agency's mission is to provide forecasts to its residents; that is, over a limited territory. However, any region's weather is a part of the global weather system. Therefore, someone has to run global weather models operationally, though it would be wasteful to do so by every country.

This reality is resulted in a kind of a two-tier weather centers arrangement. Several top centers have the high-end HPC systems required for running daily global weather forecasts. Most other countries consume the global forecast product and use it to run more detailed and short-term local model, or just tailor the global output to their local needs. These major centers are in U.S.A., the U.K., Germany, France, Japan, China.[1] Arguably, the premier center in the world is a center run as a consortium of European states. Other large facilities for forecasting are in Australia, Italy, Russia, and a few other countries.

Producing the daily weather forecast is a process that involves much more than running the computational model. The major steps are illustrated in Fig. 4.2.

Before the model can run its starting values have to be determined. This involves access to the most recent observational data of all the dynamical and physical

---

[1]Several of these centers are mentioned in later chapters.

variables in the model. The raw data comes in different formats, depending on its source – from satellite to transport vehicles to weather stations to sea buoys. Each piece of data has a location and time stamps. These are not the locations of the grid points of the model, nor taken at the exact time of the start of the simulation. Therefore, there is much extrapolation and projection to be done so the initial values of the grid points are established, and the simulation can start. This stage of mapping the observed data to the grid is called *data assimilation*. It turned out to be one of the most critical factors for a high quality forecast.

The output of the model is a set of values for each grid point and for each time step for which it was saved. It now has to be interpreted before a forecast can be produced. The output is subjected to statistical models and methods, including the use of Kalman filters for variables for which only indirect measurements or values are available. The later may apply to quantities such as wind energy or the common "temperature feels like" estimate. The graphics of forecasts for rain, winds, and heat that we see when forecasts are broadcast are also created at this stage. Finally, meteorologists add certain amount of subjective interpretation when they deliver the forecast in a layman language.

We have come to expect weather forecasts to be presented regularly like a clock-work. Industries such as travel, transportation, agriculture and others rely on it for their operations. The top-tier centers take on the responsibility to provide the results of global models simulations to all the other countries and to countless organizations and agencies. They have to have fail-safe systems in place, that not only guarantee a forecast once a day, but several a day, to be run in specific time-windows. To achieve that, these centers often acquire dual systems. One is designated the prime operational system. The other, an identical system, is a standby in case a failure occurs to the first one. They typically share a file system so the running job can resume on the second system almost instantaneously. A dual system of this size is an expensive proposition, and the standby is not left idle. It is sometimes called the development system. Revisions and updates to the model are developed there, as well as general research. In addition, it is now used for climate modeling, which, of course, is paused whenever the system needs to fill in for the operational system.

Weather forecasting centers work to a tight schedule. To get a high quality forecast, the model needs to start with high quality initial values. This means acquiring the observational data as close as possible to the start of the run. And that start time has to be as close as possible to when the forecast period start is. The processing time for the data assimilation stage, the runtime of the model, and then its post-processing has to be known with high degree of confidence. For example, if the center is to deliver a fresh forecast for the 6 o'clock news, they will count back the time needed and determine to stop consuming new data at, say, 3:00 pm. The operational run is designed to take a couple of hours. This allows for 2-4 runs per 24 hours, so that the weather is tracked anew every few hours.

The multiple runs are also useful for delivering the most up-to-date forecasts to countries in different time zones.

Determining Parameters of the Model

Within the HPC world, weather and climate modeling are two of biggest consumers of compute cycles and system resources among the classes of applications. Outside of the operational weather centers, most of usage is for climate modeling. On the face of it, weather and climate modelers can use any capacity available, subjected only to their budget of compute resources.

A higher resolution model produces more accurate results. It is closer to the continuum on which the equations are defined. However, the amount of computations goes up exponentially with the reduction of grid points spacing. When the grid interval is halved – both horizontally and vertically (though the number of vertical layers is generally independent of the spatial resolution) – 8 times more points were added. In addition, for numerical stability the time step also needs to be about halved. Which means that twice as many simulation steps are taken for the same forecast period. Overall, the computation 'cost' has gone up by at least a factor of 10. Note that whereas the added computations due to the spatial reduction of the grid interval can be, in principle, mitigated by increasing concurrency (more nodes, for instance), the additional time steps have to processed sequentially. Only faster processors can help.

The other consideration is the 'physics' content. The basic entities, such as temperatures, humidity, and precipitation, are always included. Longer forecast periods are affected by small-scale phenomena such as cloud formation. So that increasing the forecast span from, say, 3 days to 10 days, is not a linear increase.

Finding the right balance requires fine-tuning of financial and computational factors. For a researcher it is relatively simple. The compute system is what it is. The computer budget the researcher gets may be somewhat negotiated, but job queue times are less flexible. Of course, there is the pressure of papers to publish. All of that will dictate how to construct the model's parameters that will optimize the balance of runs' wait time, the number of runs, and the quality of the model, so a timely meaningful research project is achieved.

For operational weather centers, the decision-making tree is more complex and more critical. They have to look ahead a few years – the time it takes to define requirements, benchmark a future system in a competitive bid, having it installed, and bring up operationally. Given the efforts and resources needed to replace a supercomputer in an operational environment, combined with rate of technology innovations, the typical time between two generations of systems is about 5 years.

The centers keep track of the model's accuracy and present it as what is known as a *verification graph*, that shows the percent accuracy of the totality of several attributes, over some area and for a period of time, which constitute what we consume as forecast. Their performance is judged by that metric, and their mission is to increase both the accuracy and the range of high quality forecasts.

It should be noted here that one of the consequences of the exponential growth of errors for non-linear equations is that increasing the forecast period requires increasing the accuracy of the initial values. Otherwise, the size of the computational error is bound to become unacceptable unexpectedly.

As an exercise, let us assume that in 5 years, for the same budget, the center procure a system 5 times more powerful than its present one. Perhaps a factor of 3 is due to increased parallelism (each processor chip performance is advancing at a slower pace). The operational runtime has to be constant. The center may want to increase the forecast range by a day or two. For that, it will need to allocate more processing time to the data assimilation phase. The extra speed *not* due to parallelism can be applied to extending the forecast range. What is left can be applied to shorten the grid interval, and possibly add more physics content. It is a calculation that needs to be done ahead of the available technology, based on projections. A benchmark model is constructed and the competing vendors test it and attempt to reach the performance target. Expectations have to be calibrated as the evaluation process progresses.

Brushing over the many details, people have come up with rough rules of thumb, or a simple constraint, for the model's simulation time. If a center produces 10-day forecast, and it is to do it in a 2-hour run, then the simulation has to run 120 times faster than real time. So, if the model's timestep corresponds to, say, 6 minutes in real time, it needs to complete in three seconds. If this is not achieved, there may be some leeway in the value of the timestep, and it can be made a little larger.

For climate models there are no daily constraints that dictate model resolution and runtime. In practice, both for research and for policy planning, people found they need to simulate at least 10 years per day of computing.

Regional Models and Downscaling

One way to afford high resolution without giving up physics content is to reduce the area the model covers. As an example, consider that the Earth's surface, about $200Mmi^2$, is about 12, 300 times larger than the area of Switzerland. This easily allows $4x$ higher resolution, more physics (measure air quality, for instance) and topographical details, all of which adds up to the ability to describe a very localized weather, such as separating the forecast of conditions in a valley from those in higher elevation of neighboring mountains. Similarly for any other region or country.

Figure 4.3 illustrates the nested grids scheme used by the U.K. Met Office. In recent years they have run a global model with grid points separation of $17\,km$, a regional-European model at $4\,km$ intervals, and, finally, a fine-resolution grid over the United Kingdom at just 1.5 km between grid points.

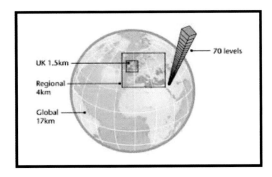

Figure 4.3: Nested Grids: Global-Europe-U.K.
Source: U.K. Met Office. ©Met Office.

The problem with this simplistic view is that any region on the globe is *not* a closed box. The weather systems know no human-made boundaries. With fixed boundary values a relatively short distance outside the region, a simulation can run for just a short simulated period before it cannot be trusted, because by that time affects from outside the grid would reach the inside of the region.

For that reason, a regional weather forecasting run has to rely on a global run over the period in question. It takes the coarser grid values over the regional grid and interpolates them onto the finer regional grid. The regional model then runs for a short period, simulated a few minutes perhaps, before pausing to refresh with the later values from the global model. This two-step process makes it possible to extract local, and finer, forecast features from the global model's output.

Climate Modeling

Weather forecasting and climate modeling simulate the same environment – our atmosphere and land, ice, and oceans, though each measures different aspects of that environment. Still, the underlying processes are the same, so the digital models are similar. Some of the prominent models can serve both for weather and climate runs.

Some would say "If the weather forecast is accurate enough only for next few days, how can I trust climate predictions of 30 years from now?." Well, the underlying premise of the question is wrong. A climate prediction is not that on January 1, 2051 the high temperature will be 55° and partly cloud, with 40% chance of rain. That would be a weather prediction in the far future. A climate prediction might be that, in the region of my state where my town is, the winter in question is likely to be 30% dryer and 5° warmer than the average winter of present time.

Climate simulation does not attempt to extend weather forecast into the future. It predicts trends of averages over sizable areas of environmental variables such as temperature and precipitation. Whereas a weather model simulates several days, a climate model simulates (many) years. Understandably, because it deals with averages and simulates long periods of time, climate models, mostly, use lower resolution grids than weather models.

Climate modeling requires the addition of some guardrails and phenomena that are not necessary in weather models, because they have no significant impact over periods of several days. The simplest climate models add the constraint of conservation laws. These are called *energy balance* models, and track the solar radiation and the reflected radiation from Earth. That's how global warming is tracked. The more complex models, known as *General Circulation Models*, or GCM, include ocean circulation and its layer inversions and currents. These are processes whose impact over a few days can be ignored.

Mathematically, one can think of weather forecasting model as an *initial values* problem. Given the values at the grid points of the quantities of interest, the model proceeds according to the governing equations. Climate model is a *boundary conditions* problem – certain conditions are specified and are imposed during the computations. They include conservation of energy (solar radiation) and volcanic aerosols (that can trap heat). They also include assumptions about human activities such as deforestation and gas emissions that affect moisture, heat exchange, and absorption of carbon dioxide etc. For a short description of *weather* vs. *climate* see [12].

Coming back to the question posed above: What gives us confidence in climate models prediction? Weather forecasts can be verified within days. We observe the real weather and, if mostly accurate, we gain trust in the model. Clearly, this is not a practical approach for climate predictions of what our world will look like, climate-wise, in 50 or 100 years. But, there is a way around this dilemma. There is detailed data on climate in the last 50 years, at least. There is documented information corresponding to the last several centuries, and geological clues that go back thousands of years. This allows the modelers to take these future-facing models and run them forward from some point in the past, say, the last 50 years or so. The output can be compared to known observed findings. These tests confirmed, for example, that the inclusion of human contribution, or anthropogenic change, to gas emission into the atmosphere is necessary in order to explain the climate changes we have experienced since the Industrial Revolution. While it is not a direct validation, when the model can reproduce the known history, we are justified in placing a reasonably high level of confidence in its prediction of the future.

The idea of looking for finer features at the region level, as described above for weather models, exists in climate modeling too. Beyond the global features, there is much interest and need to study climate aspects over, for example, large mountain ranges such the Andes. Or large valleys and bodies of water. The aim is to get more than what is provided by the statistical nature of output from global climate models. This requires adding topographical details, and, of course, the use of a finer grid.

The process of zooming in for climate studies is referred to as *downscaling*. As described by NCAR (National Center for Atmospheric Research) at [13], it is the procedure of taking large-scale, low-resolution, statistical information to make predictions at local scales. There are two main approaches to accomplishing downscaling. *Dynamical downscaling* is the process of feeding the statistically-based output from the GCM to physically-based regional model. The other approach is known as *statistical downscaling*. It involves using statistical relationships between large-scale predictors (pressure fields, for example) and local climate variables such as temperature and precipitation, and applying these relationships to the output of the global climate model.

Establishing trust in climate models is much more important than just satisfying academic standards. It allows modeling and *measuring* the effect of changes in human behavior in terms of life style, food production, use of energy sources and technologies, and more. It allows the impact of government policies and personal choices to be quantified.

Reaching consensus among mainstream climate models about how to model these effects would be of enormous benefit to humanity and the world in addressing climate change.

A summary of the origins of the use of computers for weather and climate modeling and the people who pioneered the field can be found in [14].

CHAPTER 5

Vector Processors for Weather

The Battle Between the Vendors

I N some sense the late '70s was the golden period for supercomputing and weather prediction. It was the period when top weather centers acquired the most powerful supercomputing of the time. We will see that some 30-40 year later, these centers used systems that were no more than half the peak performance of the top HPC system available.

This chapter is dedicated to the introduction of vector processors into two of the most prominent weather centers, only a few miles apart geographically, in quick succession. It is also a personal story of my early days in HPC.

ECMWF

The European Centre for Medium-Range Weather Forecasts (ECMWF) was created in 1975 by 18 European countries – the "Founding States." It would become the top-ranking weather forecasting center in the world. There are now 23 member states and another 12 countries with the status of *Co-operating States* (more about its history and links to ECMWF's progress over the years can be found in [15]). The United Kingdom was chosen to host ECMWF. It was chosen, to a large extent, because the proposed site was in the vicinity of the U.K. Met Office (known simply as the "Met Office"), the existing premier weather center.

Back in 1975, the U.K. Met was running its forecasts on an IBM mainframe, and was already recognized for its quality of forecasts (relative to other centers; not necessarily in absolute terms). ECMWF just opened their offices in Bracknell, a small town west of Heathrow airport and Windsor, where the U.K. Met also resided. ECMWF would soon establish their permanent facilities in nearby Reading, while the U.K. Met would relocate years later, in 2003, to Exeter in south-west England.

ECMWF's founders were building up their organization, hiring top scientists and technologists from the member states. They needed to develop the operational model, and they needed a computer system. Some early vector processors were already in existence – such as Control Data's (CDC) STAR-100, and the Cray Research's Cray-1 was in its final stages of development. CDC was talking about a follow-on vector processors – what would come out as the CDC Cyber 203, followed by CDC Cyber 205. It soon became clear that the two main competitors for providing ECMWF with a supercomputer are Cray Research and CDC.

It would take a few months for ECMWF's staff to come on board and put together the *Request for Proposal*, or RFP, for their first computational system. The time from the release of the RFP until the new system becomes operational was expected to take, as it did, two to three years. But development work on the model and supporting utilities had to start, so when the new system is installed the Center will be ready to produce forecasts. Unlike Cray Research, CDC had computers to offer immediately, and ECMWF leased an older system from CDC, the CDC 6600 – an earlier Seymour Cray design (1965), and installed it at their initial, and temporary, location at Bracknell. It may have looked as if this would give CDC a leg up in the procurement process. It turned out not to be so.

The Cray-1 was the computer most talked about at that time ('75-'76), as a revolutionary design by the venerated Seymour Cray. Its serial number 1 yet to be delivered to Los Alamos National Lab (for a trial period, and without an operating system). CDC had the STAR-100 installed at Lawrence Livermore National Lab, and customer satisfaction can be described generously as "low." It was the system on which early evaluations were done for ECMWF. The CDC Cyber 203, an interim system between the STAR and the much improved Cyber 205, was still too far out to be credible.

It soon became clear that the ECMWF team were leaning toward selecting the Cray-1. Not for reasons of reputation, but for technical reasons. What was considered a large memory for both Cray and CDC was 8 Mbytes. That, combined with a processing speed in the $100s$ of megaflops range, meant the model's grid had to be coarse – leading to short vectors as defined by the horizontal dimensions, and much shorter in the vertical dimension. And, as was explained previously, the Cray vector architecture of utilizing vector registers performed better than CDC's memory-to-memory vector operations on short vectors. In addition, the CDC architecture was inferior to Cray's in scalar performance.

Already in 1976 it became clear to the vendors involved that Cray Research would be selected if early performance measurements of computational kernels were to hold. They did. It would be later, in '77-'78, when a full-scale application, such as a weather model, could be run on the Cray-1. But given the information at hand, ECMWF acquired one of the early Cray-1 systems, and in August 1979 produced its first operational weather forecast. True to the organization's name that includes "medium range", it was a forecast for the next 10 days, the first operational forecast for that many days, while allowing for declining accuracy beyond the first few days. Nevertheless, an historic milestone it was.

The U.K. Met Office

I was hired by Control Data Ltd., the CDC U.K. subsidiary, to start in January 1976. In fact, I found myself on a flight to CDC's headquarters in Minneapolis the day before my official start date.[1] So was the sense of urgency on the part of CDC to build vector programming expertise in Europe. My immediate task was to become proficient with the STAR-100 vectorization style, and to evaluate what can be done for the ECMWF procurement. However, soon after, when it became clear ECMWF will procure from Cray Research, our attention turned to the expected Invitation to Tender (ITT)[2] from the U.K. Met Office.

The U.K. Met Office is probably the oldest weather forecasting organization in the world. It was established way back in 1854. Long before there was numerical modeling, and long before applying computers for forecasting. In the beginning it was for studying marine climate trends and features. Its storm warning service, now called the *shipping forecast*, is considered the world's longest running forecasting service. The Met Office produced forecasts for the country's defense toward the end of World War I. In fact, the organization was under the Ministry of Defense until recently (2011). It was a Met Office scientist, Lewis Richardson, who laid the foundations for numerical weather prediction (see Chapter 4), which started at the Met Office in 1922 – hand-calculated, of course. The first computer arrived at the Met Office in 1959, but it took until 1965 to have the first computer-aided operational weather forecast. For more on the history of the Met Office see [16].

Having the ECMWF choice for their supercomputer settled, both leading competing vendors – Cray Research and CDC – focused on the Met Office's plans. We, at CDC, thought there are a few circumstances that should improve our chances relative to the ECMWF procurement. In essence, the timing of the Met Office procurement was more advantageous to CDC – just as the ECMWF procurement timing was favorable for Cray. The Cray-1 would have been about 5 year-old design by the time the Met Office was to install the new system (1980-81). The successor Cray system, the Cray X-MP, was due around 1982. And CDC's system to bid, the Cyber 205, could be demonstrated, in factory settings, from around 1978, with the interim Cyber 203 serving to develop codes and model the Cyber 205 performance.

Being an operational weather center, performance was a crucial component of future systems evaluation, and in that area the Met Office has made a decision (some may call it a concession) that turned up to be very significant to the outcome of

---

[1]It was my first visit to the U.S. I had no appreciation of the Minnesota winter, and the first thing we did upon my arrival was to purchase a warm coat.

[2]This is equivalent to Request for Proposal (RFP), as it is known in the U.S.

the procurement. It had to do with the rules of benchmarking the weather model provided.

The benchmark rule was about accuracy, or, more precisely, how to determine that the benchmark output is acceptable. It was common to require that all the output bits match all the bits provided as reference – those produced by the current system. The Met Office, instead, specified a margin of acceptable deviation from the reference. I do not call it "acceptable error", because there is no reason to call a result that differs in the last bit or two any less "correct." It is even likely that a more modern system executes arithmetic more accurately.

A consequence of the above was that it opened the door to more robust vectorization and other optimizations of the code. For example, the vendors had to worry less about the order in which computations were done (depending on the operands, change of the order of operations could easily result in some of the last bits being different). This aspect was helpful to both Cray and CDC, of course.

However, the CDC team had another architectural feature to exploit. One that proved pivotal to the outcome. The CDC hardware – from the STAR-100 onwards, could perform in single-precision (32-bit) floating point arithmetic. The Cyber 205 could address 32-bit words, though memory access would fetch a "super-word" of 516 bits.[3] The kicker was that the compute rate of single-precision was double that of 64-bit (double-precision) arithmetic.[4] The hardware included logic that enabled a floating-point functional unit to act as two units when presented with 32-bit operands. The instruction set supported the designation of these half-size variables. Another very significant benefit of working with the smaller data types was that the memory footprint also halved, and every memory access affected twice as many variables.

The Cray architecture supported only 64-bit arithmetic.

Our (CDC's) big break was that the Met Office agreed to consider looking at the benchmarked model's output using 32-bit arithmetic, and evaluate it with respect to how the weather forecasts are delivered. In 32-bit arithmetic there are only 24 bits for the mantissa (the part that holds the displayed digits of the number). They cannot possibly be a close duplication of the 48-bit mantissa one gets in 64-bit processing. The Met Office scientists concluded, though, that given the margin of error in the initialization data from the observational input, the numerical errors due to grid spacing, and the granularity of delivered forecasts, that the 32-bit output was adequate. They arrived at that conclusion by comparing 64-bit runs with 32-bit runs against the same initial conditions (a test they could do on their IBM mainframe).

---

[3]In technical documentation "super-word" was abbreviated to SWORD. Later models access double that size – to 1028 bits. This larger data item was called, somewhat whimsically, PEN. Why? – because "the pen is mightier than the sword."

[4]I am using here today's terminology for single and double precision. At the time, 64-bit word was the standard precision. 32-bit word was referred to as half-precision.

That is not to say that we, at CDC benchmarking the Cyber 205, could just run the model at the 32-bit mode. The code was, of course, written in Fortran. And the Fortran compiler at that time did not support multiple floating point formats. Resorting to 32-bit arithmetic could be done in assembly language. However, there was another way, and it was just a little easier than assembly coding. The Cyber 205 compiler included a mechanism for calling hardware instructions directly. It was presented as a subroutine call for each instruction and was identified by a special naming convention. Each name started with "Q8" appended to a mnemonic of the instruction to be activated. The parameters in the "subroutine" call were the fields of the instruction, such as pointers to input and output operands. The first field of a vector instruction was 8-bit long for various options of the instruction. One of these bits was to indicate if this was a 32-bit or a 64-bit operation. In order to point to a vector (remember, this was in the '70s, the very early days of C and pointers, and well before it was considered for HPC applications) the compiler had a data type called "descriptor." It was a 64-bit quantity with 16 bits assigned to the vector length (allowing for lengths up to 64K elements), and the remaining 48 bits were the bit address of where the vector starts.[5] The vector instructions took *descriptors* as the operands to operate on. For example, the Fortran code for adding two 32-bit vectors into a third one using a Q8 call would look like this:

```
Descriptor desa, desb, desc

Assign desa, a(1:n)
Assign desb, b(1:n)
Assign desc, c(1:n)

Call Q8ADDNV(X'80', , desa, , desb, , desc)
```

Armed with these tools we, a small team from the U.K. offices of Control Data, set out to vectorize the Met Office model, then convert it to a 32-bit version. In those days we had to physically be where the computer was. That meant weeks on end spent in the Arden Hills (a suburb of Minneapolis-St. Paul) facility where the Cyber 205 was built. Vectorizing the dynamics part of the model was pretty straightforward. The original Fortran's loops were easily translated to vector code. The physics part is more difficult. Vectors have to be constructed since the physical processes occur only under certain conditions. This part, accounting for about half the original processing time, required much work. The availability of instructions such as *gather, scatter, compress, mask* etc. made it possible to get much of the physics vectorized, albeit with the overhead of moving data around.

Having the "CALL Q8" feature in Fortran meant that a high-level language structure could be preserved. This applied to declarations of data types, data blocks, I/O operations, and even scalar operations on operands that were not declared to be

---

[5]The bit level addressing was necessary because the architecture allowed for bit-type vectors. They were necessary for masking operations and logical instructions.

32-bit long. The resulting code still looked strange and unusual. The computational segments of the code, mostly within nested loops constructs, were a series of lines all starting with "CALL Q8", each line for one low level operation. There were several thousands of them in the model. The resulting program listing did not look pretty, nor was it easily readable. But it allowed optimum performance of the computationally intensive and most time-consuming part of an operational applications suite.

Creating this 32-bit version of the code from the standard 64-bit Fortran was not as tedious as writing assembly code. The process could be automated. And that is, indeed, what the late Bob Carruthers, who went on to work for Cray Research and ECMWF, contributed: A program to 'translate' Fortran floating point arithmetic to Q8 calls of 32-bit operations. Running the 'translator' with the vectorized Fortran source code as input, required only minor tweaks for some rare end-cases to get the error-free 32-bit source code. Another member of the team was George Mozdzynski, who would go on to work for many years in senior positions at ECMWF.

As the reader might have expected by now, Control Data won that U.K. Met procurement. Serial 1 of the Cyber 205 model was installed at the Met Office and became operational in 1982. It served the center for the next 9 years.

Of course, by this time the implementation has given up any pretense and any hope of portability, but for the possibility that someone could have written a 'reverse translator'. Turned out, that did happen about 10 years later. I got the story from Deborah Salmond who worked for Cray Research in the U.K. from the mid '80s as benchmarking/performance engineer and at ECMWF from the late '90s maintaining and optimizing the operational model. Cray bid for the replacement of the Met Office's Cyber 205 around 1989, and needed to run the model which was written in the Q8 special calls style. To achieve that the cray engineers developed a convertor utility they called *Lifeline*, which is believed to have started by Alan Dickinson from the Met Office, who went on to become the institute's director of science and technology. They estimated that the tool automated about 80% of the conversion from Q8 calls to regular Fortran. This may have been the only major use of the convertor, and it was not sufficient for Cray to win the bid. Though, as the twisted story goes, when ETA Systems went out of business shortly after, Control Data pulled out the ETA-10 at the Met Office and replaced it with a Cray Y-MP.

Given the short lifetime of the Cyber 200 and its successor ETA Systems, the moral of the story is that having to resort to extraordinary efforts to win a benchmark may, to use a military analogy, win a battle, but not the war. More on performance metrics and benchmarking in Chapter 31, which is dedicated to the topic.

Fast Forward 40 years

The Cyber 205 was rated as having theoretical peak performance of 200 megaflops (at 64-bit arithmetic; 400 megaflops at single-precision). Installing it at the Met

Office delivered a peak performance jump of $15x$ over its predecessor, an IBM System/360 195. The U.K. global weather model's grid spacing was only halved (from 300 km to 150 km) and a 50% increase in number of vertical levers – to 15. This matches well with the estimate of computational increase when halving the horizontal grid, since more physics content was also added to the model. It was replaced by a series of systems from Cray Research, NEC, and back to IBM, and Cray.

The latest (as of 2021) system at the Met Office, are three Cray XC40 installed in 2015. Together they rate at 16 petaflops theoretical peak performance. This figure is 80 million times the 1982 Cyber 205's peak performance. The current Met Office's global model runs now at grid spacing of 10 km and 70 vertical levels. That would account for about 10,000 times more computations if the same 1982 model was used now. Of course, there are more computationally intensive physics and chemistry processes now, more forecasts per day, more regional and local runs. And only one of the three systems is dedicated to operational runs. The other two are used for research, development, and climate modeling. For more see the "High performance computing" section in [17].

ECMWF stayed with Cray Research as their supercomputer provider through six generation of Cray systems for close to 20 years, until 1996.[6] Fujitsu was the supplier of the next four system, followed by a long stretch of IBM systems from 2002 till 2014, when a dual Cray system (the Cray XC-30) became, again, the flagship compute engine for the organization.

ECMWF documents the progress made over the years by the increase in the number of the model's grid points relative to the more relevant system's *sustained* performance. In 1979 the Cray-1 was rated at about 10 megaflops for the model that had 200,000 grid points. 35 years later, the ECMWF system performs at 300 teraflops on a model with over 200 million grid points. The sustained performance went up by a factor of 30 million, while the grid's resolution went up "only" by a factor in the order of a couple of thousands. The difference can be explained as a combination of factors: More feature-rich model, higher fidelity of forecasts, more time spent on data assimilation, more operational runs per day. For a good summary of ECMWF's history of their supercomputers progression see [18].

At the time of this writing the future plans of both the Met Office and ECMWF regarding their computer systems are known. ECMWF will switch their vendor again. For the first time it would a European supplier – Atos is contracted for a Bull system with a price tag of just a little under $100M$. The organization is also moving its data center from Reading in the U.K. to Bologna in Italy. And it will open an additional office facility in Bonn, Germany.

The U.K. Met Office went further with a larger departure from past norms for a weather center. It signed an astonishing 10-year $1B$ deal with Microsoft to provide its operational forecasts via multi-site, cloud-base computing using Cray systems.

---

[6]Coincidentally (or not), in 1996 SGI acquired Cray Research.

Both these esteemed centers have come a long way since the late '70s when the largest supercomputers available went for under $10M$.

We will return to discuss the current state of weather and climate modeling, in the context of HPC, in Chapter 27.

II

The Epoch of Multiprocessors

Dawn of Parallel Programming

CHAPTER 6

Macro Parallelism
Multi-Vector Processors

As vector processor technology was pushed to the limits possible at the late 70s, another approach is being superimposed. Now that it has been established how to stream, or pipeline, operations that are independent of each other, why not take a large set of such variables and divide them among several processors? The idea is that we can multiply the performance derived out of the system by placing multiple processors within the system. They would run under a single control and a single operating system. Thus we entered the age of *coarse grain parallel processing*. Of course, the system can also run multiple independent jobs.

Early Work

For the sake of historical accuracy, it should be pointed out that the idea of using multiple processing units for high performance computing originated earlier. Here are two cases:

The better known of the two is the ILLIAC IV.[1] Obviously, the "IV" means there were earlier versions. For HPC, however, the ILLIAC IV is the relevant reference. It is the design that had 64-bit processing elements (previous designs were based on 1-bit elements). This is a good place to describe this one-off design since it was a stand-alone system (even if accessed through a front-end computer) like a vector processor, and it had multiple processing elements like an array processor (discussed in Chapter 8). Even though the design line did not continue beyond the single machine that was built, some of its architectural and programming concepts survived and contributed to what would become mainstream HPC.

The development of the ILLIAC IV started in the mid 60s by a group led by Daniel Slotnick. The version that was eventually built, by Burroughs Corporation,[2] had 64 processing elements (PEs) and was delivered to the University of Illinois at

---

[1]ILLIAC stands for Illinois Integrated Automatic Computer.

[2]Burroughs later merged with Sperry Univac to form the company we know today as Unisys.

DOI: 10.1201/9781003038054-6   **59**

Urbana-Champaign in 1970. It was later moved to NASA Ames Research Center in Mountain View, California [19].

While visiting Lawrence Berkeley National Lab (LBL) in early 2020, a fews weeks before the start of the COVID-19 lockdowns, I was fortunate to meet Kenichi (Ken) Miura. Ken had a long career in HPC. He is Professor Emeritus of the National Institute of Informatics in Tokyo, and was a Fellow at Fujitsu Laboratories until a few years ago. I have known Ken since the early '80s and during his many years at Fujitsu America. What I found out at the Berkeley meeting is that Ken was also a member of the team that developed the ILLIAC IV. He recalled the reason the system was moved from the university campus in Urbana-Champaign to California. These were the Vietnam era days of great upheaval and student unrest on campuses. The ILLIAC project was funded by DARPA. It was felt the machine was not safe on campus.

This one-off system was not considered a great success. It was plagued by massive cost overruns and delays. The result was a machine a quarter the size of what was originally planned for. It was used by only a small number of people. Nevertheless, lessons learnt in those early days laid the ground for parallel processing as we know it today. A small example: Red-Black ordering technique was mentioned as very useful for vectorizing iterative solvers (more on that in the discussion of co-design teams in Chapter 23). Well, Ken Miura tells me: "This was first done on the ILLIAC IV. I wrote a program with this scheme."

Perhaps less known, an even earlier multi-processor design was an actual product (not a one-off) before the ILLIAC IV. It was called the CDC 6500, and this is how it came about: Several years after co-founding Control Data, Seymour Cray's designed the first of the "6000 Series" – the CDC 6600, launched in 1964. At 3 megaflops, it was the fastest computer of its time. The '6600' was also a revolutionary design in that it had 10 Peripheral Processors (PP) supporting the central processor that itself had additional parallelism through its 10 functional units that could execute concurrently. Cray, known for his lone, single-handed, later designs, collaborated very closely with Jim Thornton in those years[3] Together, after finishing the '6600', they engineered a variant, called the CDC 6400 that was completed in 1966. It was a slower version of the '6600', with a slower clock and a unified functional unit for arithmetics (no parallel execution of instructions). Of course, this made the '6400' considerably cheaper.

Now comes the interesting part: It was mostly Thornton (Cray was already busy designing the CDC 7600) who took the '6400' and architected a dual-CPU version. This was the CDC 6500. With its 10 PP processors, it consisted of 12 processors, and was liquid cooled, unlike the single CPU '6400' that was air cooled. One of the CDC 6500 systems was installed in 1967 at the U.S. Navy weather center, Fleet Numerical Weather Central (FNWC), known today as Fleet Numerical Meteorology

---

[3]The Computer History Archive keeps an online version of a wonderful book by Thornton about the design and applications of the 6600 ([20]).

and Oceanography Center (FNMOC). They added a second one, and we return to what FNWC did with the dual 6500 in Chapter 7. An actual CDC 6500 can be viewed at the Living Computers Museum in Seattle, Washington.

Cray and Thornton went further and created a dual-processor system with one '6600' CPU and the second a '6400' CPU. This was the CDC 6700 that came out in the late '60s and was the most powerful in the 6000 Series.

The Multiprocessors of the '80s

While the ILLIAC IV and the CDC 6500 and 6700 were still in operation but not much in the public eye, the world of supercomputing turned to vector processors. We had to wait till the 80s for the return of multi-processors. This concept is with us since.

The best known extension of vector processing to multi-processors is the Cray X-MP. Computer architect Steve Chen, then at Cray Research, built on Seymour Cray's Cray-1 design to create the Cray X-MP. There were numerous improvements to the vector processor itself and its memory system, but the more impactful innovation was architecting two, and later on four, such vector processors into a single system. A single operating system controlled all the processors. Each processor had a (theoretical) peak performance of 200 MFLOPS. The maximum configuration of 4 processors reached, at 800 megaflops, close to what we can call the gigaflops era.[4] The Cray X-MP was the most powerful computer when it was launched. It was also a great commercial success for Cray Research. In the tradition of both Cray and Chen's designs, and unlike the dreary square boxes of the past and most present days designs, the Cray X-MP looked beautiful (Fig. 6.1).

The Cray X-MP was just the first of a series of systems made up of multiple vector processors. A somewhat competing design came from Seymour Cray. It took about 10 years from finishing the Cray-1, twice as long as the 5-year cadence of Cray's previous designs, for the launch of the Cray-2. This was not entirely due to technical difficulties, though, there certainly were those too. In 1979 Cray resigned his official position at Cray Research and became an independent consultant to the company and moved his lab to Colorado. The Cray-2 can be seen as Seymour Cray's successful 4-CPU design years after his first attempt, while still at CDC, at 4-CPU design to succeed the CDC 7600. It was dubbed the CDC 8600, but the component technology at the time was not up to the task of packaging necessary for the job. It was then, in 1972, that Cray left CDC and founded Cray Research. The Cray-2 was launched in 1985, and at 1.9 gigaflops peak performance was the first truly over-gigaflops system. It was a very compact design, necessary to achieve what was the fastest cycle time of its time – 125 MHz (or 8 nanoseconds). And this required a special liquid cooling of all the modules to extract the heat generated in

---

[4]The CDC Cyber 205 with 4 vector pipes (not 4 processors) reached the level of 800 megaflops earlier – but only at the reduced precision of 32-bit operands.

Figure 6.1: A Cray X-MP that was installed at NCAR. Source: NCAR.

the dense circuitry. The design was also esthetically pleasing as can be seen in Fig. 6.2.

The Cray-2 was not as successful commercially as the Cray X-MP. About 25 units were installed at customer sites. Its strength was having a huge memory for its time. Its size was counted in 'words' as was the custom then: 256 megawords. That is 2 gigabytes. The size made the latency of memory access longer, but for applications with very large datasets it was still far superior than having to swap data in and out of storage devices.

While Seymour Cray's Cray-2 was a departure from the Cray-1 design, Steve Chen continued to build on the Cray-1 when he designed the Cray X-MP. This

Figure 6.2: A Cray-2 that was installed at NASA Ames Research Center. Source: NASA.

continued when Chen developed the Cray Y-MP after finishing with the X-MP. The Y-MP had an improved processor, with faster clock. And it could be configured with up to 8 vector processors. Thus, its peak performance, at 2.67 gigaflops, was more than 3 times that of the X-MP. When the Y-MP came to market, in 1988, Steve Chen was no longer with Cray Research. This part of the story will be told in Chapter 11.

The trend of integrating ever more vector processors into a single system continued at a faster pace. The Cray C90, based on the Y-MP architecture, with a 35% faster clock, could be configured with up to 16 processors. It was put on the market in 1991 – a mere three years after the Y-MP.

Another less known example of multiple vector processors, but no less interesting, is an IBM project. Enrico Clementi, an IBM Fellow at the time, needed more compute power than was possible on a single IBM 3090 with vector facility for his computational chemistry research. So, he found a way to connect together four such systems to jointly work on a single computational job. The system was known as "lCAP 3090" (lCAP or LCAP, for short), for Loosely Coupled Array of Processors. This was a one-off project with no further product impact, but was a noteworthy application of multi-processors. I tell its story in Chapter 7.

The focus here is on computers developed in the U.S. But, contemporarily, there were three Japanese companies that produced multiple generations of vector processors. They are Fujitsu, NEC, and Hitachi. They produced competitive systems compared to the ones from the U.S. In particular, the technological foundation they build, and the culture of government-business collaboration in Japan, produced the world's top systems at least four times: The *Numerical Wind Tunnel* (1993, see also page 112), the *Earth Simulator* (2002), the *K computer* (2011), and *Fugaku* (2020). Amazingly, these top systems came out at the constant cadence of nine years. All these systems where developed in cooperation between government agencies and labs and computer companies – Fujitsu for NWT, K, and Fugaku; NEC for the ES. The scant mention of Japan's considerable contribution in the HPC space is due to this book's focus on the U.S. experience (I certainly skip over important work done in Europe).

Macro Parallelism

The multi-processor design was the HPC computer architects' way of adding performance faster than what was enabled by component technology. Or, in other words, much faster than the pace according to Moore's Law. That was the hardware's answer. To realize the new-found potential the system software needed to be revamped so that the operating system can manage several processors simultaneously. The application had to be provided with the tools for breaking it up across processors, if the performance gain was to apply to a single application and not just for increasing throughput.

The software techniques that were developed in that period of the '80s for running a single application across small number of processors are the foundation for future software tools and techniques in use in today's clusters on a much larger scale. The 'standard' software tools were created in the '90s, and continue to evolve to this day (see also Chapter 17). Beyond vectorization, the application had to be spread over multiple CPUs that, generally, shared a single memory system. That is, each processor could access the whole memory directly (that was not the case for the lCAP system). This required a new way of thinking about the parallelization process.

Vectorization involves identifying *independent* identical operations that can be performed on elements of arrays. This a process that examines the lowest-level details of the program – the innermost loops, or the 'kernels' of the code. When considering how to divide an application between processors we need to think in terms of big chunks, taking a top-down look at the code. Now we don't deal with a single instructions stream. Each processor manages its own instructions stream. Another significant change, and one that turned out to be troublesome at times, was the addition of inter-processor communications.

Two main methodologies emerged: *Task Parallelism* and *Data Parallelism.*

Let us dispense with task parallelism first. Spoiler alert: It is rarely effective. It is reasonable to think of this "divide and conquer" idea of multi-processing in terms of assigning different parts of the program's routines to different processors. Basically, this is dividing the *code* into several parts. The pieces of code assigned to different processors have to be independent of each other in order to perform in parallel. Think of a weather model where we can split the dynamics part from the physical processes. It can be set up so the processor doing the dynamics computes the state of the next time step, while the physics part of the current time step is computed on another processor. This time-staggered procedure can be extended when the model is coupled with an ocean model (that often has a longer duration for the time step), and possibly land-ice and sea-ice components. It is also possible to find some pieces of code that can be done simultaneously within a single step, but likely with frequent stops for synchronization and passing on new values to be used. It is a complex process to execute.

And there are other practical issues with task parallelism. For me, the top four are:

- *Identifying parallel sections.* Even in codes organized by routines that reflect separate aspects of what is simulated or computed, it is difficult to group such routines such that all the routines in one group will be independent of all the routines in all the other groups. After all, then, in the '80s, the tens or hundreds of routines had to be grouped into small number of groups (for the, typically, 4 or 8 processors of the system).

- *Load balancing.* Arguably the strongest objection to task parallelism is the fact it is next to impossible to divide the amount of computational work equally among the processors. Every task, or part of the code, is doing something different than the others. In a typical multi-step simulation much of the resources available would be idle some (or much) of the time, waiting for the longest-running task to complete its apportioned work.

- *Tasks accessing the same data.* All the program's data is available to any of the separate tasks, so tasks may compete for access, which would cause a performance bottleneck.

- *Scaling to more processors.* Once the segments of code are divided to a fixed number of tasks – equal to the number of processors on which to run, it is not possible to change that number without starting the parallelization process from scratch. This makes for an inflexible implementation.

While in task-based parallelism the code is broken among the processors, in *data parallelism* the data sets used by the code are divided. each processor gets the entirety of the code, or at least all the code that can be parallelized, but only a fraction of each of the data arrays. This approach overcomes the issues encountered in the task-based method. The arrays on whose elements the same operations are done are simply partitioned equally between the processors. The load on each processor is perfectly balanced. Each processor touched only its own data – there is no congestion. And the procedure for dividing the data scales naturally.

In both *task* and *data* methods there may be serial sequences that can be run on one processor and communicated to the others or run on each processor(redundantly, but saving some communication overhead). Synchronization instances occur in both methods, but less frequently for task-partitioned codes. A perfect task-parallelized code will fork out after the starting setup, and join back for the final output (or at the end of a time step).

Synchronization is more complex for data parallelism. When an array is split up among processors, the boundaries of each sub-array are determined by values at the edges of other sub-arrays. This means the code in each processor has to 'know' which processor holds which side of the 'neighbors' to its own arrays' boundaries. And at some point (or points) of each iteration the processors have to be in lockstep and exchanges the values of some of the elements.

Expressing the inter-processor communications and exchange of data was central to the evolution of tools for parallel processing. I return to the topic, with the help of several contributors, throughout the book.

CHAPTER 7

Making Use of Multi-Processors

Examples of Applications on MP Systems

I n the previous chapter we saw some instances of configuring a small number of powerful vector processors to operate as a single system. Another approach to applying multiple processors was to attach coprocessors of many small processing elements to host computers. Both approaches meant the user has to find and develop software and coding techniques that enable all the system's processors to work in tandem on a single demanding computational task.

Partly paving the way, and partly a demonstration of an approach different than that of the integrated multi-processor systems, were a couple of one-off configurations that deserve a mention. One is a weather modeling system from the '60s, the other a computational chemistry project in the '80s.

Fleet Numerical Weather Central of the Late '60s

In my early days with Control Data, at the start of my work on the U.K. Met Office future procurement, I was taken to a visit at what we called *Fleet Numerical* in Monterey, California. We went there to talk about their future Cyber 203 and 205, but while there were told about their past CDC's systems. The one that stuck with me was that of a dual CDC 6500. The subject of early applications of multi-processing brought back that fuzzy memory.

As is the case in other countries, the U.S. military has its own weather centers for tactical and strategic reasons. The navy used to have its main computing at FNMOC (Fleet Numerical Meteorology and Oceanography Center, as it is known now). It has been there, previously called Fleet Numerical Weather Facility, for about 60 years. Though my focus in this book is on the civilian weather and climate

DOI: 10.1201/9781003038054-7

organizations, I bring up FNMOC for its unique, and little known, distinction in the history of weather models.

In Chapter 6 we described the CDC 6500 and mentioned that the organization now known as FNMOC acquired first one (in 1967), then a second '6500' (in 1969). Recall that each of these systems had two processors. What the scientists and programmers did with these systems was remarkable. They developed the *first ever production weather model utilizing multi-processors*. But this was not only a dual processor implementation. They hooked up the two independent systems into a *four-processor* configuration and ran a 4-processor model in 1970.

I reached out to find out more and found a historical perspective that mentioned the *6500 episode* ([21]), and through intermediaries was introduced to its author – Tom Rosmond. Rosmond worked as a civilian research meteorologist for the Marine Meteorology Division of the Naval Research Laboratory (NRL) in Monterey for 30 years, retiring as an NRL senior scientist. The NRL scientists in Monterey worked closely with the FNMOC staff, and though Rosmond arrived in Monterey a few years after the CDC 6500 period he worked with people who were directly involved in that episode. He told me more about this very early MP application:

"Philip ("Pete") Kesel was a young naval lieutenant and a graduate student at the Naval Postgraduate School in Monterey at the time. His master's degree project was to develop and code a northern hemisphere forecast model. When he finished he was assigned to Fleet Numerical and he brought his model with him. He put the model on the CDC 6500 but, running on a single processor, it was too slow. It took more than an hour for each day of forecast. It was too slow for any sort of operational schedules – the required three-day forecasts, twice a day, would have taken between three and four hours.

"At Fleet Numerical, Pete teamed up with Frank Winninghoff who was a civilian member of the IT staff, and said, *We have to come up with a way of using all four of these processors so we can run this model operationally in a timely fashion.* And so they set about seeing what was going to be required to do that. Obviously, there was a lot of things that were not available in those days to do this sort of thing. They realized they were going to have to come up with some sort of a file locking mechanism so processors could share a file system. Some sort of synchronization so that the processors could march in lockstep. And they were going to have to basically design all that from scratch because there was nothing in the Fortran language or the operating system that would allow it. They concluded they had to make some operating system modifications. And that's what they set out to do.

"The CDC 6500 had an Extended Core Storage (ECS) on which resided a shared file system that all four of the processors could access. However, they had to come up with a way of locking parts of that file system so processors didn't compete for the same space at the same time. And it was basically set up as a master-slave arrangement where one processor was the gatekeeper that told all the other three processors when they could have the file they requested. When permitted, the other processors would write to it or read from it, and when done would send a message back that said, *Okay. The file is open for other users.* They accomplished the design by inserting some changes to the operating system where they set bits at various places."

Kesel and Winninghoff worked on this very early parallel version of a weather model in 1970. It took them 3 to 4 months to deliver a parallel production model. They describe the model and its parallelization in a 1971 paper ([22]). Three aspects had to be addressed: How to partition the model; how to get programs on different processors to communicate; and how to device a synchronization mechanism among the processors. (The second and third items are described in another paper that includes additional authors involved in the implementation – see [23].) It is interesting to look at the approach to parallelization taken back then.

As mentioned by Rosmond, a pivotal component of the system was the extended core storage. Firstly, because it was directly accessible by all four processors. And secondly, as important, this is where all the data resided even as it was manipulated by the processors because the local fast memory was too small and was able to hold only the immediate slice of data being computed.

Given that the device holding all data was also the only path of communication between the processors, it is, in hindsight at least, surprising that the partitioning scheme chosen was not one of simply splitting the data and letting each processor executes the same code – what we would call today *domain decomposition* or *data parallel*. That is not the strategy taken. Though, in [22] Kesel and Winninghoff list the data parallel approach as a future action item. Interestingly, forced to consider it only as a way to increase the model's resolution. Without partitioning the data the 6500's local memory would be insufficient.[1]

The partitioning strategy taken was what would be later called *task parallel*. Kesel and Winninghoff divided the code among the processors; not the data. The operational job flow has three phases: input and initialization, running typical time steps, and the output phase. The first and last were partitioned onto 3 processors, but the time-consuming part of advancing the forecast by repeatedly computing the next step was divided 4-ways. Two processors took on the east-west and north-south momentum equations, the third did the thermodynamics equation, and the fourth computed the moisture equation. They could do it all in parallel because the difference scheme used required only reading the previous time step values.

Remember, there was no MPI (or any other messaging library), nor processor interconnect or a parallel file system. The scheme above requires that at all times all four processors work on values from the same time step. The synchronization mechanism required added functionality to the operating system. This was done via a couple of routines that were added to the 6500's peripheral processors which

---

[1]I could not find out whether a data parallel implementation was done before the 6500s were decommissioned and replaced by more powerful single-processor systems.

handled the I/O of the system. The changes allow one processor to establish a "master" status. It kept track of which block of data was assigned to which processor, and when it was released. It held advancing to the next time step until all the processors finished their task on the current one. In [23] the authors offer details of their creative way of devising a method for inter-processor communication and a fail-safe synchronization mechanism using what they call *Buffer File Method*. The buffer file was a set of blocks organized as a uni-directional ring. When a program finished processing a block of data it would place it in the buffer file, and only then another program, waiting on this data block, could read and manipulate that data. (There is much more to the scheme, that can be found in [23].)

The resulting performance improvement was impressive. The 3-day forecast took 3 hours (184 minutes) on a single processor. The parallel 4-way version completed tree-fold faster – in 60 minutes! Think of it as a speedup of 3 out of possible 4. Or, as 75% parallelism efficiency. This is a remarkable achievement given the overhead of the synchronization scheme, and the use of task-based partitioning where careful load balancing between the processors is not possible.

A possible explanation of the effectiveness of the project is, ironically, the small local memory of the system. It forced the data to reside on an external device that was directly accessible by all the processors. In other words, data was not exchanged. It just had to be accessed in order. The I/O time was significant for the single processor version – it also had the data on the ECS, and the different tasks accessed different arrays much of the time, so the I/O time was also cut down appropriately.

The methods the scientists at Fleet Numerical had to use to expose parallelism do not resemble today's tools. For one, the systems' architecture is so different. But it is interesting to observe that we didn't get to where we are today in our first attempt.

The Enrico Clementi Project

A later example of connecting several host-coprocessor pairs together was the project led by IBM Fellow Enrico Clementi that was mentioned before (see page 63). This work was done in the mid '80s and was contemporary with the early days of Cray multi vector processors. It resulted in what was known as the lCAP system. Logically, we can think of it a precursor to a cluster of servers, each with a GPU accelerator. The lCAP story is entwined with that of Floating Point Systems (FPS), the story of which is told in Chapter 8, and also includes Prof. Ken Wilson from Cornell University, the recipient of the 1982 Nobel prize in physics. I talked to Shahin Khan, who is today an HPC technology analyst and commentator after a long career in HPC in several companies, and then a student at Cornell. He recalls:

"When the FPS-164[2] came out and Prof. Ken Wilson got one at Cornell, Enrico Clementi, an IBM Fellow, was also looking at buying an FPS-164. And because Cornell had one, he was looking to recruit scientist types as summer interns from Cornell. He had written to the head of Cornell Academic Computing. I managed to see that letter and contacted Clementi, and after attending an FPS programming training session at Clementi's lab got the job. Clementi's vision was to assemble a bunch of FPS-164s together and split the codes into parallel chunks. He himself had been behind a code called, IBMOL which was an ab-initio quantum chemistry code. IBMOL was clearly a target code for this parallel system.

"A computationally intensive component of the code, the Self Consistent Field (SCF) calculation, was a candidate routine that could be carved out. The SCF part was migrated to the attached array processor (the FPS system) which then became a workhorse for the department. Soon after, Clementi bought more FPS-164s until we had ten of them. Each of the attached processors was connected to one of two mid-range IBM mainframes. However, the IBM 3090 was just around the corner.

"That was the beginning of writing a parallel code. We needed some systems programming work on the mainframe so we can parcel out pieces of the app to different virtual machines each of which would then communicate with their corresponding attached processor, and then reassemble the results back after receiving them. Some library routines were written, which were akin of some to what MPI (the Message Passing Interface library) does these days."

Khan reminisces on how things were more 'primitive' on those days and how the system got its name:

"The success of that initial activity was becoming known in computational chemistry circles, and gradually in other parts of the HPC market. Clementi was invited to talk about the system at some conference, and we had to figure out how to use it to make presentation slides. This was before PowerPoint apps on PCs, so we used a Tektronix graphics terminal (a storage tube) that was sitting in the lab, with a joystick, a plotter, and a thermal printer that would capture what was on the screen. Someone had written the code for that system to give it access to fonts etc. Clementi would bring his sketches and slides and I would put it on the Tektronix and he would stand behind me and direct the process. It was a very rudimentary slide system but beat having to wait for professional graphics and camera work which would at best take a few days for each iteration. And we needed a name for the project and system. *Coupled processors* became *loosely coupled processors* and then *loosely coupled array processors*. And I said: 'you know, coupled array processors doesn't include all the mainframes and other processors that are out there. What if we call it *loosely coupled array OF processors* and that became the formal name, with lCAP as an acronym with a nod to El Capitan in Yosemite."

And the project evolved and grew:

---

[2]The FPS-164 is an array processor from FPS.

"lCAP was used initially for computational chemistry because the lab already had strength in that area. But then Clementi started a visitors program where he had scientists come from around the world and bring their code. We would port the code to the system and see how it ran. That became an avenue for getting additional types of code, and before long we had the geophysics folks and others. And so the code base expanded well beyond computational chemistry. This also fit well with Clementi's very ambitious and pioneering "Global Simulation" vision that would look at the whole scientific and engineering workflow.

"Clementi's lab was in Poughkeepsie, New York in one of the original old IBM buildings, in a beautiful setting. By the time we got the 3rd FPS-164, we were running out of space. FPS had just announced the follow-on product, the FPS-264 which more than tripled the performance to a peak 38 MFlops, and new mainframes were coming. At that time, IBM was just building a brand new campus in Kingston, New York, about 30 minutes north of Poughkeepsie. There we could get a very large data center with enough room to add more systems and people.

"With the pending arrival of additional systems, including the FPS-264s, everything moved to Kingston. In short order, there were first 10, and then 20 attached processors connected to 3 mainframes with lots and lots of disk drives and it became a significant supercomputer center for its time. The system in Kingston ended up being 23 attached processors and 2–3 large mainframes, a pretty significant supercomputing capability. The system architecture also became a blueprint for the Cornell national supercomputing facility in the late '80s, one of the original 5 NSF national supercomputing centers."

The lCAP project continued to expand and evolve into the late '80s. There were ambitious designs that did not get approved because of a combination of budgetary constraints and some inside turf infighting. One of the design casualties was a way to connect multiple systems via a shared memory system. The final version – called lCAP-3090 – had a different set of building blocks: A set of 4 IBM 3090-400, each made up of 4 vector processors, making it an interesting 16-way parallel system with 4 distributed memories. The lCAP experiment is captured by Clementi and his colleague Gina Corongiu in an article written a few years after their departure from IBM (see [24]).

It seems appropriate to summarize with Khan's appraisal of Clementi's vision as it relates to how HPC should be applied:

"Parallel processing was a big component of the lab. When I first joined there were about a handful of people and then it grew up to about 40-50, mostly research scientists from around the world. The topics of research also expanded from ab-initio quantum chemistry to include Monte Carlo and molecular dynamics simulations as well as fluid mechanics. Clementi had a grand vision, definitely ahead of the times, to create a complete scientific and engineering workflow. He called this "global simulation" and he wrote books and integrated software to pursue it. We used to joke that at one end of the simulation you just entered the atomic number of the elements and out came a car at the other end. His ideas are just starting to become more common, but, in fact, part of Clementi's vision was that we can start with the initial calculations and come up with the materials needed, simulate the materials into what needs to happen downstream, and then add fluid dynamics and structural analysis to it, and so on. The idea was that we can devise a pipeline or workflow of computations that would start with basic science and would become progressively more engineering-based, in stages that are all connected, with compatible data formats and such that the codes would hand each other data in the proper way. This vision is starting to look real as people now see the full workflow and can focus on overall productivity with a business metric attached."

One of the consequences of IBM's difficulties and restructuring of the early '90s was the closure of the IBM Kingston facility, and Clementi's lab with it (another consequence relevant to HPC history is mentioned in Chapter 11).

Multitasking on the Cray Multiprocessors

The most common, and enduring, use of multiprocessors (MP) in HPC in the '80s was on Cray systems. Cray Research had to develop tools that will assist the application programmer in using multiple processors by a single application code. In those days the hardware and software stacks were still propriety and the system vendor had the sole responsibility of providing such tools (with occasional assistance from early users from within government labs).

The problem in the abstract was one of exposing parallelism in scientific computing codes. The Cray terminology for this in those days was *multitasking*. This was not entirely new. Exposing fine-grain parallelism was a known practice at that time, and was done to create vectorized codes – pipelined, but made up of independent operations, as if done concurrently.

The MP architecture added another layer to consider. This one of actual concurrency between distinct processors, which means between separate instruction streams. Chapter 6 concludes with a discussion of *macro parallelism*, and here we refer to it with its Cray naming conventions. Breaking up the code so that each processor computes a *different* task was called *macrotasking*. Just as we saw with the CDC 6500 at Fleet Numerical some 15 years earlier, this was the initial approach taken on the MP Cray systems. Soon after, the focus turned to partitioning the data among the processors with all executing the same code. This is really an extension of vectorization. The vectors are divided, not the code. Given that for this approach one looks at the finer details of the code, Cray aptly called it *microtasking*. Of course, serial ('scalar') parts of the code were done, somewhat redundantly, on each processor. Both methods – macrotasking and microtasking, require identifying and inserting breakpoints for synchronization. These are the places in execution all the processors had to reach before proceeding. There were, typically, more such synch points in micro-tasked code than when macro-tasked.

The compiler capability that was developed to automatically vectorize parts of the code, often with the help of user-supplied directives, was now extended to automatically micro-task codes. It would rely on the same analysis as for vectorization, but would have to add the synchronization aspect. This capability was referred to as *autotasking*. A review of Cray's multitasking and some results can be found in paper by a team from the Pittsburgh Supercomputing Center (PSC) that was presented at the very first ACM/IEEE Supercomputing Conference in 1988 ([25]).

The idea of directives-assisted multitasking in Fortran, and later in C, in a shared memory system would be taken up by the community in the '90s as part of the move away from the proprietary silos. It would become the OpenMP application programming interface (more on that in Chapter 17).

CHAPTER 8

Attached Processors, Microprocessors, and Mini-Supers

Precursors to GPUs and Challenging Large Vector Processors' Hegemony

THE '80s saw another architectural development that was adapted to, and adopted by, scientific computing. In addition to the use of multiple general-purpose processors, there was the idea of providing more specialized devices that can accelerate compute-intensive tasks. In particular, floating-point operations. Such a device can be attached to, and controlled by, a host processor. Fortunately, it did not have to be invented. The Signal Processing sector invented this technique already. A board was populated with processing elements and interconnected in a way that favors FFT (Fast Fourier Transform) computations. The thought was that this approach can be applied to numerical calculations in general.

These kinds of accelerators were referred to as *Array Processors* or *math* or *attached* coprocessors. The more correct term is *Attached Array Processors*, since later on there were systems of array processors that were considered stand-alone systems.

The story of attached array processors is similar to that of vector technologies in an important aspect: Both impacted HPC for a period of time (though, array processors had a lesser impact compared to vector systems), and then were left behind, only to re-emerge later on.[1] Vector instructions were added to cores of single chip server microprocessors, and array processors reappeared as today's robust General-Purpose Graphic Processing Units (GPGPUs). As Shahin Khan, who we

---

[1]A thorough discussion and comparisons of vector processors and array processors of the '80s can be found in Roger Hockney's and Chris Jesshope's book[26].

DOI: 10.1201/9781003038054-8

met in Chapter 7, has said: "The attached array processors were basically set up as an I/O device where a computing task would be 'written' to them by the host computer, and then on the receiving end the task would be executed after which the data would be 'read' back by the host. The math-coprocessor/attached processor approach was a prelude to GPUs and the whole accelerator era that we see now." We will return to these topics in later parts of this book.

The following narrative captures the volatility of HPC companies in the '80s and the '90s, accompanied by the changes to HPC system architectures.

The Story of Floating Point Systems

Floating Point Systems Inc. (FPS) was, arguably, the most prominent builder of coprocessors (or attached processors) for fast floating-point numerical computation for HPC applications. Founded in 1970 by ex-Tektronix engineer/executive Norm Winningstad, it was in the '80s that it became a player in HPC (until then FPS was focused on signal and image processing). After great success in the attached processors markets, the company betted its future on a product that was not successful. FPS nearly went under. Recovering, it bought a minisupercomputer company (Celerity), and shortly after was acquired by Cray Research. The latter was acquired by SGI, which sold the division that was the continuation of FPS to Sun Microsystems. It made Sun a player in the server market. Some years later Sun itself was acquired by Oracle, and soon enough the FPS remnants were gone.

Let me take a small detour from the attached processor narrative to tell the story of how and why FPS was nearly destroyed prematurely. It turned out to be a 'teachable moment' for me.

The FPS story intersects with my own. I joined FPS in 1985 following the spinoff of ETA Systems from Control Data. In the 18 months I was there, I witnessed FPS riding high with a successful 64-bit array processor, and falling hard as a result of placing too much of the company's future on a new product, called the "T Series" ("T" stood for Tesseract), which was a huge departure from its other products. It was an interesting and innovative design. A kind of a hypercube such that its nth degree consisted of 2^n nodes, each node a processor. Its connectivity allowed for many useful computational topologies – rings, meshes, and even those for all the stages of FFT. The processor was proprietary but used VLSI from third-parties Inmos (the transputer), Weitek, and Texas Instruments. The memory technology was borrowed from the video game industry. Each node had its own memory, which had to be accessed in chunks of 1024 bytes with predetermined boundaries. It was nicely suited for problems where all dimensions were a power of 2. In the general case, the control unit had to move data around – one element at a time. The chosen high level language for programming the machine was Occam, a somewhat obscure language that was co-developed with the transputer in the early '80s. John

Gustafson, who was a senior staff scientist at FPS at that time, described the T Series and how to program it in a 1986 article ([27]).

Gustafson, who would later become known for his formulation of Gustafson's Law and the idea of weak scaling (see page 255), told me about the origin of the T Series project: Ken Wilson (at Cornell university) managed to secure a generous, DARPA-sponsored, earmark from Congress for an experimental parallel machine. It was to support his interest in large scale QCD computations (see also Chapter 9). That was in 1983, when Wilson and Larry Smarr were lobbying for the creation of national academic supercomputer centers. And so it happened, as Gustafson recalls, that "Well into the project, Cornell pulled the rug out from underneath us by saying they would not accept a Congressional earmark. Their policy was always to compete for funding through the traditional peer-reviewed proposals and thought that by accepting an earmark they were telling the world they would instead start using cronyism to get money for Cornell."[2]

As pointed out, the T Series did not capture the market – all of its four customers acquired small configurations. I doubt many people remember it today. The signs were there. The vast majority of real-life applications do not fit the very specific data structure that matches the architecture. They also contain serial code that can not be handled well. Only artificially tailored tests performed well on the T Series. The design was too specialized and restricting. Then there is the use of Occam. The community was not going to switch from Fortran or C because of one company's choice.

Gustafson, who was intimately involved in the T Series project captures its failings thus: "We were assured by the engineers that the scalar performance of the T Series will only be six times slower than the vector performance. The actual product delivered was 1,000 times slower at scalar. Unbelievably bad balance. that was probably the biggest single reason for the failure of the T Series." And "There was a terrible delay in the transputer. It was actually two and a half years late. It meant that we couldn't develop much software and have it ready with the hardware. We had to just give raw hardware out to the customers that have been waiting for it."

But there is more to the story beyond the technical aspects. The architect of the T Series was not an FPS employee, but an external consultant brought in by the CEO. Its development received the highest priority at the time, at the expense of other, revenue producing, projects. Gustafson explains the use of non-company resources: "The reason for going to outsiders to do the design was largely for secrecy. The CEO was afraid people would find out we were building a hypercube of array processors and that it was so easy to do that they'd beat us to the market if they knew. As it turned out, Intel did beat FPS to the market, and they knew about

---

[2]This turned out to be a major blow to FPS, as it was seen to have misled its investors by declaring the roughly $10M deal as done. Prof. Wilson did not see it coming, according to Gustafson.

hypercubes from the same place we did: Geoffrey Fox and Chuck Seitz who built the first hypercube system – the Caltech Cosmic Cube."

The background story here is that the Caltech team first looked at the FPS array processors as the building blocks for their design, but eventually went with Intel boards (which were offered at great discount). The idea of constructing parallel systems with hypercube topologies was a sound one, as was demonstrated by several companies. More on that in Chapter 12.

Open criticism of the T Series project was not tolerated, as I learnt from personal experience. Immediately upon being hired by FPS, and before moving physically to its facility, I was asked to review a Tesseract's architecture design document and write a report about my findings. The requestor, a senior manager, wanted an opinion untainted by the company's prejudices. I delivered a few-page report that highlighted concerns about the design. The memory system, the reliance on close to 100% vector-parallel code, and the programming language support were found to be the major weaknesses. The report was read by only a handful of people, to the best of my knowledge. The VP at the department I was in decided to quash the report. Showing it to the CEO would have not changed plans and investments in a project in its final development stages. But it would have ended my short tenure at FPS. The internal hype was such that the company even claimed the T Series to be the world's fastest supercomputer in 1986 ([28]) if at its maximum 16,384-processor configuration. In fact, only very small n-cubes were ever built (16, 32, or 64 processors), with mostly synthetic kernel benchmarks for claimed performance.

This short experience and the following 18 months of watching a company deteriorate because an esoteric project was suffocating its successful products taught me and reenforced several lessons:

■ A product design needs to be based on real-life constraints, not an idealized application and parameters.

■ A successful product has to include an ecosystem that allows its integration into an existing market.

■ A product better come from within the company to match its experience and skillset. An idea imported from outside that deviates from the current company's trajectory and imposed by forceful management is rarely successful.

■ A company where criticism and open debate are silenced is incapable of course correction and doomed to fail.

Engineering resources for the follow-up products of the money-making array processors were starved. FPS was losing money (and people), their revenue and market evaluation plummeted. It was only the quality of FPS's in-house engineering, change of management, and a fortunate constellation of market forces, which allowed the entity to continue its existence in different forms.

I have witnessed a similar situation once more, this time at Intel. It was considerably less extreme for the company, but did cause some damage. I refer to the Itanium saga some 10-15 years later, and discuss it in the chapter on HPC at Intel (Chapter 18).

We now return to the rest of the FPS story, starting with the topic of where it excelled: Attached array processors.

Khan followed the FPS journey from when he was an intern working on early FPS products, while at Cornell University. It was then that I came to know him when we collaborated on some projects related to FPS and Cornell. He would stay with what was the original FPS group through all its transitions and ended this journey as an executive at Sun Microsystems. Here is his story:

"My initial exposure to array processors was at Cornell University where Ken Wilson, the Nobel Prize winner, and others like him outside of Cornell were looking for more computational power. He got an AP-190 Array Processor from FPS, which was attached to one of the IBM mainframes on campus. The system was designed for signal processing so it had a 38-bit word size, and did not have much programming tools for a scientist. Ken and his team actually wrote a Fortran compiler and job management software for it, called APEMAN, within the Cornell Computer and Information Technology group. That project was very successful. Ken Wilson even mentioned the work with Floating Point Systems in his Nobel prize speech. In fact, as far as I know, FPS is the only computer company that has been mentioned in a Nobel Prize acceptance speech.

"Such efforts showed that there was a market for these attached array processors beyond signal and image processing. But scientific work required higher precision and much better tools. This is very similar to what GPUs went through decades later.

"Either the success of the Cornell effort by itself, or possibly combined with other data that FPS had, caused them to say: *okay, there is a market for scientific engineering computation; but it needs to be 64-bit arithmetic.* FPS then developed the FPS-164 attached processor which was released in 1981. Whereas the previous array processors were rack mountable and designed so they could be ruggedized and installed through a hatch of a submarine or a tank, the FPS-164 was really a mainframe size machine. In fact, its backplane was too large for one big box, so the system came in 2 refrigerator-sized boxes that got bolted together on site to form a pretty wide box. Its theoretical peak was 11 megaflops in 64-bit arithmetic."

For the history buffs, there is still a readable specification document of the FPS-164 posted online. See [29].

The '80s and into the early '90s was a period of much experimentation and turmoil in the HPC marketplace. The 'Big Iron' vector processors were challenged by coprocessors accelerators. They were challenged by 'mini-supercomputers' that were 'Cray-like', less powerful, but with much better price-performance ratio.

Parallel programming on a small number of processors opened the doors to systems with larger number of less powerful processors, the server market, and the use of commodity CPUs. The microprocessors would take over as the processors building blocks in just a few years later (see Chapter 17).

It is interesting to document how the transitions described above are reflected in FPS' history. This story shows the turmoil and reshaping of the HPC market and its players. Khan continues:

"In 1985 FPS announced the FPS-264. The product and the timing of the announcement were brilliant because the 264 was way faster and significantly less expensive than people expected it to be. It really propelled FPS to the heights that it achieved. The attached processors of the 64-bit variety started becoming a bigger and bigger portion of the revenue stream. FPS was the darling of Wall Street for a while and was doing better and better until new stand-alone mini-supercomputers came along.

"Meanwhile, FPS products needed to hang off of a mainframe or a minicomputer which themselves were under attack by microprocessors. FPS needed a new strategy. They were dabbling with massively parallel systems in the form of the T-series. They had experimented with data flow architectures, played with systolic arrays, etc. So while all that hands-on experience made their engineering team one of the best in the industry, it wasn't leading to a successful system in the market.

"The solution came when FPS acquired Celerity Computing in 1988. Celerity had a modern Unix based system with a good vector processor. It was one of the early pioneers of the RISC model and was focused on number crunching. Celerity and a number of other companies were going after users of the DEC VAX that was the standard in minicomputers. All had their own custom CPU, and all fell on hard times because of the microprocessor. At the time of the acquisition, Celerity had a strong engineering team and a product that was ready to be launched, but had cut most of the rest of the functions. FPS, on the other hand, had a pretty impressive worldwide organization and market presence but its product line was weak. So, it was a marriage made in heaven. The Celerity product was launched as the FPS 500 series.

"The FPS 500 became a contender to compete with Convex and Alliant. The FPS engineering team already knew that they were not going to be in the CPU business and had to use an existing microprocessor. The choices for a server class microprocessor back then were really either HP-PA, Sun-SPARC, Intel i860, DEC Alpha, or SGI MIPS. Intel x86 and Motorola 68k were squarely PC class chips. They chose the Sun SPARC, adopting it for the scalar portion of the FPS 500. But the real motivation was to use Sun operating system Solaris and all that came with it including a very large application base. It would now become the FPS 500S ('S' for SPARC), run Solaris, and continue to offer vector coprocessors for the HPC market driven by FPS compilers and libraries.

"FPS 500S used an ECL [Emitter-Coupled Lgic] implementation of SPARC. ECL is complicated and the parts were late and FPS continued to get weaker and there was an economic downturn that made things hard, so by 1991 FPS needed help. Along came Cray whose own systems were too big and specialized for traditional commercial database applications but was being approached by large commercial customers looking for a 'business supercomputer'. FPS seemed like a great fit, a company that understood supercomputing but had a system that could run all the commercial apps you wanted. So from a Cray standpoint the acquisition of FPS was going to fill that gap and allow them to expand into the traditional enterprise IT market.

"The actual acquisition of FPS by Cray was dramatic, however, since Cray did not commit in time and FPS filed for bankruptcy, laying off everybody and preparing to deal with the aftermath before Cray came back and acquired 'selected assets' and rehired the re-assembled teams, forming what became the Cray Business Systems Division (BSD).

"Soon after the Cray acquisition the team started collaborating with Sun on both companies' next-generation systems. They licensed a new parallel packet-switched interconnect technology from Xerox PARC. Sun was having no interest in bigger or more complex systems, but Cray built a 64-processor system with the next-generation SuperSPARC CPU, and Solaris OS. And this time, there was no vector attachments to complicate matters any more since the HPC market was to be handled by Cray's existing and expanding product line. That SPARC system became the Cray Superserver 6400, or CS6400.

"Cray was moderately successful with the CS6400 despite go-to-market and channel challenges. Work was starting on the follow-on to the CS6400, again as a collaborative project with Sun. This time using what would become UltraSPARC and with new interconnect built by the companies themselves. In early 1996, that system was getting ready to be launched as the Cray Starfire.

"But then, quite unexpectedly, SGI announced that it would acquire Cray. SGI was a serious competitor of Sun. It had a brand new highly scalable product line, the Origin series, that it was just about to announce. It had no interest or use for the SPARC-Solaris (the FPS) part of Cray. Despite a lot of advice to the contrary SGI decided to sell that business to Sun.

"I believe the view from SGI was that this was not going to be a threat to them because they had their own Origin system about to be launched and selling the business would have economic benefits without any perceptible market impact. To this day there is a controversy about whether it would have been better for SGI to just 'kill' that SPARC product. One day after the Cray-SGI deal closed, it was in the middle of 1996, that system and essentially the FPS component of Cray moved over to Sun.

"The transformations of FPS – acquiring Celerity, then becoming a part of Cray, SGI, and Sun – is a pretty unique piece of computing history. What was Cray BSD quickly became the high end systems group at Sun. The Cray product became the Sun Ultra Enterprise 10000, code-named Starfire. And because the team that came from Cray had intimate understanding of HPC they became the lead group for Sun's HPC activity.

"Armed with a large global sales force and fueled by the industry rush to adopt SAP R3 and Oracle applications, which the system ran so well, Starfire did extremely well. *100 days, 100 systems, $100m* was a line that my group came up with that was used in Sun's earnings call, making Starfire a barometer for how Sun was doing as a company and the only Sun product whose sales were discussed publicly. The system went on to a $1B run rate and ended up with an installed base of over 5,000 units."

And on this high note, the remnants of a small pioneer engineering-driven HPC company from Oregon, having survived numerous upheavals, departed the HPC world with a server product supporting database environments.

Enter the Mini-Supercomputers and the MPPs

While the MP vector processors ruled the supercomputing world, a quiet revolution was taking place. Advances in process technology that offered higher degree of integrated circuitry enabled faster designs and higher performance than that of the one-chip personal computer processors. It was at that period, circa 1990, that the phrase "The attack of the killer micros" became in vogue. Coincidentally, the Unix operating system, written in the high-level language C and therefore easily ported, was there to quickly offer a full software stack on new architectures. As a result, the period from the mid '80s to the early '90s became either the "golden age" for innovation in scientific computer systems or a period of great confusion, volatility, and uncertainty in the HPC market, depending on one's perspective. Many start-up companies, and some established ones, engaged in offering more affordable scientific-numerical computing. They used the new hardware and software technologies to build systems that are a significant fraction of the performance of better-known vector processors, but with a smaller fraction of the cost. Roughly speaking, the 'sweet spot' of these minisupers was at about a quarter or a third the performance for about a tenth of the cost relative to the 'big iron' supercomputers. By one count ([30]), 35 companies offered what came to be known as *minisupercomputer* systems in a period shorter than 10 years. Not many survived beyond the '90s.

To understand the place of the minisupers in the HPC landscape we need to look at the low end of technical computing: the workstations. The early machines that can be considered the forebearers of the workstation concept were systems, introduced in the '60s, such as the IBM 1130 and DEC's PDP-8. They were too big to be placed on a desktop, but could be installed in an office environment and support a single user sitting by their console. In reality, these early systems were used as a departmental or even as a campus-level resource. That is, a shared resource in the form of one-user-at-a-time. The advent of the microprocessor and denser memory technology allowed both higher performance and the shrinking in size of workstations. The workstations of the '80s and '90s consisted of a display monitor and a keyboard on the desk and a modest size box, often called *tower*, at the side of the desk. They became truly personal workstations, and an identified market segment of technical computing served by most computer companies who built PCs, and some who did not (e.g., SGI and Sun).

The technical computing market of the '80s, then, had the supercomputers, with performance in the gigaflops range, at one end, and the workstations, of megaflops range, at the other end. With this performance gap of three orders of magnitude between the categories of products, there was a definite performance and price range that was not served. And the technologies for filling up the gap were present.

Two classes of architectural solutions emerged: the minisupers and the *massively parallel processing* (MPP) systems. Some lump both types of systems into a single category and call it *minisupercomputers*. In my view it is useful to differentiate between them. The MPPs had a greater impact on the evolution toward today's clusters, as we will see (Chapter 12).

The minisupers were a scaled-down version of the MP vector processors. Not scaled down in the number of processors, but in the power of each CPU. Arguably, the most commercially successful of the minisuper start-ups was Convex - founded in 1982 with Steve Wallach one of its two co-founders. Convex's central idea was to use mainstream semiconductor tools to produce an architecture similar to that of Cray. Their line of products, with up to eight processors, had lower performance than the Crays, but better price-performance. HP acquired Convex in 1995 and shortly after terminated the minisuper product line. About as successful as Convex was Alliant, which existed from 1982 to 1992. Its systems allowed for up to 32 processors, also with vector instructions capability.

Digital Equipment Corporation (DEC) has been a presence in technical computing, notably in academia, since the '70s. Its products were not in the supercomputing class and initially competed with IBM's mid-range systems. By the late '80s DEC produced the VAX product line that included vector instructions and multi-processors option. By that time, DEC were looking to position their offerings' capabilities well above those of the technical workstations while having a price-performance advantage over supercomputers. That put the VAX systems within the minisupers class, albeit with a much larger share of the market.

What is common to the minisupers of mid '80s to mid '90s is the use of microprocessing technologies to develop systems with 64-bit custom design CPUs and modest level of parallelism via multi-processing.

CISC vs. RISC

Propriety processors have their own instruction set. But they each tend to correspond to one of two instruction set architecture (ISA) methodologies; some would even say *philosophies*:

- CISC – Complex Instruction Set Computing

- RISC – Reduced Instruction Set Computing

This topic is of interest since it still applies today, though with a twist (see Chapter 17).

The idea of CISC is to define hardware instructions that perform a series of operations with the issuing of a single instruction. For example, a CISC *add* instruction would include fetching of the operands, performing the arithmetic, and storing the result. In addition, the repertoire included compound operations combining things such as loop control, mixing of addressing modes, recognizing data structures, and more. Clearly, the goal was for the hardware expression to closely resemble statements in high-level programming languages. The approach also resulted in compact assembly language codes, not much more difficult to code than the high-level languages. This all seems like a good thing, and was implemented early on by IBM on mainframes, and later in the x86 architecture of personal computer microchips. However, CISC also meant that instructions took longer to complete with varying execution times, and their implementation in hardware required more circuitry. Providing complex instructions also meant catering to a large number of possible useful combinations of basic operations and operand types. This, in turn, generated a very large number of instructions for CISC implementations.[3] An important drawback of CISC is that it made pipelining (vectorization) and instruction-level parallelism more difficult, if not impossible.

RISC ISA is based on breaking down high-level operations to their smallest logical parts. The *add* operation now would require two load instructions, followed by the performing of the addition out of the registers holding the operands, and then storing the result register to memory. It looks more cumbersome from a programming perspective, but not from that of resource utilization. The complex instruction would reserve and hold the *load* and *store* ports, as well as the registers and the functional unit for the duration of the instruction. In the RISC way, the *load* is freed as soon as the operands are in their registers. So is the *adder* unit as soon as it is done. This allows new *loads* to start in preparation for the next operation.

---

[3]Take, for example, the x86-64 chip, which is very common in today's servers. People came up with different counts, as treating instruction variants is somewhat subjective. That said, by one count (see page 3 in [31]) the ISA has 981 distinct mnemonics and 3,684 instruction variants, many of which exist there for maintaining backward compatibility and are rarely, if ever, used.

With RISC there is a smaller number of instructions, the machine language foot-print is larger, and the compiler has more instruction scheduling work to do. But more resources are used in parallel, with likely higher performance.

The term RISC was popularized in the '80s with several microprocessors developed in the Bay Area: MIPS, SPARC, HP-PA, and more. Interestingly, it was at IBM, in the era of CISC-based mainframes, that the foundations for the RISC architectures of the '80s were laid, by the Turing award winner and highly regarded computer architect John Cocke. Cocke's early work in the '70s led to IBM's RS/6000 workstations and the PowerPC processor. But in hindsight we can credit Seymour Cray as perhaps the first architect to 'go RISC'. Against the backdrop of the IBM CISC mainframes, he designed the early Control Data machines in the '60s as true RISC implementations (though before the days of the microprocessor).

A Parting Thought

The refresh rate and the cost base made microprocessors a lethal competitive option to custom CPUs. But systems based on them would require many more processors to match in performance. There is a saying attributed to Seymour Cray: "If you were plowing a field, which would you rather use? Two strong oxen or 1024 chickens?." This was the case for the specialized systems for supercomputing. It hit home the idea that breaking up a problem into too many small parts results in such a loss of efficiency that the distributed solution is rendered useless.

However, after several generations of microprocessors, the Cray's analogy would have been more suitably phrased as: "To plough a field every year, would you rather have two oxen that you can replace every five years, or ten donkeys that you replace every two years?." Now the answer is not all that clear cut. It is true that systems based on microprocessors were not upgraded every two years, but a five-year upgrade cycle now was an advance of two technology generations. Extrapolating forward it is pretty obvious the microprocessor will win this race.

The answer to the "ten donkeys" phrasing regarding the minisupers is unclear. Both the high-end big-iron supercomputers and the minisupers departed the HPC scene at about the same time – by the mid '90s. At least, as far as U.S. manufacturers are concerned (NEC and Fujitsu in Japan continued to produce vector processor class systems). Cray went through a period of producing specialized MPP systems, before switching to commodity server boards connected by a proprietary high-performance interconnect network.

Eventually, as it turned out, the great Cray was wrong even in the "thousand chickens" sense. The high-end systems of today are made up of hundred of thousands of processing units, as we will see in Part IV.

CHAPTER 9

Studying the Standard Model

QCD on supercomputers

Q UANTUM Chromodynamics (QCD) is an area of computational physics I was
personally involved in. This allows me to relate, from personal experience, to
the role computations play as the third leg of science by complementing Theory
and Experiment. By the early '80s I was no longer a practicing physicist. But I
had access, in my role as a resident scientist on behalf of Control Data at Colorado
State University, to a supercomputer – the CDC Cyber 205. A physicist friend
and past colleague, the late Kevin Moriarty from Dalhousie University in Halifax,
Nova Scotia, kept in touch. An ad-hoc research project ensued in the general field
of lattice gauge theories for quantum chromodynamics – the theory of the strong
interaction, the force that applies to quarks and gluons, or Lattice QCD ([32]).
Other physicists got involved at times, including Claudio Rebbi now at Boston
University and Michael Creutz from the DOE's Brookhaven National Laboratory.
Of course, this activity was just one of many such projects at universities and
research labs around the world. We were not the only ones, nor the most prominent,
carrying on these computations.

The Standard Model of particle physics describes three fundamental forces –
the electromagnetic, and the *weak* and the *strong* forces (but it does not include
gravity). As scientific theories go, it is a very successful theory. Success of a scientific
theory is determined by its predictive power, and its correspondence to experimental data. For example, the Standard Model predicted the existence of elementary
particles that were later found by experimental physicists. The part of comparing quantities calculated from the theory to those measured in the lab is where
computations come in. The quantum field theory equations that express the model
are very complicated. They don't have, in most cases, an analytic expression that
can be easily calculated. Instead, researchers use a method known as *perturbation*

DOI: 10.1201/9781003038054-9

theory, where it works. One starts with a simpler, approximate, system for which there is a solution, and adds corrective terms, that can be calculated numerically. These can be power expansions, for example. The approach works extremely well for Quantum Electrodynamics (QED) through the application of Feynman diagrams.

QCD is the equivalent to QED, where electrodynamics – the study of electron-photon interaction, is replaced by *chromodynamics*[1] – the study of the *strong* force, or the quark-gluon interaction. Unfortunately, the QCD equations that arise for even simple cases of this strong-force interactions cannot be computed, let alone solved, with perturbation methods. The simple explanation for this is that the *strong force* is too strong. The perturbation technique involves successive terms of exponentials of a constant proportional to the strength of the force. in QED, that of the electromagnetic force, this constant is small enough so that the series of terms converges. For the *strong force* it is too large.[2]

Therefore, non-perturbative methods have to be deployed. QCD is a case of a gauge theory over an infinite-dimension path integral (a term familiar to those who studied particle physics). To make it possible to compute, the continuum problem has to be discretized (see the *Short Introduction to Scientific Computing* chapter). Thus, Lattice QCD.

A short (but relevant) digression: The physicist who introduced this approach to QCD computations in the mid '70s is Ken Wilson, who was at Cornell University at the time, and was a recipient of the Nobel Prize in physics in 1982 for analysis of critical phenomena. Wilson also played a central role in the HPC world. He was the founder director of the Cornell Theory Center (CTC) in 1985 (now called the Cornell University Center for Advanced Computing, or CAC). It was then that Shahin Khan and I, working for FPS at the time, got involved with the creation of the center. The CTC was unique among the National Science Foundation's (NSF) five original supercomputer centers in choosing array processors as the main building blocks of its compute capability (see also Chapter 8). They created a system of some 10 FPS array processors configured together, making a novel form of a supercomputer.

Back to the computational problem: The equations for the values at the grid points have to be such that when the interval between points approaches zero, they become the continuum equations of quantum field theory. As noted above, perturbation methods don't work for strong force interaction.

The method, that became popular and successful for Lattice QCD, is a random sampling algorithm called Monte Carlo (MC). It is used in several branches of science, including a number of physics applications – of which QCD is one. The method, developed in the early '50s, is attributed to Nicholas Metropolis, who led

---

[1]The term *chromo* is used because the attribute that distinguishes the different types of quarks is called, whimsically, *color*.

[2]Do not confuse the strength of a force with its range. The strong force is many times stronger than electromagnetism, but its action is felt only at distances many scales smaller. Gravity, by the way, that acts at universe-scale distances, is also the weakest of the four fundamental forces.

a group in Los Alamos National Lab that developed it. The group included John von Neumann and Stanislaw Ulam, the famous scientists of that era. It is said that the name, Monte Carlo, a reference to the casinos there, was not only appropriate due to the randomness nature of the method, but also in honor of Ulam's love of these casinos.

And so it was that in the early '80s we began to apply the MC method to Lattice QCD. The procedure is very compute intensive. It involves generating pseudorandom numbers for all the lattice links (4 links for each grid point), but the essence of the MC method is that each random value (sort of a probability between 0 and 1) is multiplied by a weight factor. And this weight factor is an exponential with an exponent proportional to the strength of the *action* at that point. It also contains a factor related to the lattice spacing. The latter makes it possible to repeat the computations with different spacing value and extrapolate its length to close-to-zero, that is the real-life continuum. Calculating exponentials is computationally expensive. The grid, lattice, is 4-dimensional (as it needs to be for relativistic physics) and each pass, or iteration, is repeated many times, so that the final average is more accurate. Every step requires a sweep through all the 4D lattice points, which makes the problem also memory bandwidth intensive.

This was the period when the top supercomputers of the day were closing on the one-gigaflops peak performance. The Cyber 205 we used had a theoretical peak of 800 megaflops at 32-bit arithmetic, resulting from 16 vector pipelines (of single precision) running at 20 nanosecond clock rate, or 50 million cycles per second. It had 64 MegaBytes (accessed as 8M words) memory. By comparison, a MacBook Pro today runs at a rate 40-50 times faster than the Cyber 205. Its single chip processor has up to 8 cores (in 2021). So, a present-day laptop has a nominal operation rate some 20–25 faster than the early '80s supercomputer. Todays laptops even come with hardware supporting vector operations, or streaming of results. Even more impressive is that a laptop today can have a memory capacity that is 1,000 to 2,000 times larger than the 64MB of the 80s supercomputer.

Back then, about 40 years ago, we were able to run with a grid of around 20^4 points (range of 20 points in each of the 4 dimensions). A rare "hero" run was when we got the lattice up to 32 points at each dimension. I talked to Claudio Rebbi about Lattice QCD then and now:

"Now I'm involved in a study using grids of 96^3 by 192 points. Runs today are of order of 100^4 in size, which is a huge increase over our runs in the distant past. Computers help in two areas: One is due to the sheer size of problem we can run. It allows us to investigate theories that go beyond the Standard Model. That, combined with advances of computational methods, allow us to get results with astounding accuracy. That's where computers have helped enormously. Another area is in what we could study. In the '80s we studied Fermions and pseudo-Fermions where we let the gauge fields indicate quarks propagation. Now we can solve for the quarks motion. It's a set of linear equations that have to be repeatedly solved as the gauge field evolves. The equations are sparse, luckily. Still, we have a system made of 10^8 points (and links) multiplied by the number of the internal degrees of freedom, which results in over a billion components."

Rebbi adds that today each step, or iteration, can be completed in seconds. This is when the calculations are done on some of top HPC systems in the world. He refers to the large DOE national labs systems, such as those in Oak Ridge, TN and in Argonne, ILL. The application makes use of the GPUs of these systems. Still, he says, "Running these jobs is not straightforward. It take quite a lot of expertise to make an optimal use of the compute systems. Crucial aspects of the programming include how the data is arranged and managed."

The MC method had some successes in validating the theory of the Standard Model by calculating quantities that can be compared to experimental results. It had been used to calculate masses of particles without introducing any arbitrary constants. The proton's mass, for example, came out very close to that measured in the lab. Another achievement was the energy (temperature) value when confined quarks transition to a quark-gluon plasma.

Our work on using computations to validate the Standard Model forced us to overcome the limitations of supercomputers of that time. The follow-up to applying the MC method highlights the central role that numerical methods play in bridging theory and experiment in science. This is a tale that, in many different forms, was played out in many other instances confronted by HPC practitioners.

The output of the Monte Carlo method computation is a set of 'configurations'. A configuration is a collection of $SU(3)$ matrices, one for each of the 4 links of every lattice site. An $SU(3)$ matrix has 9 complex values, or 18 real values. So, a lattice of 16^3 by 32 is a dataset of over 9.4 million variables. These values are numerical representation of the gauge-preserving dynamical variables of the quantum space. This is the 'raw data' upon which we can perform the computational measurements of physical observables. In our case, we 'placed' a quark on the lattice and computed at each lattice site the 'propagator' – an operator that describes the propagation of the quark. What this entails is solving a very large set of linear equations. The vector to be solved is made of the propagator values on the lattice sites. The coefficients matrix is made of the configuration values.

For the purpose of this discussion it is sufficient to note that the system to be solved is so large as to warrant an iterative solver. And because it has a 'simple' coefficient matrix, the Conjugate Matrix (CG) was a suitable algorithm. The choice made, we faced implementation challenges.

The following episode is an opportunity to describe an example of the dynamics between numerical algorithms and the compute system.

Supercomputers are often ranked by the potential speed of arithmetic operations they can perform – the number of floating-point operations per second. It is also well known that the vast majority of HPC applications perform at a rate that is only a small fraction of that peak potential performance. Memory bandwidth is, arguably, a more appropriate metric. Moving a data element takes an order of magnitude longer than performing an operation on it. But in our CG-solver case the situation was far worse. The problem's dataset did not fit in memory. That is, much of the data was kept on a disk in the computer room. Therefore, even when several operations are done on the data slice when in memory, the execution time was I/O-bound. The solver implementation had to be what is termed an *out-of-core solver*.

The optimizations we did for the computation stage – vectorization and the use of the *gather* instruction instead of multiple transpositions of the large matrix, did not address the I/O-time issue. The problem was aggravated by the fact the regular CG algorithm steps through the data twice each iteration, while the data could not be kept in memory. Data slices had to be rotated in and out. The hardware we were given was fixed, but there was another aspect to consider – the algorithm itself. The basic thought was *can we do more computations that advance getting to a solution on each piece of data brought in?*

Rebbi came up with an approach that may apply to other iterative algorithms too (and I implemented it). His idea was to write down the arithmetic steps for doing two iterations. Doing so allows the use of calculations for the first iteration as immediate input for computing the second. We still needed to go through all the data twice, but now we performed the work of two iterations. And because the I/O time was so dominant in getting to a converged solution, we essentially halved the time to solution. For detailed derivation and explanation see [33].

I bring up the modified CG solver story as an example of the interplay between algorithm and hardware system architecture. It is a recurring theme in the lives of HPC application programmers. Such examples provided insights to co-design teams that became popular years later.

CHAPTER 10

HPC for the Automotive Design: Early Days

How HPC Saves Time and Resources

O NE of the very early industry adopters of HPC was the automotive design industry. Designing a vehicle involves materials' strength and weight, airflow over frame, noise, stability, and more. It involves aesthetics and comfort, and regulatory safety measures. Computationally, these call for finite-elements and finite-volume solvers, for fluid dynamics codes, and for visualization. It all falls under the heading of Computer Aided Engineering (CAE).

One of the early pioneering HPC users in the automotive industry was Sharan Kalwani who worked at General Motors for 10 years. Kalwani has a long history within the HPC community with stints at both end-user sites from the private and the public sectors, and with system vendors. When we talked he narrated how computing entered and grown in Automotive. Kalwani opened with:

"The automotive computing story began sometime in 1983. Initially the focus was on crash analysis, and software engineers had to construct the digital models. Later they worked with suppliers and partners – CAD[1] people, in order to generate the meshes for the crash safety analysis. The General Motors (GM) research labs were the pioneers in this. The initial crash models contained less than 1000 elements. Today they run on 10 to 20 billion elements. It has grown not by simple orders of magnitude but by leaps and bounds."

The use of HPC for Automotive grew beyond the crash safety aspect:

"One of the members at the GM Research Lab was a thermal aerodynamic specialist, therefore familiar with Computational Fluid Dynamics (CFD). CFD in Automotive grew from that. I still have the original published catalog of early

---

[1]CAD stands for Computer Aided Design.

DOI: 10.1201/9781003038054-10

90

software for engineering for automotive and aerospace design that was published in the late '80s as a thick book by Cray Research. It includes all the codes available then and who you can get them from. It describes a lot of the early history of HPC at GM. From there it grew to other automotive companies. Ford picked it up one or two years later, and then in the same time frame the European automakers also got interested in it. They all started with the same baby steps that GM did. Trying out very small models, given the power and size of the Crays of those days. The codes had to be tuned by hand for performance. Then the European software ecosystem started growing, and for the next 20 years there was a healthy competition between North American and European software for CAD and CFD. And then the Japanese were also there on the hardware front. We, at Cray Research, had fierce competition from NEC, Fujitsu, and Hitachi and their vector machines."

Crash analysis on computers involves creating a structural model, described as a 3 dimensional mesh shaped like a car, with physical attributes assigned at the grid points. These will include elasticity, resistance to external forces, and other material properties. The simulation proceeds by subjecting the car being tested to simulated frontal collisions at varying speeds against other cars in motion or into a wall, as well as side collisions. The output describes the damage done to the car and allows design corrections where needed to meet safety standards. A computer run instead of material cars driven into a wall. Done multiple times.

The CFD type of computations are applied well beyond modeling the airflow around the vehicle for improving its aerodynamic properties. They include diverse aspects such as engine combustion analysis, internal cooling and heating, ventilation, fuel flow, and lubricant engine cooling. The many digitized simulations save plenty of real-world experimentation that is more costly and time consuming.

But that is not all. HPC is applied to other areas too: NVH – Noise, Vibrations, and Harshness. Computational Electromagnetics (CEM). Battery analysis via electrochemical simulations. Durability analysis. Material sciences – finding potential compounds that are both light and strong.

What has been happening in the Automotive industry is an example of the major shift, a revolution indeed, of how digital modeling entered manufacturing, largely replacing mock and material-built models. It has dramatically changed the design, development, and testing phases. Getting there was not always easy. Engineers exercising imagination and creating a vision had to convince business managers of the feasibility and benefits of such a process transition. Kalwani was there in those early days and remembers:

"We had to convince the management of the use of, and investment in, HPC for designing production and cars. And it was not just the proof of concept. 'we can do safety car crash with HPC.' They would say 'that's fine, but tell me how will it benefit the company. How will I save money? How will I accelerate the design? What is the benefit to the car consumer?' And they threw at us lots of questions. And we had to come back with actual numbers and business analysis. Management would often, at the last minute, put a high-level executive in front of us and say, 'Convince him.' And we would have to give them answers that would survive a very hard probing by business people or economists or marketing people. They would try to tear our proposals apart. And then we would respond back, 'Here is the benefit.' And express the benefit in terms of time, money, or market share or whatever is the metric they threw out. HPC benefits in different ways. And by demonstrating the different ways, it actually works as a multiplier. Think of it as a cube. So there is the top face of the cube, front face of the cube, side face of the cube. And those are the visible parts and you can show concrete benefits in three dimensions. And we can also show benefits on the faces of the cube that are not visible to you. Then it becomes even a more persuasive argument about the benefits of HPC."

Elaborating on the actual benefits, Kalwani continues:

"With the increasing power of HPC we are able to evaluate literally thousands of designs in a very short time. This allows us to select a design that optimizes a given criteria. We can actually generate or analyze several different designs. Either the design is optimized for material cost, or the design is optimized for manufacturing assembly cost. For example, you may have created an excellent design, but one that is very hard to maintain. For instance, you have to remove 20 parts to get to one part. This would make the after-sales service cost very high. And that's a negative. So with the help of HPC, we could actually create many, literally a thousand, different designs in one design period of three months. And then executives and product managers can choose any particular one depending upon market conditions and the sale value of the car. If it is a $20,000 car, then you don't want to have high aftermarket maintenance cost. If it is a higher-priced car then you don't mind the after-sale maintenance cost but then you want the reliability to be extraordinarily high. So giving the planners a fantastic range of options at their fingertips was a real benefit of HPC. They saw that they could have a spectrum of decisions they did not have before."

The role of HPC in manufacturing, that started in earnest in the '80s, continued to grow. Digital, numerical, modeling became the norm. We return to this theme in later chapters.

CHAPTER 11

End of an Era

The last gasps of proprietary vector CPUs

W HEN I was approached, toward the end of 1986, by Bob Korsch, the venerated past manager of benchmarking at Control Data and now (1986) at Cray Research, about joining Steve Chen's new project, I thought this could be both my dream job and a lucky escape from the turmoil at FPS, where I was at that time (see Chapter 8). Instead, after joining Cray Research, I was treated to the proverbial front-row seat of the events that marked the end of vector processors in the U.S.

The microprocessor, enabled by the development of integrated circuit technology, was becoming an overwhelming competing approach to the specialized big-iron vector processors.

IBM's mainframe IBM 3090 had a Vector Facility and was used for scientific computing well into the late '80s. However, IBM was already developing the POWER (Performance Optimization With Enhanced RISC) architecture as its microprocessor product. The resulting product line would replace the mainframe-plus-vector as IBM's offering for technical computing.

Control Data went out of the vector processors business in 1983, and shrunk and disintegrated over the next decade or so, when it spun out ETA Systems with CDC's financial backing. It brought to market the ETA-10, a successor to the Cyber-205, which was far from a commercial success.[1] In 1989, CDC shut down ETA and folded some functions back into CDC, mostly for continuation of contractual support to existing customers. It was out of the supercomputer business, and soon settled with customers and eliminated all remnants of involvement with vector processors. (We saw an example of that in the U.K. Met Office story on page 55.)

---

[1]Though never at ETA myself, and at FPS at the time, I collaborated with fellow physicists on a followup to our work on the Cyber-205 aptly titled *Applications Development on the ETA-10* (*[34]*).

DOI: 10.1201/9781003038054-11

The demise of the mini-supers and the emergence of MPP systems, left Cray Research as the sole bearer of the old-style vector processors in the U.S.[2] What transpired is the story of Cray's two most celebrated architects – Seymour Cray and Steve Chen, and two short-lived supercomputer companies.

Cray Research makes hard choices

Seymour Cray finished the Cray-2 by 1985, and was designing the Cray-3 since. Steve Chen was done with the Cray Y-MP around 1986 (though it was sometime in 1988 before the Y-MP started shipping to customers), and started a design dubbed *Cray MP* (the project I was hired into).

Meanwhile, in that period of late '80s and early '90s, there were two other development projects that were critical to the company's survival.

Chen's MP Project was a departure from his Cray Y-MP line, so another team was engineering a continuation of the Y-MP – involving more processors (up to 16) and a faster clock. It resulted in the Cray C90 models that provided much of the revenue in the early '90s.

At the same time, very aware of the 'killer micros', Cray Research started yet another major project based on microprocessors and what would be Cray Research's first MPP product. A protege of Seymour Cray, Steve Nelson, was the lead architect of what would be called the Cray T3D. It was quickly followed by the very successful Cray T3E. Both used a processor from another company – the Alpha chip from DEC.

The company, under financial pressure mostly due to competition from microprocessor-based products, could not afford to support all these projects at the same time. Something had to give. The Y-MP and the C90 generated revenue. The microprocessor revolution was the future. The Cray-2, which arrived nearly 10 years after the Cray-1, attracted very few customers. Both the Cray-3 and the MP projects were technologically very ambitious and were considered risky if not questionable. History proved the skeptics right.

Between '87 and '89 both Cray (the person) and Chen were pushed out of Cray Research.

Cray Research attempted to adjust to the changing HPC environment. Challenged by low-cost vector minisupers Cray-like competitors and the fast advances of the microprocessors it tried to stay relevant and competitive. Its vector architecture strategy was to build on the Y-MP, itself based on the X-MP. Cray's answer to the minisupers was the air-cooled scaled-down Cray Y-MP EL. It was followed

---

[2]Again, vector processors continued to be developed by three Japanese companies – NEC, Fujitsu, and Hitachi, for years. Some marketed globally beyond the '90s, other for Japan's domestic market.

by the Cray J90, the minisuper version of the C90. In 1990 Cray even acquired a minisuper company – Supertek that 'cloned' the X-MP design in CMOS.

At the high-end the C90 was followed by the T90 (1996), considered the last of the old-style shared-memory vector processors. Later systems would have vector capabilities, but with non-proprietary microprocessors and distributed memory systems. More on that in the later parts of the book.

Cray even acquired FPS as a division for superscalar servers (see Chapter 8). All of which did not alleviate the company's financial hardships, and it would undergo several incarnations. Acquired by SGI in 1996, then by Tera (2000) that changed its name to Cray, and finally, in 2019, as a subsidiary of HPE.

The remarkable fact is that throughout the turmoil and transitions Cray remained a major player in HPC, and a much admired company.

Seymour Cray and Cray Computer Corporation

Seymour Cray's relationships with Cray Research was as a consultant, having resigned his CEO role so he can focus on designing supercomputers. Of course, a consultant with much leverage and influence. Even though, he needed more distance from the center of activities so, as he put it, he will not be "bothered" so much. The Cray-2 was designed in Boulder, Co. where the company set up a lab for that project. It was shut down in 1982 (though the Cray-2 was not shipped as a product until 1985). After another stint in Chippewa Fall, WI. Seymour set up shop (or lab, more correctly) in Colorado Springs, CO. and worked there on the Cray-3.

It was then, sometime in 1988, that the Cray Research management made the hard decision to cut funding for the Cray-3 project – entirely or drastically. It would have resulted in delays or a lesser system. Seymour declined the offer and left Cray Research with his team. In 1989 the Colorado Springs lab became the main facility of a new company – Cray Computer Corporation.

And the development of the Cray-3 continued. In a departure from his tradition of using well-established semiconductor material and processes, Cray built the Cray-3 with gallium arsenide. One prototype was built by 1993 and installed at NCAR while still being debugged. No other customers were found for the Cray-3.

Ignoring the market realities of the microprocessors and the MPP architecture, accompanied by advances in tools for parallel programming, that combined to produce HPC solutions with far better price-performance ratios, Seymour Cray went ahead and embarked on a Cray-4 design. He was hoping, so the thinking goes, that the machine will be so powerful that the economics of pricing will not enter the equation, as can be inferred from his *oxen vs. chickens* comment (see page 84).

This did not end well. Cray Computer filed for bankruptcy in 1995.

Lesson learnt, Seymour Cray started another company – SRC Computers (SRC being his initials). This time opting for the MPP style architecture. Tragically, Cray died in 1996 as a result of injuries from a car accident.

SRC still exists. It develops reconfigurable computing solutions, but is not involved with HPC.

Steve Chen and Supercomputer Systems Inc.

The following episode is one where I can share first-hand experience. This allows for a somewhat more detailed telling of the story.

I joined Cray Research and started working for Chen on the first working of 1987 (not the most hospitable season to move to Wisconsin), as head of the newly formed applications group of the MP Project. The other groups were Technology (process), Hardware (design), and Software. All not much more than boxes on the org chart. The focus in the first half of the year was on staffing, from within Cray Research and externally. Then, in late August 1987, we were notified that funding a project of the scale and ambitious goals as planned cannot be afforded. The project is to be shut down, and the company will do its best to find the staff alternative positions in other divisions.

By then there were some 25-30 of us working for the MP Project. Quite a few have just relocated, with their families, from as far as the East and West coasts. Anxiety and emotions were high. Chen gathered the project's principals – 6 in total, and stated that he did not want to abandon the project, but to continue it outside of Cray Research. Form a new company, and 'we'll figure it out from there..'. And, to us: 'please let me know in a couple of days if you're joining; and who from your group wishes to join'.

All but one of the principals decided to join and be co-founders of a new venture.

The separation from Cray was amicable. Our new entity was given temporary use of a small facility with some history. It was the old lab Control Data built for Seymour Cray when he sought some distance from the company's bureaucracy and moved to Chippewa Falls, where he grew up. Located in a nondescript wood building, in a northern section of neighboring Eau Claire known as Hallie, with a few offices and some lab space, it was later given to Seymour when he left Control Data and formed Cray Research. It seemed, at the time, a sign for hopeful future, and a generous gesture from Cray Research.

More than half the engineers from the original project joined the startup immediately, and we barely squeezed into the old lab building. A larger facility, and funding had to be found quickly.

The termination of Chen's project was a fairly big news in the computer industry and attracted attention from other companies. After all, he was then the supercomputer system architect whose products were the most powerful and the most commercially successful. Executives from the high-end computing at IBM were quick to start a series of conversations with Chen. IBM wished to be a major player in HPC. IBM had a most advanced process and material technologies. The new startup needed investors before it had anything concrete to offer, besides track record and reputation.

The IBM executives also understood that IBM was too large and structured to be able to let a somewhat risky entrepreneurial project thrive, or even survive, within the company. Chen wanted to keep his creative independence. Within days both sides agreed on a somewhat novel framework for a cooperative relationship. The new company will remain an independent startup company. IBM will be a major investor, but not exclusive. Its funding of the new company will be incremental and subject to progress and meeting of milestones.

The announcement that Chen founded a new company came out in mid-September 1987. The founders invested some seed money, as did a couple of early-hired executives in administrative and marketing capacities. The mood among the employees was one of great excitement and optimism. In a true startup frame of mind everyone was accepting of the rough initial accommodations and administrative support.

With the yet unannounced agreement with IBM at hand, we were able to move to a larger facility in south Eau Claire. There we had adequate lab space for the various aspects of the hardware, office space for design and software development, and room for adding personnel.

Somewhere along the line, the company's name was chosen: Supercomputer Systems Inc. (SSI). A big sign on the new location carried the name.

Neither IBM, nor SSI wanted to have IBM be the sole provider of financial support. The goal of SSI was to stay and prosper as an independent entity. SSI did not want to relinquish control to venture capital providers, either. But we wanted to find a way to involve more of the HPC eco-system in the project. In particular, we were looking for end-user input to some design ideas.

The result was a scheme that combined end-user input and financial investment. We signed 5-year cooperative agreements with major corporations that involved annual contributions from them, regular review and feedback meetings, and 'dibs' on early systems from SSI. To avoid conflicts and to preserve open discussions, we chose each partner from a different industry. They were Aerospace (Boeing), Automotive (Ford), Energy (Electricity De France – EDF), and Chemical (DuPont).

In addition, SSI engaged an agency of the U.S. Department of Defense that was interested in early access to the eventual system.

All that meant that, while operating in stealth mode, we were under pretty extensive scrutiny from early on. Delegations of the three entities – IBM, the commercial partners, and the government agency – visited us in Eau Claire, WI at regular intervals of about every three months to receive progress reports and offer feedback and critique. The process served as a very constructive form of oversight.

The administrative and the marketing-sales functions were streamlined and minimal in size. The technical development was composed of four groups, not differently than while still at Cray: Design (called 'Hardware'), Process (called 'Technology'), System software ('Software') and Applications and Libraries ('Applications').

Starting in 1988 we recruited engineers at a fast pace (possibly too fast at this early phase). The hiring spree was largely successful due to its timing. It was not only Cray Research that was in a certain amount of turmoil. Minisuper companies were failing. IBM was contracting its in-house HPC activities. The Technology group benefitted from nearby Cray engineers looking for a new challenge. The Hardware group added several architects from technical computing companies that collapsed or were struggling. The Software group added, among others, a complete team of ex-Cray compiler writers located in Livermore, CA. For the Applications group we had plenty of candidates from minisuper companies and even from academia.

And then, from the end of 1988 to early 1989, when Control Data pulled ETA back in while terminating most of its positions, SSI gained from an experienced pool of engineers in all the required disciplines. And all from neighboring Twin Cities, MN.

Within a couple of years SSI's headcount grew to some 300 employees. Growth that exceeded the rate and progress of the product development.

It seems that things have lined up very favorably for SSI. It only had to deliver on its promise; and do so on time.

And the promise was for the world's fastest supercomputer of that time. The plan was for a multiprocessor system targeted initially at 64 CPUs, each at performance range of multiple gigaflops. The key for such an aggressive target was a most advanced and innovative multi-layer integrated circuit, multi chip module, and novel packaging. This was supported by high density switching power supply. The circuits density and a clock rate at the nanosecond range required liquid cooling by immersion of the whole unit.

The technology aspect was challenging, and at the same time the plan was to develop a complete software stack – OS (Unix), compiler (Fortran), and libraries (numerical, utilities). The idea was to be ready for production applications to run on the system upon completion of testing the hardware.

This was a great experiment of what would later be referred to as *co-design*. There were two dynamics in action: The Hardware and Technology dynamics, where the architects test the limit the process technology would allow in terms of gates available, chip size, pin count etc., and clock speed as a function of power and cooling. And concurrently there was the Hardware-Software-Application interaction. The designers proposed architecture details – instruction set, cache sizes, functional units, and the Application group followed with an instruction-level simulator with which Software started writing and testing a compiler, library routines and utilities, while application engineers tested computational kernels and technical writers were creating the documentation. It was truly a marvelous environment.

Alas, all these checked boxes of *required* or *nice-to-have* items were not sufficient for success. Partly, because of matters within our sphere. Within a couple of years the technology readiness was at least a year behind schedule. Achieving its design goals proved as hard as was predicted, and not all were achieved. In general, only in hindsight we can tell if our technology and design goals were pushing the proverbial envelope or bursting it. It appears some technical targets were beyond reach for the period and the time allowed.

Nevertheless, the technical obstacles notwithstanding, toward the end of SSI's fifth year – 1992, a prototype processor was powered on and functioning. A sealed cube of cooling liquid surrounding a very fast processor.

Powering on this single processor was, of course, most gratifying and a moment to celebrate. But there were already reasons to be concerned about the future of SSI. Being about one or two years late in a five-year project is not unheard of, and not unexpected given the complexity and the technical challenges. However, our early fast growth meant higher rate of spending, and we have exhausted IBM's committed 5-year investment. It was a difficult period for IBM (the same that affected the Clementi project – see page 72). A sharp decline of mainframes sales, and tough competition in storage, networking, and the printer markets, caused IBM to lose billions of dollars for several years. About half its employees lost their jobs. IBM was about to start a major shift toward applying microprocessors and emphasis on software and services. In that environment it could no longer continue to support SSI.

Late 1992 saw frantic efforts by the SSI management to find other sources of funding to complete the project, and go from a prototype to manufacturing a product. These efforts failed, and in January 1993, unceremoniously, SSI's facility doors were locked and entry prevented to its employees.[3] There was equipment and IP (intellectual property) to capture and divide among the entities that supported SSI. Other than being locked out of the building initially and very controlled entry later on, the closure process was smooth and generous. Most, if not all, employees found employment elsewhere pretty quickly. This was, after all, a collection of highly skilled computer professionals. Most of us, did not have to travel for interviews. Recruiters for computer and high tech organizations came over to Eau Claire to talk to us.[4]

As much as many of us would have loved the SSI story to have ended differently, the reality is that even a successful product would have been too late for the HPC

---

[3]John Markoff of the NYT captured the moment in a 1993 article ([35])

[4]That was how I ended up at NASA Ames Research Center in Mountain View, CA.

systems of the '90s. Microprocessors, commodity components, MPPs, and clusters would soon doom any chance for commercial success of propriety big-iron systems.

Cray Research would hold on for a few more years offering vector processors. And they survived by transitioning to microprocessors and many-processors systems.

And so the glorious era of the 'classic' vector processors came to an end. Some of us remember it fondly.

III

The Epoch of Microprocessors

The Age of Massive Parallelism

CHAPTER 12

Toward Massive Parallelism

Microprocessors Establish Presence in Top HPC Systems

W HERE the '80s saw the emergence of parallel computing at the high end, it was, by and large, on a small scale. Starting at two processors and getting to under ten processors by the end of the decade. Of course, array processors had many more processing elements, but they operated under the control of a host and executed a single instruction together. The '90s saw the arrival in mainstream HPC of systems made up of hundreds and thousands of processors each running its own copy of the operating system, or at least a 'light' version of it.

The MPPs

This is the architectural school known as *Massively Parallel Processing* (MPP). The choice of the term MPP is an example of present-focused marketing influence. Yes, the MPP systems of the '90s had about two orders of magnitude more processors compared with the minisupers and the MP vector processors, but their concurrency levels pale relative to today's large clusters. The term "massively parallel" has been used later for very large clusters and in describing accelerators with many processing elements. Here MPP is referred to a class of systems built in the late '80s and into the '90s by companies that no longer exist or exist but are out of the MPP systems business.

As with any major change, this did not happen abruptly. The origins of Massively Parallel Processing (MPP) can be found years earlier. A prominent example of the early development was Goodyear Aerospace's Massively Parallel Processor (MPP). Yes, this was its name. It had 16,384 (16K) *1-bit* processing elements (PEs) arranged as 128x128 two-dimensional array. It operated at NASA Goddard Space Flight Center from 1985 to 1991. Though each PE executed the same instruction – in a Single Instruction, Multiple Data (SIMD) fashion, just like an array processor, it is a worth noting milestone. This is because of its scale, albeit with very simple processors, and it being a stand-alone system.

DOI: 10.1201/9781003038054-12

The move from proprietary specialized supercomputers to systems made of microprocessors came at a cost. Many more components were needed to equal or exceed the custom systems' performance. The additional cost was not in terms of price (the micro-based systems were less expensive), but due to the higher level of parallelism needed in order to use the whole system, or most of it, on a single job. The challenge of parallelism is that when more parts participate efficiency is lost. This efficiency loss puts limits on the degree of scaling that is practical. At least a limit on how much we can scale a single job. Of course, many-processor systems are useful for multi-job environments since they allow flexible partitioning of the system depending on the scheduled jobs and the resources they require.

Upwards scaling of a computing system involves two attributes, one positive and one undesirable. Of course, the aim of scaling is to increase the *capability* of the system. But, invariably, it comes with at a price: added *complexity*. Shahin Khan observes: "Scalability is the ability to add more capability than complexity." That is, when scaling results in a positive outcome. And he adds a caution: "And there comes a time when complexity exceeds the incremental benefit of scaling and, unfortunately, you usually don't hit that wall until you've built a pretty big system and it becomes too expensive to fix."

The MPP sector took a different tack than the minisupers, architecturally . The latter attempted to be a scaled-down, and more cost-effective, versions of the classic big-iron supers. The MPP architects started with low-end microprocessors and devised schemes for connecting a large number of them tightly. It was a departure from the traditional high-end systems, and it was quite successful for a few years. The MPP concept was a step toward the idea of clusters.

The period from the mid '80s to the mid '90s saw a number of MPP companies come and go. They include MasPar, nCUBE, Meiko Scientific, Kendall Square Research, ICL DAP, Parsytec, SUPRENUM, and others.

The most successful of them was Thinking Machines Corporation (TMC). It was in existence from 1983 to 1994, when it filed for bankruptcy and its assets were acquired by Sun Microsystems. TMC was the darling of the U.S. government for a few years, and marketed 5 generations of its Connection Machine (CM) designs. Starting with the use of simple processors and hypercube topology for the early implementations, the CM-5 was built from SPARC processors on a *fat tree* network.

It is a testament to its success, and to some extent the power of the U.S. government's purse, that the CM-5 took the top 4 places in the first edition of the TOP500 list in June of 1993. And a fifth entry within the top 10.

Enter Intel

By the early '90s Intel has had over 20 years of experience in designing and producing microprocessors. It became Intel's main product line after it exited the memory chips business in the mid '80s (realizing it cannot compete with Asian chip manufacturing). What is relevant to HPC history is that ten years after its first microprocessor, in 1981, it produced a 32-bit microprocessor – the Intel iAPX 432. It was not the x86 architecture used in the chips IBM was getting from Intel for its PCs. The product was not successful. A young Pat Gelsinger, today's Intel's CEO, was a lead engineer of the 432. Fortunately, he quickly got involved with another design, the i486 – an x86 design, in production from 1989, that was more successful and provided a basis for a long line of x86 CPUs (the i386 that preceded the i486 was, in fact, the first 32-bit x86 chip).

But Intel did more than just designing and manufacturing chips. By the late '80s it had a division that developed systems. Rather than being exclusively a technology provider to companies that are 'system houses', it chose to play in the systems market too. That would last about 10 years. Having a systems division broke the clean separation between technology and systems providers that is the basis for the *commodity business model* (a theme we discuss in Chapter 17). The technical lead and the visionary behind the idea of Intel building HPC systems was Justin Rattner, who would later run Intel Labs and be Intel's CTO. It was a shrewd move that would greatly contribute to the emergence of x86 servers as the most popular components in today's clusters.

First were the MPP products in the early '90s. The Paragon product line, launched in 1992, did not use an x86 CPU. It was based on a RISC processor (the i860). Following a product called iPSC/860 and a one-off from Caltech (call Touchstone Delta), the Paragon could have up to 4K i860s interconnected in a two-dimensional grid. As did other MPPs, the OS on each node was what was known as *light-weight kernel* (LWK) – a stripped down OS that contained only what was needed for running a computational task and communicating with other nodes. Connecting to the outside world, including storage and file systems, was left to a host or special nodes known as *head nodes*. In the case of the Paragon, Intel used an LWK developed at the Sandia National Lab (SNL). This turned out to be a prelude to another, and more significant collaboration between Intel and Sandia Lab.

The Sandia Lab is one of three DOE national labs with responsibility for the nuclear weapons stewardship (the other two are Los Alamos National Lab and Livermore National Lab). As such, in the '90s, these labs were the recipients of the most powerful supercomputers DOE procured, under a program named *Accelerated Strategic Computing Initiative* (ASCI). The Sandia system selected in 1996 was called ASCI Red built by Rattner's Intel supercomputer division, with overall system design by Sandia computer architect Jim Tomkins.

ASCI Red was based on concepts taken from Paragon, but went much further – and not just in scale. The Paragon's i860 RISC CPU was replaced by the Intel Pentium Pro - an x86 CISC chip. A main aspect of the innovative design was that this large-scale distributed memory system was constructed as the sum of four task-oriented partitions. The nodes in each partition were optimized for their partition's function: Compute, Service, I/O, and System.

The ASCI Red system was an inspiring example of a close collaboration between Industry and an enduser. It was very successful in a couple of ways. First, its performance: When it became operational in 1997 it was ranked the top entry on the TOP500 list. It was the first to achieve the symbolic milestone of surpassing teraflops on the HPL (High Performance LINPACK) test. ASCI Red retained the no. 1 position on the list until 2000, with upgrades along the way. Second, and arguably more significant, ASCI Red's longevity was unprecedented. it was decommissioned in 2006, after nearly 10 years of operation. It is a testament to its robustness and reliability that the system was still operating reliably at the time and was taken out only so that power and space can be given to newer systems using technologies 10-year more advanced ([36]).

Effective Parallelism in Distributed Memory Systems

The previous decade's MP processors ushered the beginning of parallelism being at the center of applications' performance. The computational work had to be split into pieces that can be executed concurrently. But the programmer did not have to worry about where the data is. The memory was shared. Each processor had equal access to all of it.

That has changed with the introduction of MPPs. The sheer number of processors made it impractical for all the CPUs to be connected to a single memory system. The solution was that each CPU had its own local memory. Distributed memory architectures were becoming the norm.

Now the application programmer, and the compiler, had to worry not only what computations can be done in parallel, but also where the data is, and how to make data items in one node available to other nodes. Parallelization became a lot more challenging.

Reasonable performance dictated that data required by a node be in its memory when needed. And when a chunk of data is placed there, the more computations scheduled on it, the better. Devising a parallel algorithm, more often than not, started with planning the distribution of the data, followed by working out the sequence of the arithmetic operations. Data locality is a key to effective parallelism.

Apart from a class of problems known as "embarrassingly parallel", nodes need to know values computed at other nodes. During the execution of a code the nodes need to transfer data to and from each other. Interconnect schemes vary among MPPs and among the future clusters, but for any of them the application has to synchronize exchanges of data – letting other nodes read one node's data before it is overwritten, but not before it is computed, for example.

Then there are what is known as *global operations*. Such an operation is one that requires computations that span multiple, or all, nodes. For example, finding the sum of the elements of an array that is spread out over the system. This would require doing some local work (e.g., a partial sum), then not proceeding until all the pieces were combined for the final result.

Data locality and data movement, or managing memory and interconnect, became a major consideration for applications' performance and the focus of parallelization. What started with small number of large chunks of datasets has evolved to managing thousands of servers, often with attached accelerators. This is today's world of clusters and other high-end systems.

CHAPTER 13

Engineering with HPC

How Engineering Apps Responded to Changing Architectures

T HE development of a material product involves the interplay between *design* and *modeling*. When computing is applied the modeling phase is a *simulation*. The *design* can be thought of as a set of constraints or rules. They may specify shape (geometry), dimensions, weight, functionalities, maintainability, cost and more. Think of a car: The designer draws a shape. Size is determined according to the number of passengers, storage volume, roads' lane width, etc. A limit could be set on the weight. Or it could be derived by functionality requirements such as fuel economy and engine acceleration goal. And a mechanic has to be able to reach various parts with some quantified degree of ease, and so on. The design rules and constraints, or *design goals*, constitute an envelope within which an engineering solution is to be found; preferably, an optimal solution.

The physical modeling phase has become digital simulation. A model is defined mathematically based on the design rules and derivative quantities are computed. For a vehicle these may be attributes related to weight, airflow, noise, resistance to impact at some speed range, smoothness of ride, etc. If no solution is found within the given 'envelope' then the design has to be modified. More typically, parameters within the digital model are adjusted to avoid 'overkill' while improving a factor that fell short of the design goal. This ability of quickly and easily test variants of the model is the big gift of computing for engineering-based product development.

The '90s saw the transition from big-iron multiprocessors to massively parallel distributed memory systems. This has greatly affected the engineering segment of the codes running on supercomputers. More than other application segments, computer aided engineering (CAE) and computational fluid dynamics (CFD) relied on complex third-party software, also referred to as ISVs (for Independent Software Vendor). The industry needed high-quality and reliable applications to comply with regulations and standards, and home-grown codes were not a feasible solution. The

DOI: 10.1201/9781003038054-13

ISVs were not large corporations, and the shift to parallel systems did not come easy.

Sharan Kalwani, who we met in Chapter 10, describes how the automotive industry adjusted to the changing compute environment in the '90s:

"Sometime in the early '90s we started having systems set up as constellations. These were not clusters, but SMP[1] machines put together in a fashion similar to that of a network of workstations. And then around mid to late '90s, clusters started making in-roads, and constellations disappearing. And there were the MPPs, such as the Cray T3D and T3E, that several car manufacturers, perhaps 4 or 5 of them, experimented with."

But, continues Kalwani, "It turned out that most of the CAE software did not scale well on those MPP machines. CFD codes did, but not CAE. And they could not afford to have two different architectures. When clusters started becoming more popular, the third-party CFD and CAE vendors, by necessity, migrated toward that architecture. And naturally, the automotive, engineering, and aerospace industries followed."

The parallelization of commercial engineering applications came about relatively slowly. There appeared to be some resistance by the vendors to migrate to parallel systems and a wish that simulations would continue on powerful workstations. It wasn't stubbornness, necessarily. The codes were old and would have to undergo major restructuring to expose parallelism. Kalwani explains:

"I wouldn't call it resistance. It was the difficulty in making their codes work in parallel. Take MacNeal-Schwendler, for example. Several of us that looked at the source code and the problem was what we call *dusty deck*. Those codes were written for small memory machines. They involved a lot of matrix computations written as *out-of-core solvers*. The matrices resided in storage, brought into memory piece by piece, manipulated there, and written back to storage. It was very difficult to render this process in parallel. A complete rewrite of the base code was required, an effort that ISVs such as Ansys or MacNeal-Schwendler or many of the other CAE vendors found difficult to do."

Software vendors that did not have the 'baggage' of the past and with smaller presence in the market faired better:

"For example, HKS was a vendor that wrote a finite element analysis software suite called Abaqus. It didn't have much success earlier and was better positioned for a rewrite that was successful in scaling on multiprocessors. Another company that migrated to parallel implementation early on was LSTC (Livermore Software Technology Corp.). Their best-known code is LS-DYNA. They had success in running on multicore hardware far better than MacNeal-Schwendler, and a few of the other ISVs. The latter were forced into rewriting portions of their codes so that they

---

[1]SMP stands for either Symmetric Multi-Processing or Shared-Memory Processors. The Crays and the minisupers of the '90s were SMPs.

could take advantage of parallel systems. Still, CAE codes, when compared CFD, continue to fall short in scaling. Today CAE apps can show good scalability up to a few hundred cores. Beyond that, it's a little bit of a struggle and of diminishing returns."

Crash Analysis (or Simulation) was mentioned earlier (Chapter 10) as an early CAE application for the automotive industry. In the early days of the '70s the modeling was more rudimentary and involved input data of pressure and distortions of panels from physical experiments. The advances in computation capabilities achieved by the '90s allowed simulations from 'first principles', meaning relying on the mathematical expression of the impact of a collision between two bodies – one of them a car, defined with all the details of shape, geometry, strength etc.

To appreciate the impact of such an ability I turned to Robert Lucas who is a Division Director at USC's Information Science Institute. He is also a research consultant and contributor to the development of LS-DYNA. Lucas put it this way:

"A General Motors speaker at a conference said that crash simulation reduced the time from concept to production from 5.5 years to 1.5 years. There are companies that are so confident in their engineers and their simulation tools, that they don't build prototypes for crash testing before going into production."

The use of HPC allows the design time to be cut by close to four times by having digital models that can be quickly modified and tried again. Until the desired safety is verified computationally. Only then, when deemed production ready, is a physical car tested for crash safety worthiness.

CHAPTER 14

HPC for the Aero Industry

How HPC Saves Time and Resources

A IRPLANES and motor vehicles have some similarities. They transport people and cargo. They both strive to be safe, comfortable, efficient, and fast. However, they differ much in the medium in which they operate. This dictates much different shapes and design considerations for safety and comfort. Nevertheless, the computational tools used by the two industries are very much the same. They include design and simulation, concerned with shapes, and strength and flexibility of materials, the flow over a frame, crush resistance, etc. In short, both aviation and automotive use of computers is in the areas of CAE and CFD, with some material modeling.

The aircraft industry, commercial and military, was an early user of technical computing. Since the days of vector processors designers and developers would run fluid dynamics codes (CFD) to simulate airflow and turbulence over the proposed frame of the aircraft. And finite-elements or finite-volume CAE applications to test the properties of the structure. The governing equations for the computations were well understood, but the accuracy of the results depended on the level of resolution of the discretized form of the equations. And that was limited by practical limitations of compute power and memory size and bandwidth before the '90s.

As a result, when a design model appeared to show promise in achieving structural goals and flow properties, a physical mock-up model would be built. It was often just a component of an airplane – a wing, the tail, or the fuselage. The model was then tested in a wind tunnel to verify (or not) the CFD simulation. It was also used to figure out the layout of electric cables, air filtration inside the cabin, access to parts for maintenance, seats arrangements, galleys and more.

At that stage, computers provided guidance toward an optimal design and the ability to quickly test new design ideas. But physical models were needed to verify and tweak the design, and, most importantly, for the engineers and the public to have confidence in the resulting product. The trust in simulated models grew with the increase of compute power and advances in numerical techniques, and as computed results compared well with physical evidence and experiments. This was an incremental process.

As the reader might imagine, the 'holy grail' of aircraft design was to get to an optimal design by digital simulations alone. Indeed, Japan's National Aerospace Laboratory (NAL) collaborated with Fujitsu to build a one-off parallel system of vector processors they named *Numerical Wind Tunnel*. When completed, in 1993, it toppled the CM-5 from the top of the TOP500 list, and retained the #1 position until 1996.

Jet Engine Simulation

Before looking at the whole aircraft, let us consider how computers assist in the design of a very crucial component: The jet engine. To understand this subject I talked to James Ong, who is a technical specialist at Rolls-Royce U.S.A. Ong's expertise is in the structure aspects of turbine jet engines, and specifically in their CAE simulations on HPC systems. He starts with differentiating the aviation challenges from those of the automotive:

"The significant difference between crash simulation for the auto industry and jet turbine application is that the energy involved in aero events is significantly higher than that of the auto industry. For us, the model is not sophisticated enough to get close enough to the physical behavior. It is getting better. The gap between the simulation versus the physical test given to us is becoming smaller. But there still is a good gap there. For the auto industry the gap for crash analysis is significantly smaller, and the physics for automotive is well-understood. But for turbine jet engines, there is still quite a bit of work to do."

Ong also sees challenges that were introduced by the transition to parallelism over distributed memory systems. Challenges that need addressing by both software engineering and the choice of numerical methods: "We introduced a lot of numerical issues which we did not experience 30 years ago under the shared memory environment." A particular area of difficulty is the turbulence aspect of the airflow: "Turbulence exhibits strong non-linear behavior. Of the many models being proposed I did not see any robust enough to truly help the design process. Though they improve over time, there's still need for further developments."

That said, simulations for turbine engines play a major role in the design. Ong again: "Simulation is a big part of our design process. We do a lot of the pre-wind tunnel simulations to guide us. Early in my career we didn't have much information, and did tests on the NASA wind tunnel based on guesswork extracted from past projects. Today we certainly do a lot more simulation and minimize the number of iterations. But I don't think we are close enough to totally getting away from the wind tunnel."

Ong concludes: "The simulation capability in our industry is not as good as it is in the auto industry."

Full Digital Design of a Complete Aircraft

It turned out that the story of the first commercial airplane that was designed completely electronically, using computer models, before the first prototype was ever built, belongs to Boeing. It was the Boeing 777. The project that started in 1990, saw a commercial flight of the 777 in mid 1995. Throughout the duration of the project, the detailed design and the experimentation of changes were done digitally. Components, such as the wings or the tail, were modeled separately, but even their assembly was modeled digitally.

One of the best people to talk about that period at Boeing is Albert ("Al") Erisman, who I have known from the days of Boeing's partnering with SSI (see page 97). Erisman was the director of technology for computing and mathematics for the Boeing Company in the '90s and a Boeing Senior Technical Fellow. It was a 300-person organization that included some first-rate mathematicians that innovated some numerical methods and produced a highly regarded math library (among other things).

Getting to an all-digital design was the culmination of many steps in the use of computing for aircraft design. Back in 1988 Scientific American published a special issue on *Trends in Computing*.[1] Among other wonderful articles there is one by Erisman and Ken Neves.[2] The article, "*Advanced Computing for Manufacturing*" ([37]), is about the use of supercomputers in airplane design (as well as in the automotive and the petrochemical industries). Here I choose two examples from the pre-777 era:

The Boeing 737 was around since the late '60s. A new version of it, the 737-300, was launched in 1984. Its main change was a more efficient engine. These new engines are more efficient by being enclosed in larger diameter nacelles. That meant that, due to ground clearance constraint, these engines could not be hung under the wings. The problem was that placing them in front of the wings was known, by wind tunnel experiments, to result in unacceptable levels of drag. This is where HPC enters the picture. Simulations showed the details of the airflow and where drag was generated that could not be observed in the wind tunnel. It showed the designers where the problem arose. Careful and numerous computer simulations produced a modified top of the nacelle that resulted in a smooth airflow over it. An engineering solution that allowed the use of an efficient engine without adding drag.

---

[1] A fun fact: I managed to acquire a hard copy of this historical issue. Its 1988 face value was $3.95. Scientific American issue today costs $25.99.

[2] Ken Neves managed the research and development programs for the Engineering Scientific Services Division at that time, and later became Senior Technical Fellow and Director of Computer Science Research at Boeing.

The second example involved the 757 and 767 models. They were to be completed at about the same time with the 767 a few months ahead of the 757. Boeing decided to use the same cockpit design for both models, mostly so that pilots can take a single certification training and be certified for both. The cockpit of the 767, a wide-body plane, was already designed, but the 757 is not wide-body. The aerodynamics around the area where the cab and the body intersect is critical to flight efficiency. Time was short and did not allow for the trial-and-error method of building mockups. Instead, the designers ran many designs of the 757 cab, which had to be wide enough to hold the 767 layout of the cockpit instruments, and, in a matter of weeks, honed in on the aerodynamically best solution. In the words of the article: "Boeing engineers were so confident in their results that they recommended fabrication of the new plane begin prior to wind-tunnel verification of the results."

As Erisman describes the evolution toward a digital design he links it to the transitions in system architecture: "The progress actually started in the '80s and even in the '70s. The new architecture created opportunities for applications that hadn't existed before. And those applications then drove the need for more robust architecture. It is sort of a cyclical process where these feed on each other. That was true for the all-digital design as well. At first it was not about doing away with a physical mock-up, but the idea of reversing the order. It always used to be that a mock-up was built, and then they would do analysis on what they thought they could do with such a design. What happened with the 777 is they would do the design first, and a mock-up or a first prototype or whatever would become the test of it. So, in fact, you always tested it physically, but the difference is whether it is in the inner loop or the outer loop of the design process. And, of course, a mock-up is much more expensive than computer runs.

"Thinking about this idea of architecture giving rise to new algorithms, one of the big breakthroughs that started happening around the time of the 777 design was this idea that you could set the parameters for the design and then run the analysis, ever more complex analysis, to be able to say, '*Okay, it meets these specs,*' or, '*It is weak here,*' or, '*It doesn't accomplish all that we wanted here.*' What you can then get to is reversing that loop as well by saying, '*Which parameters will give me the optimal design?*' What that means is that you basically parameterize the space of designs and use optimization to select the best parameters to achieve a certain design. It is one more level of abstraction where, if you wanted to optimize for performance, instead of, '*I have a design. How does it perform?*', you can ask the opposite question, '*Which parameters will give me the best performance?*'. While, obviously, taking into account that there are constraints on these parameters."

There are many constraints upon which ranges of parameters can be examined. They can be grouped into several major categories, or *systems* that make up a modern aircraft. Starting with *structures* that can be broken into its main components: fuselage, wings, empennage (the 'tail', made up of a rudder and an elevator), landing gear. Then there are the *electric* and *electronic* systems, *air filtration* system, the *layout* inside the fuselage, the *cockpit*, and more. The design process has to consider the *aerodynamics* of the structure and the expected *payloads*. And, of course, the *propulsion* generated by the engines, that become a part of the structure and also greatly affects the aerodynamics.

There is much interplay among the pieces that make up the aircraft, and the correct way to look at the design process is as a very complex optimization problem. And that is not all. Beyond the engineering challenges there are other factors, just as critical: *Manufacturing* – level of difficulty, tooling, maintainability. And the business considerations of *finance* and *market*. To which the planners and designers add a *performance* design goal, which is subject to the interplay between the structure and its aerodynamic profile, as well as a likely constraint the defines desired fuel efficiency level. In short, a complex problem for a digital simulation. Especially, given that a flying object has to be particularly reliable and safe.

The numerical simulations are indispensable but doing them by component or for different aspects (material strength, aerodynamically etc.) brings up the human factors. We have the aerodynamicist, the payloads designer, the structures designer. Each may believe their part the is most important part of the airplane. Each advocates for their piece of the optimum solution, which may impact negatively another aspect. In reality, the final configuration may require each proponent to give up on something dear to them in order to get the best airplane.

The quantities of interest in the simulation framework can be thought of as a set of inputs from which other quantities are computed. Their interplay forms the design process cycle. One of Erisman's colleagues was Thomas Grandine, now retired, who was a Senior Technical Fellow at the Research and Technology organization of the Boeing Company. He explains:

"There are two types of variables to keep track of. One set of variables forms the independent variables of the problem. These will be the design variables, and these quantities will include things like length of the airplane, wingspan, engine thrust, wing sweep angle, wing thicknesses, and other variables which fully describe a potential airplane that one could build. Other variables are dependent variables, or engineering analysis quantities. These variables describe the performance of an airplane: drag, takeoff noise level, fuel efficiency, weight, etc. These variables are typically quantities that can be measured or computed and which depend on the values of the independent, or design variables. The main idea of design optimization is to choose the design variables to produce an airplane whose dependent, or analysis quantity variables, have the more desirable values.

"This problem must be solved in the presence of constraints: The airplane has to fit at a standard gate at an airport, noise level must be at or below some prescribed level, the airplane has to be able to fly all of the routes it is designed to fly, etc."

Digital design does not mean *automated* design. Human judgement comes into play constantly. Suppose there is a wing design with a shape or a curve that gives additional performance, but it adds cost to manufacture. Someone has to decide how much more cost is worth it for what additional performance. Or, perhaps more importantly, how much uncertainty of crash safety is tolerable. Erisman explains:

"There' is a lot of judgment that goes into the process. But what the high-performance computing did is it enabled you to explore the design space, including things that you could not have ever explored before. And we added another feature for the 777 design: We used CATIA, the software suite from the French company Dassault. The design system had a 3D design capability, and you have thousands of models from CATIA that come together to form the whole airplane. And the question is, if they're designed in different departments by different people, how do you make sure that everything fits when it comes back together? And so, again, using high-performance computing, we developed a visualization system called Fly-Thru, that allowed you to bring all these electronic design pieces together into one airplane and then walk, virtually, through the airplane. And you could walk down the aisle, you could drop through the floor, you could look at the systems, you could go up in the cockpit, and you could just visually inspect how all these things came together.

"You could take a journey through the airplane and look at manufacturability issues. And then we extended that to haptics (the simulation of the senses of touch and motion). You want the airplane to be designed for maintainability. If you have a subsystem in the airplane and you need to be able to remove it, is there enough clearance to be able to allow you to take this out and repair it? We actually build a robotic system that allows someone to grab a hold of the part and twist it and pull it out (digitally). As they pull it out, they are pulling it through a robot. And if it encountered a barrier, it would stop and it would make a noise. This idea of being able to think beyond the design of the airplane to its maintainability, that was a part of the 777 project too."

The digital design process allows a robust exploration of the solution space subject to the requirements and constraints. It benefits the manufacturer in getting a product to the market faster and cheaper, while closer to the margins set by the specs. Is there also a benefit to the consumer – the passengers, in this case? In addition to getting to our destination faster and at a lesser cost (due to efficiencies derived from lighter materials, aerodynamic shape, and engine design), there is the safety aspect. According to Erisman:

"A better design improves the analysis and the resulting performance, with improved confidence in the margins, by affording a new insight into the design. Similarly, with safety. I remember saying in the 1990s, based on a graph that we had, that if you look at the increase in passenger miles traveled, and if the safety performance in the '90s had been what it was in the '50s, there would be a major crash every day based on just the increase in the number and the amount of air travel. These more sophisticated designs are much more robust with respect to things like lightning strikes, bird strikes, turbulence, and so on. The structures we have now are stronger and much more capable of flying in adverse conditions.

"In the 777 case in particular the idea of eliminating the mock-up as a design tool was really the key. There are two things that you save. One is you get an improved performance, and the other is you improve development cycle time."

Grandine offers a further insight into the use of HPC for aircraft design and the benefits it provides. Turns out it improves the management of the supply chain.

"With the models of the airplane prior to the 777, you had paper drawings, so you would have 2D sections of some of the surfaces. The tool and die maker would interpret these 2D sections and handcraft some surface on the tooling and this physical tool would then really become the real-life definition of the surface that you are modeling. And sometimes we would want to change suppliers for a particular part. Then you can't just send the part definition, the drawings, to them, because you'll get a different part than the first supplier was creating. The only way you could actually move the production of one of those parts from one supplier to another was to buy the tooling from the first supplier and give it to the second supplier so that they would be using the same tooling. Once you had a full 3D definition, now you can say this is the official definition of the part, and you can send that to suppliers. And so that made a huge difference up and down the supply chain for the company to be able to send to the network of some 35,000 suppliers exact definitions of what it is you are expecting them to produce."

In addition to structure and aerodynamics, there is another very important aspect to the design of aircraft: The materials from which to build the plane. There is a constant search for compounds that will be light, but strong and flexible. HPC is an indispensable tool in that search. Computational chemistry codes for molecular dynamics can be applied to simulate materials and compute their properties. Boeing likes to use an application called LAMMPS, an open source code developed at the DOE Sandia Lab in the mid '90s. LAMMPS stands for Large-scale Atomic/Molecular Massively Parallel Simulator, which explicitly expresses that the application was designed for large problems and to be run on highly parallel systems, fitting nicely to the emerging high-performance computer architectures of the time.

Jumping ahead in time, these days LAMMPS accounts for a large amount of compute cycles. Composites have to go through a cure cycle in autoclaves. Experience shows that with LAMMPS engineers can model quite accurately the chemical structural interactions. Again, allowing for quick explorations of materials. This aspect is becoming an integral part of the design optimization process. Grandine explains:

"People are starting to take a serious look at simultaneous size, shape, and material optimization. In the past, the structural engineers will first prove out some material. The materials engineers will say *this looks like a really good material*, and then the structures folks will try to put together some design using that material. But where things are certainly headed, and this has got HPC written all over it, is to be able to design the material at the same time you are designing the structure, and to be able to alter the material to accommodate load paths and structural needs in various places of the design. Much of this very customizable material is enabled by 3D printing, where you can vary material composition as part of the printing or deposition process."

Why Not a Single Coupled Model?

What triggered the question is the analogy to climate modeling. There, the model is made up of several components that represent different parts of the Earth System: Atmosphere, Ocean, Sea Ice, Land Ice. The software model has a *coupler* that manages the time-progression of the simulation by invoking each component in turn for each time step.

The automotive and aerospace models have two types of applications to support the design and digitizing of their products. CAE for the structural aspects, and CFD for the airflow around the object. Could the structure and the flow models be coupled? Should they?

The differences in context between climate and vehicle modeling explain why the latter isn't coupled. Weather and climate change over time and the model is tracking these changes. In the case of a car or a plane we model what is essentially a *steady-state* environment. For instance, while the 'structure' is fixed the simulation is of its reactions to certain configurations of speed, pressure, and temperatures. This is a CFD code applied to a given shape and surface and material properties. The design process may be iterative with a change of geometry or material followed by CFD test of the external impact of the change.

The CAE and the CFD models don't share the same space on which they operate (unlike coupled earth models). CAE operates on the structure from its outer surface inwards, while CFD is applied from the outer surface outwards. Numerically, CAE model uses finite-elements or finite-volume methods, whereas CFD is calculated over a mesh type of a grid.

Therefore, for vehicle digital design one can think of a scripted runs where, say, a wing angle may be tested at a range of values, but a coupled CAE-CFD model is hard to contemplate or justify.

For aircraft design it goes beyond the separation of structure and fluid dynamics. The presence of multiple subsystems makes the modeling more complex. Erisman talks about the difficulty in attempting a single model:

"You have a system that is so interconnected with various factors that it would be infeasible to operate. Then you say, 'Let me break this down into components, and then let's figure out how we can assemble these components.' In an airplane design, you do as much as you can with these pieces, but then you ultimately have to pull the pieces together. We found that it does not make sense to put them all in at the same time. doing so is what I called the *push button airplane*. And that doesn't make sense because in a lot of cases, there are judgments to be made. If you build a model that puts all the pieces together, you have to parameterize this in such a way that gives different weights to changes in different parameters. But if you do them separately and then you look at how they come together, you can actually identify places where a small change here doesn't make much difference, or where a small change here makes a lot of difference. And that helps understand intellectually how to put the pieces together."

Erisman explains that the sensitivity to small variations in the parameters is of concern since the manufacturing process cannot represent the simulation results with absolute accuracy. The matter of assigning weights to changes through judgement calls is also a source of variation since these are likely to differ from one person to another. Erisman points out that looking at a family of solutions allows for better judgement calls. Then there are the *hidden assumptions*. Here is an example: In one case a rippled wing design was aerodynamically marginally better than a clean leading edge. But that benefit "was completely washed out by the cost to manufacture a rippled wing, and likely both ignore the later cost of maintenance."

To emphasize the folly of ignoring *hidden assumptions* Erisman tells a story: "We were doing a study for the city of Tacoma, trying to cover all of their routes and schedules with fewer buses. When we showed the dispatcher the way they could save three buses he responded, *You have bus 12 on route 47. There is a hill on route 47 and bus 12 can't get up that hill.*"

The moral of the story: Human judgement is still needed after design studies are done. The studies aid in making better judgement calls.

So, the structure is simulated by a CAE applications and the aerodynamic behavior is computed with a CFD application. And then there is another facet to the design that is computer-aided: Product definition. Boeing has been using CATIA to describe all the parts *as-built*. Grandine gives an example of a difference between what a part looks like in flight and the as-built definition: "When the plane is sitting on the ground, the landing gear and the fuselage are supporting the wings and the wings will hang from the fuselage. But as soon as the plane takes off, the situation is reversed, and all of the fuselage is now being carried by the wings. For example, on the 787 you can have a 25-foot difference in location of the wingtips relative to your seat on the airplane. When the plane is on the ground, the wing tips will be hanging low. Once the plane takes off and you are cruising, that wing tip will be 25-foot higher than it was when you were on the ground. There is a lot of structural deformation. It is all by design. We want the wing to flex and support that load. Therefore, early on in the design phase, the aero engineers will be designing the cruise shape of the wing. It is the intended shape after the plane has taken off and is flying. But now the trick is to figure out what shape you need to build so that when the plane is in flight, it achieves the shape that you actually design. This is a very difficult inverse problem that needs to be solved."

Another such *inverse problem* to be resolved before CATIA creates the parts' models that are sent to manufacturing comes from the propulsion parts. As grandine describes it: "Jet engines can run very hot. Around the pieces on the inside of the engine combustor and the nozzle of the engine there are built-in gaps that are designed so that as the engine heats up the thermal expansion will close and seal those gaps, and you will get this very smooth nozzle surface. The hot condition is the design condition, and you need to figure out what to manufacture that will fit that."

There are many of these inverse problems that need to be solved. Boeing uses different codes to design the *in-operation* shapes. They are then passed on to CATIA operators that create the actual part definitions.

We can appreciate why, at the time of the 777 design, the Boeing team decided that the assembly of the components will not be automated. But, Erisman says: "You could go a lot further than you did before. You can take steps from confirmation of the design to getting to an optimal design. Each one of those is a big step because each one puts a design function in yet another inner loop, and that is what requires high-performance computing."

CHAPTER 15

The WRF Story

Multi-use Community Model

A LREADY back in the '70s it became clear that much can be learnt from a higher resolution weather models. They would give more accurate prediction. And, more importantly, such models allow the inclusion and capture of phenomena that occur on smaller scales. These include topographical effects by the presence of mountains and valleys, and the impact of local bodies of water and forests. Higher resolution makes it possible to model physical and chemical processes – even the formation of clouds and cloud-resolving processes. Of course, higher resolution meant many more grid points to describe the model, and a vast increase in the amount of computation for a given period of time (as was discussed earlier in Chapter 4). Therefore, initially at least, high-resolution global models were out of the question, and the developers and researchers had to settle for regional and local models.

One such project was a regional *mesoscale model* (MM) that started in the '70s as a community development project, coordinated and maintained by a joint team from Penn State University and the National Center for Atmospheric Research (NCAR) (we return to NCAR and its models in a later chapter). The model, whose latest generation is called MM5, has evolved over the decades, until 2005. It gave rise to what became a very successful open-source community model that is the subject of this chapter: The Weather Research and Forecasting model (WRF).

The person to tell the WRF story, John Michalakes,[1] got his start with numerical weather prediction working for the MM project. He went on to become the lead software engineer for WRF. Having known John for a number of years we met (virtually) and I asked him to tell me about the origins of WRF.

---

[1] At the time of writing John Michalakes is Consultant to Naval Research Laboratory Marine Meteorology Division in Monterey, CA on behalf of the UCAR Visiting Scientist Program.

DOI: 10.1201/9781003038054-15

"I was involved with WRF before there was WRF. First there was the MM5 a community model that was both microscale and mesoscale.[2] That was in '96-'97. I got involved with NCAR through a DOE climate program while working at the Argonne Lab at the time. It started as a 3-way project with ORNL, ANL, and NCAR for parallel implementation of CCM – the Community Climate Model. While working on a parallel version of Penn State/NCAR Mesoscale Model – MM4 and the subsequent MM5 work, the NCAR team started talking to NWS (The NOAA National Weather Service) about a mesoscale model. They came up with the idea of the WRF project – a new model developed from scratch. Since I knew HPC and software engineering I was asked, around '98, to lead the software engineering and architecture aspects. First, for 3 years as an Argonne employee, before joining NCAR in 2001, and working on WRF until 2010. After 10 years WRF was essentially done as a HPC/software engineering project. The scientific staff was adding packages and features, and I was the only HPC guy there, and thought 'it's time to move'."

Michalakes didn't quite leave WRF. He moved to nearby NREL – the National Renewable Energy Lab (also one of the DOE labs). There he worked on applying WRF to farms of wind turbines used as a source for renewable energy. This is but one example of the versatility of WRF. Here it is serving as a simulation tools for what is essentially an engineering project. We will return to this theme.

WRF enjoys a huge user community – according to its website ([38]) it is used in over 160 countries by some 48,000 users. In addition to it being accessible to all, the model owes its success and popularity to the solid software engineering it was built on. In the words of Michalakes:

"When I started at NCAR I said '*before we write a single line of code, let's put down requirements.*' The software working group did the requirements. They were both functional and non-functional. 'Functional' meaning those requirements about how the users wanted to use the model, and 'Operational' requirements were for the weather forecasting services, such as resolution, specific physics, etc. The parameters that would be useful for a particular type of application – severe storm, forecasting, research, etc. And there were other 'non-functional' requirements: performance (to run at certain speed), maintenance, single source – to support both monolithic and clusters, MPI and threading, and modularity. It had to be maintainable across these platforms within a single source. These requirements were expressed as design specs and first presented at the ECMWF workshop in 1998 ([39]). The point is, no code was written until the design and the architecture are spelled out and agreed on. And this was a major reason for the success of WRF. Not only was it well architected, but as often happens, people would come with other requests, and we could always go back to the agreed upon requirements and use it to not deviate from the original plan. Made it a lot easier to herd all the cats..."

---

[2]*Mesoscale*, also called *synoptic scale*, indicates a coarser resolution than *microscale*, but a finer resolution than *macroscale*.

Being a mesoscale model – one with a high resolution grid – meant the WRF was to be a regional model. Michalakes again:

"WRF was never meant to be a global model (though eventually there was one such version). It was, however, put to a broad range of purposes: atmospheric, climate, chemistry, NWP (Numerical Weather Prediction), basic atmospheric research, high resolution convective modeling for severe weather etc."

Most of the WRF users are researchers in hundreds of universities and research institutes. But WRF is also used operationally by national weather centers to produce local and regional weather forecasts. For example, at different times it was the operational model for the South Korea Meteorological Administration, the National Centre for Medium Range Weather Forecasting in India, the Central Weather Bureau in Taiwan, the National Centre for Hydro-Meteorological Forecasting in Vietnam, and more. The China Meteorological Administration adapted the WRF software architecture to create its own version – GRAPES (Global/Regional Assimilation and Prediction System). The U.S. Air Force ran WRF operationally for a number of years[3] and still use it for high-resolution US-only forecasts. NOAA NCEP currently runs WRF as part of the High Resolution Rapid Refresh nowcasting application (to be phased out).

The WRF project is an example of how the combination of adhering to design decisions of a software undertaking and embracing an open source community of developers and users can result in a very successful product. WRF turned out to be a useful and popular tool for researchers to explore and refine features that can be added or improved in earth system models.

Michalakes: "WRF has always been open and freely available. There's typically one release per year, and one workshop annually. 2-3 tutorials a year. Those are very popular and always fill up, with wait-listed hopefuls."

Having said that, there is a cautionary tale here too. Referring to the idea that the original specs will resolve all future disputes, Michalakes notes: "That's not to say we didn't have problems along the way. There were what we called the 'core wars'. NCEP (NOAA's National Centers for Environmental Prediction) never abandoned their dynamic core used in their ETA model (called NMM for non-hydrostatic mesoscale model). At the end there were two divergent developments – ARW (Advanced Research WRF) by the NCAR team, and NCEP's NMM. Both use the same software architecture."

NCAR is still the center and the driver for WRF-related activities. Though WRF's infrastructure is in maintenance mode, components and subgrids are still being developed by researchers. The project is moving to an open development paradigm where the NCAR team are the gatekeepers as to what enters future releases.

---

[3]Both the U.S. Air Force and the Korean center switched from WRF to the U.K. Met Office's Unified Model a few years ago

We will return to WRF (Chapter 20) and see how the use of the model impacts us outside its home at research labs. To set the stage we should mention some of the variants and components built into and on top of WRF.

One important spin-off is Hurricane-WRF (HWRF). Starting from WRF it became a separate mode and is used operationally by NCEP. Hurricane predictions is a specialty. The intensity of the storm is particularly hard to predict. Hurricane tracking is critical for warnings and preparedness and 2-3 days of advanced warnings and advisories are needed.

Another variant is WRF-CHEM. It adds chemical content to the meteorological equations of WRF, such as interactions and transformations of gases and aerosols, and is used to study and forecast regional air quality.

There are other add-ons and adjustments made for the specific applications that WRF serves. They are described in the chapter noted above.

Chapter 26, "The NCAR Models", describes a model newer than WRF – Model for Prediction Across Scales (MPAS), that is the current focus of development efforts at NCAR. Though there is some user migration from WRF to MPAS, which is a global model, it is a slow migration according to William ("Bill") Skamarock, who is a lead developer of MPAS. He says that much of the academic research is still being done on WRF. One reason for that is practical – global models require a lot more computational resources. But also, according to Skamarock, because "WRF has a huge number of capabilities. There are many options. For agriculture, solar, wind etc., with subgrids to support them. The users can apply very high resolutions to use the features properly. And to refresh the boundaries (for longer runs) people used to use the U.S. model – GFS. Now they can also get those from MPAS, and people are doing that."

Another senior NCAR researcher, Roy Rasmussen from the Research Applications Lab (see Chapter 20), is sticking with WRF for now despite acknowledging that MPAS's approach is good and promising: "WRF is a tried and true model that has been in use for over 20 years. It is really a good model." In fact, there continues to be a rich body of added components for specific research projects on WRF. Representing the user community of WRF, Rasmussen explains their relationship to the development and evolution of the model:

"We don't develop WRF. We define the parameterization – resolution and variables for what we are looking for. We provide parameterization for things like precipitation, aerosols, cloud formation, hydrology, the micro-physics and the land surfaces. Those are some of the most heavily used parameterizations in the model. Lots of people develop parts of the code, especially on the parameterization part. The current weather models were helped by people who put in the content over the last 25-30 years, and that feeds into the nightly weather forecasts by the media."

WRF enjoyed much loyal following. Good indicators of WRF's relevance and durability are its user base size and publications rate.

Skamarock checked the stats as of the end of 2019: The user registration count was just short of 50,000 (49,925 to be precise). What's even more impressive is that new user registration annual rate has averaged 4,150 over the last five years. That is, it is being constantly 'refreshed' even if some of the older registrants are inactive now. During the same period there have been 813 WRF-related publication per year on average. With 960 in 2019 alone. The cumulative publication count reached 6,545.

To put this in perspective, here is an anecdote from Bill Skamarock:

"I was reading an article on the Hubble space telescope and how NASA is still supporting its use, even though it is well past its expected lifetime. One of the reasons supporting this decision was that a lot of science was still coming out of the ongoing observations, and they cited a scientific publications rate of papers using Hubble data at something under 1000 per year, which interestingly enough is about what WRF use is producing. The other thing to note is that the publication rate derived from WRF studies is at its high-water mark."

Another vote of confidence comes from NCAR's Roy Rasmussen:

"Though I'm on the application side, if I were operational I'd want to use WRF. The reason for it is that there are thousands of WRF users who keep improving the model. After that many years the result is a really good model. When there are changes they run tests – that have to be as good or better than before."

A relatively recent overview of WRF's history and its capabilities can be found in a 2017 Bulletin of the American Meteorological Society article ([40]). The uniqueness of WRF is that, as is indicated by its name, it is simultaneously a research tool and an operational weather forecast model. It provides a dozen 'idealized' datasets ('scenarios') for research. For HPC practitioners, one of its most known dataset and application is known as CONUS – for CONtinental (or CONterminous) U.S. that is defined on 13 km and 3 km (very high resolution) grids. WRF on a CONUS grid also shows up frequently as a benchmark in HPC procurements in the public sector. We return to some of the many ways WRF is applied in Chapter 20.

CHAPTER 16

Planning Ahead

A remarkable workshop

T HREE years before the demonstration of the first teraflops systems (June 1997 entry on the TOP500 list), a group of HPC professionals from multiple disciplines gathered in Pasadena, CA in February of 1994 to think and confer about when and how we can get to petaflops-scale computing. It was a workshop sponsored by six federal U.S. agencies. Its principal organizers were Paul Messina (at that time at Caltech) who chaired the event, Paul H. Smith (at that time at NASA HQ), and Thomas Sterling (at that time at USRA[1]).

It was a remarkable 3-day by-invitation U.S.-centric workshop. There were 65 attendees, many of whom the who's who in the HPC community. They came from government agencies, supercomputer companies, semiconductor industry, academia, and national labs. Their individual skillsets were nicely divided among the four central focus areas of the workshop:

- Applications and Algorithms
- Device Technology
- Architecture and Systems
- Software Technology and Programming Models

The four working groups convened separately and proceeded in parallel for most of the time. There were both formal and informal interactions among the participants throughout the workshop, followed by getting together at the end to share

[1]Universities Space Research Association (USRA) is a nonprofit consortium of universities now under the auspices of the National Academy of Science. It was founded over 50 years ago at NASA's request to advance space science by fostering collaboration between academic institutes. It established distinct divisions, institutes, and centers of excellence.

the groups' findings and recommendations. The workshop is well documented in a book report that bears its title *"Enabling Technologies for Petaflops Computing"* ([41]). More on this below.

Sterling is now a Professor, Intelligent Systems Engineering, and the Director of AI Computing Systems Laboratory at Indiana University. He has a long and distinguished history in HPC (more about his pioneering work in Chapter 17). About the '94 meeting he says: "The Pasadena workshop was the finest technical forum I have ever participated in throughout my more than four decades career. I credit Paul Messina for his leadership and vision but a number of others contributed to its organization and intellectual structure. Maybe I'm just being naive, but it was exciting to the point of almost being existential. And that's just how I remember my reaction to it."

I consider myself fortunate to have been invited to participate in the workshop, representing NASA Ames Research Center. Most rewarding was to be assigned to the architecture and systems working group. It included some well-known and recognizable figures who were some of the most successful supercomputer architects of the time. See the group photo below (Fig. 16.1).

Seymour Cray gave a keynote address at the opening of the event. Cray was not a frequent attendee of conferences and similar public gatherings. He was known to prefer solitary work at his lab. It is likely he showed up at this meeting at the behest of government agencies that sponsored his development of the Cray-4. While the workshop was about petaflops computing, Cray made a prediction about the nearer-term teraflops-scale system. Based on a Cray-4 $80,000 cost for one gigaflops, projected improvement of 4x in four years, and the ballpark estimate that memory costs about 50% more than the processors, Cray stated that a teraflops system is possible at a cost of $50M within four years (in 1998). He went further, referring to another project at Cray Computer Corporation, stating that a system made of millions of bit-processors for image processing, programmed to perform floating-point ops, can also be built to be a teraflops system in four years for roughly $50M.

Sadly, both projects did not materialize. Seymour Cray died before the first teraflops machine was built. But his 'prediction' was eerily close. The June 1997 TOP500 list has ASCI Red (described in Chapter 12) as the top system and the first to exceed the teraflops mark. Its cost (before an upgrade)? – about $46M ([42]). Both the timing and the cost are within a reasonable margin of error. Curiously, the architecture of ASCI Red is neither a collection of small number of powerful processors – Cray-4-style, nor an ensemble of millions of bit-processors. It's somewhere in-between: A networked system of some 6,000 microprocessors.

Figure 16.1: Architecture Working Group, Workshop on Enabling Technologies for PetaFLOPS Computing Systems. Source: David Barkai.

Thinking ahead toward a petaflops era Cray saw this as a transition from micrometer-scale features to the nanometer realm. That is what did occur over time. But Cray, having been challenged to make a 'radical' proposal, talked about biological systems. He dismissed biological computers, but mused about 'training' bacteria to make transistors. Though an ongoing research area, with some early applications for materials and biomedicine, there is no use of it in fabs for computer chips. Maybe there'll be such application in the more distant future.

A few words about the findings and the thinking of the subgroups gleaned from the report ([41]):

The application drivers were large-scale numerical simulations. Emphasis was placed on the need for new algorithms to deal with the scale of parallelism and awareness of data locality to overcome the latency becoming a hindrance to performance. Whereas in the past people touted a rule of thumb that called for a byte of memory for a flops of computing, the Applications and Architecture subgroups concluded that, in general, 30TB of memory would be sufficient for a petaflops system. Geoffrey Fox's (then at Syracuse University) leadership of the Applications subgroup is credited for reaching this important conclusion. There was an expressed concern that the precision (word size) employed may have to be increased for petascale problem sizes.

The Device Technology group found that semiconducting silicon will continue to be used for memory and (possibly) for processor logic. They concluded that superconducting material *may* provide a much higher speed logic at very low power, and that optical technology will be essential for interconnect and storage.

Architecturally, there were at the time both vector processors and microprocessor-based distributed memory systems. The participants considered scenarios based on the level of parallelism: *Coarse Grain* – made up of hundreds of pipelined processors each teraflops-capable with shared memory. *Medium Grain* – thousands of microprocessor-built with workstations class processors in the range at 10–100 gigaflops peak performance each, in a globally addressable but largely non-local memory environment. *Fine Grain* – hundreds of thousands of microprocessors (1–10 gigaflops) with memories co-resident with the chips. That was the Processor-in-Memory (PIM) idea, and expected to cost much less than the other two approaches.

Things looked grim on the Software front. The view was that the software framework ought to provide a model for programming methodology and manage parallel resources and activities. The most commonly used programming models at that time, message-passing and data-parallel, were seen as challenging even for the scale of computing of the mid '90s. The group called for a more general and comprehensive model that will encompass the different modes of parallelism and maintain portability across hardware implementations. They highlight the lack of adequate tools for resource management and assistance to application programmers.

There were further meetings and projects as a result of the Pasadena workshop.[2] They span the next five years, and Sterling provided sketchy details: "There were several topical workshops. The first, in 1995, was chaired by Rick Stevens from Argonne National Lab and held in Bodega Bay (northern California). It was a unique experience of a two-weeks long workshop on application drivers and formulating how Petaflops capability would advance science. It even included three science fiction writers. There were two separate workshops, also in Bodega Bay, on system software. The second one of these was about distributed systems and how to collaborate. The last 'Petaflops meeting' was in 1999, dubbed Petaflops 2, and held in Santa Barbra."

Sterling recalls a major project, if controversial, that was funded after the Pasadena meeting:

"The HTMT (Hybrid-Technology, Multi-Threaded) architecture project ([43]) was a direct consequence of the 1994 Pasadena Petaflops workshop. It was determined by the associated leaders of the participating federal agencies that a significant step forward would be to inform future directions by substantive alternative models. Specifically, in the form of point design studies. The NSF (National Science Foundation) was selected to lead a set of such studies over a six month period. One motivation was to highlight the potential opportunities that may be presented by a diversity of emerging enabling technologies. Guang Gao and I submitted a proposal for HTMT that considered the synthesis of superconducting logic, optical interconnect networks, PIM smart memory, and holographic storage. The major theme was how to overcome the latency challenges with multi-threading. This was one of eight projects selected, and we undertook the paper and pencil study. We concluded that multi-threading would not be sufficient to overcome the latencies. I was happy to deliver a negative result. But we started to think about alternatives. We came up with what we called "percolation" which was a form of asynchronous message-driven computation. We presented our results at the third of the series of in-depth petaflops workshops.[3] It was not well received by several of the attending architects. However, in the back row of the room were program managers from the key agencies. And they decided that the HTMT project should be funded to go forward, with much more money. I proposed that we merge HTMT with one of the other projects: Peter Kogge's PIM-based project. They liked the idea. It lasted for more than two more years and studied the ideas in detail, delivering both positive and negative results. It launched me on an intellectual trajectory that has continued to this day."

Though the proposed architecture did not materialize as a petascale system, it serves as an example of the research that was spurred by the Pasadena workshop. A 1999 conference paper describes the scope and results of the project (see [44]).

---

[2]I did not participant in any of those. It was about a year after the Pasadena workshop that I left NASA and joined Intel.

[3]That meeting where the eight studies were reported on was chaired by Sterling with Fox as the technical program chair and held in Oxnard, California.

The Pasadena workshop's report contains a list of items under the heading of "Results" that is really a set of predictions about what a petaflops system would look like. Here is a summary:

1. It would take 20 years for a petaflops-scale computer to be feasible (that is, expected around 2014).

2. Such a system will be made of advanced versions of today's (mid '90s) multiprocessor architectures.

3. Memory will be a dominant factor. But much less that byte-per-flops will be sufficient.

4. Effective performance will require "radical departure" from current methods of memory-processor interactions.

5. The rate of progress will be determined by the mass-volume market; not the high-end.

6. Semiconductor technology will be the main medium of components, with optical devices providing interprocessor and memory access.

7. The system's part count (number of chips) will be between 100,000 and one million.

Jumping ahead of the chronological sequence in telling the HPC story: How did things turned out for petascale computing?

The first petaflops system was a one-off system from IBM called Roadrunner delivered to Los Alamos National Lab in 2008. It consisted of 6,480 of what may be referred to as *nodes* (interestingly, close to the number of processors in ASCI Red, the first teraflops system), each made up of a dual-core AMD Opteron processor and two IBM PowerXCell processors functioning as *accelerators*, each attached to one core of Opteron. Its total memory size was just over 100TB (the Pasadena workshop stipulated that 30TB is an adequate design target), and it consumed 2.3MW of power. Roadrunner was decommissioned after five years (contrast that with ASCI Red, the first teraflops system, that stayed in production for 10 years).

The second system to reach petaflops was more in line of the-then prevailing HPC architecture, and typical of most other systems of the petascale era. Jaguar, as the system was called, installed at Oak Ridge National Lab, reached the petaflops level in late 2008 after several upgrades. It was a Cray XT5 system – a cluster with proprietary interconnect. Its x86 processors were quad core Opteron CPUs from AMD. There were close to 40,000 processors (just over 150,000 cores), and the total memory size was at 360TB. Jaguar's power consumption was at what was then a staggering 7MW.

We can now consider the accuracy of the workshop's prediction. First, evaluation of the list above:

1. The first two petaflops systems appeared earlier than predicted – by about 6 years.

2. The statement is a little vague, but it is safe to assume that the term "today's multiprocessor architectures" referred to Cray vector multiprocessor style of architecture. The high end of the latest TOP500 list at the time of the workshop was a mix of multiprocessor systems and MPPs. The petaflops systems were closer to high-performance clusters made of high-end commodity microprocessors and high-performance interconnect. With this interpretation, the prediction missed the mark.

3. Correct prediction. Managing data, i.e. memory, became the challenging aspect of HPC systems. The memory size was indeed much less than byte-per-flop, though 3-10 times larger than the workshop's figure of 30TB. The concern that higher precision might be required did not materialize.

4. MPI, the message passing interface, was the main mechanism for handling the distributed memory for parallel programming on petascale systems, with OpenMP being the tool for the shared/local part of the memory. MPI existed from the early '90s, and OpenMP was formalized in late 1997. there was no "radical departure" from the known methods of the '90s.

5. The rate of progress was, indeed, determined by the mass-volume market. The processors and the memory components in the petaflops systems were those used in the volume market of the data centers, though the high-performance interconnect of the high-end systems was developed more specifically for the HPC market.

6. It was a correct prediction that semiconductor technology will remain the main medium of components, and optical cables began to be present in interprocessor networks and memories. The Jaguar system had fiber optic network, but coaxial cables were still common.

7. The predicted range of a petaflops system part count turned out right when including processors, memory, and network.

From the three parallelism-level scenarios considered (coarse, medium, fine) the early petaflops systems fall somewhere higher than *medium* (more processors, each less powerful),but not quite *fine* – perhaps at its low end.

In hindsight, it is interesting that the term "server" was not uttered in the workshop, though servers became the building blocks of HPC systems. Nor did the participants foresee the dominance of x86 processors by the time the petascale era began.

Power consumption was a critical figure in designing future systems, and at the Pasadena workshop the ceiling for a petaflops system was assumed to be around 1MW. In reality, the Roadrunner consumed more than double that figure and Jaguar a whopping 7-times more. The latter was more typical of subsequent petaflops systems.

What to me is the biggest miss of the workshop is the role accelerators (GPUs) were to play in future HPC systems. At the very high-end GPUs contributed much of the computation power in petascale systems beyond the first two. They remain a fixture of HPC systems.

The lesson: Device technologies appear amenable to forecast of 10-15 years out, but which will be adapted is not. High-end system architecture evolution seems incremental, but we often fail to predict the "winner" among competing architectures.

The coming-together of the HPC community to think ahead and work out issues, as was done when planning for the petaflops era, was not the last one. Some 15 years later, when the time came to think about exascale systems, another series of meetings took place. I was there too and describe it in Chapter 23.

The Pasadena petaflops technologies meeting is also credited with being the trigger for the DOD/DARPA High Productivity Computing Systems (HPCS) program described in Chapter 19.

IV

The Epoch of Clusters

Standardization of Coarse and Fine Parallelism

CHAPTER 17

Standardization

The Demise of propriety processors and software stacks

T HE start of the new millennium marked some 30 years since the introduction of the microprocessor concept. The Intel 4004, debuted in 1971, was the first one to be marketed. A 4-bit CPU, built with 10 micrometer feature resolution, achieving an instruction issue rate of about 10 microseconds (92K instructions per second).[1] Of course, it was not suitable for the HPC workloads of the '70s.

By 1990 we had over a decade of use and progress of microprocessors in personal computers. The microprocessor became the disruptive technology for HPC systems of that era. Starting with little ability to perform numerical computations, with memory, clock, and software so far from what was needed for HPC, that it did not seem threatening to the supercomputer designs of that time. However, the capabilities of the microprocessor improved at a pace that could not be matched by the propriety computer architects. A typical cycle for development of a next generation supercomputer was about 5 years. The microprocessor marched forward at the pace of Moore's Law – doubling the number of gates every 18 to 24 months. In those days its performance just about doubled at this rate too. By the early '90s the microprocessor was capable of floating point operations in single precision (32-bit arithmetic). It was 'only' between one or two orders of magnitude less able than the then-current supercomputer processors. It made sense now to consider systems based on price-performance. This is where the micro won hands down. After all, the microprocessor process technology was supported by the huge, and fast growing, personal computing market.

---

[1]As a reminder: Present gate density is 1,000 higher, and the clock rate O(1,000) faster. Note that Moore's Law applies to the number of gates on a chip, which is proportional to the area, and therefore progresses as the square of the feature resolution increase.

DOI: 10.1201/9781003038054-17

Workstations

We saw how the idea of using less-capable processors that can be manufactured more cheaply and can be air-cooled was gaining strength with the advent of minisupers and MPP designs. This approach can be seen as *scaled-down supercomputers*.

However, the microprocessor revolution resulted also in a different approach. One that created a new class of technical computing devices. The personal computer, the PC, while not suitable for technical computing, popularized the notion of a *personal*, desk-top or desk-side, computer. Thus was born the *workstation*.

Starting in the '80s several companies created products based on microprocessors that were considerably more expensive than PCs, but were capable of decent performance of scientific codes, while still fitting in an office environment on a desk or next to it. They used faster components and considerably larger memories, had 32-bit arithmetic functional units, used (mostly) RISC architecture (see page 83), and were running the UNIX operating system. The better known workstation vendors were Apollo Computer, Sun Microsystems, and Silicon Graphics (SGI).

SGI also added graphics accelerators that made their products particularly suitable for visualization. This is an often overlooked implementation of the *accelerator* idea in HPC. A theme that is pervasive in today's top HPC systems.

The focus of this book is on supercomputers – the high end machines, but the technical workstation turned out to be a building block on which future large systems were constructed. It served as a model for today's server board.

And, as described in the following, workstations were used in what we can be called the proof-of-concept of modern clusters.

As PCs advanced, both those based on the Intel x86 architecture and those made by Apple, the distinction between workstation capability and that of top PCs blurred. Though initially with 32-bit floating point arithmetic, when workstations from Sun and SGI supported 64-bit arithmetic, the PC-based workstations had a significant price advantage. They benefited from the high volume of the PC market, as they were mostly just the biggest configurations of the high-end components of the wider market of PCs.

The Beowulf Cluster

Teams at NASA Goddard pioneered early parallel processing with their Goodyear MPP in the mid '80s (Chapter 12). They did it again less than 10 years later with what came to be known as the *Beowulf Cluster*. The idea was to harness multiple nodes – workstations, and later servers – to work in parallel on a single job. The architectural technology for achieving it was developed in 1993–1994 by Thomas Sterling and Donald Becker.

I find the Beowulf project historically significant as an early proof-of-concept, or prototype, for the cluster architecture that dominates HPC now. Unlike the earlier and concurrent experimentations with collections of workstations, Beowulf looked more like a single system, managed from a single control point – typically, one of the workstations or servers. We need to think of Beowulf not as some specification of hardware components and software stack, but as a technology of clustering for parallel execution ([45]). One could choose from a variety of compute nodes, operating system distributions, and parallel libraries. There was a guiding principle, though. As Sterling puts it: "If I had to say what is a Beowulf, I would say it is a commodity cluster, completely COTS (Commercial Of The Shelf)." The hardware was chosen from consumer products, not systems that would go into computer centers, and the software was taken from the open-source community. He adds: "In retrospect, what we did was obvious. But at that time it was not obvious. And in fact, many of our colleagues thought it was a terrible idea."

As Sterling tells it, the goal was to produce a system such that a one gigaflops cluster was less expensive than a high-end workstation. It was accomplished, by Beowulf's three generations of clusters from 1994 to 1996, with Intel x86 microprocessors, 100 megabits/second Ethernet, and Linux operating systems with PVM for its first generation and the MPICH variant of MPI in subsequent generations. Beowulf would routinely exhibit a performance cost advantage of 10-20X for real-world applications over MPPs of similar performance, and at least 4X advantage over workstation based clusters which went for premium priced components.

Twenty years after the birth of Beowulf there was a workshop in honor of Sterling (it was his 65th birthday). Jim Fischer, a program manager for high-end computing at NASA Goddard under whom the Beowulf project was run, gave a talk titled "The Roots of Beowulf" ([46]) that provides a useful historical perspective on the origin, motivation, and the construction of Beowulf.

Turns out NASA identified a need for a workstation (in the sense of a single-user system) that will cost less than $50K and deliver gigaflops scale performance. Hidden there was also the need for portable and shared software – this is an important element of the eventual solution. The state of affairs in HPC world, as NASA saw it, was one mired with some fundamental issues: Proprietary software and hardware inhibiting portability, poor price/performance, performance bottlenecks diminishing productivity, operational instability of existing systems, incompatible architectures and programming models, and cumbersome acquisition process. Fischer's team was looking for ways to make parallel computing more accessible.

That was the environment that nurtured the new concept. Fischer recalls: "I remember well the day that Thomas Sterling and John Dorband came to my office and told me about the Linux PC cluster idea that Thomas and Don Becker had conceptualized. . . As they described the plan, I could see that the Linux cluster would be amazingly inexpensive. . . When they left my office I was onboard too."

Indeed, the Beowulf project delivered on its promise. Sterling, Becker, and Dorband managed in 3-4 years to demonstrate the concept and get it out of the lab. By 1996 there were at least two Beowulf production clusters that delivered 1 gigaflops at about $50K on real applications. There were two 16-processor systems at 1.1 (at LANL) and 1.26 gigaflops (at JPL) running two different applications with a price tag of $50K and $60K. The architecture was proven to further scale when these two clusters were brought together at the Supercomputing '96 conference to form a roughly $100K 32-processor system that executed a real application at 2.2 gigaflops.

The Beowulf concept was duly recognized for its contribution to HPC. In 2022 the Space Foundation inducted the NASA Beowulf Project, naming Sterling and Fischer into the Space Technologies Hall of Fame. A few months later, the International Conference of Parallel Processing (ICPP) at its 51st meeting gave the first Beowulf paper presented in 1995 that year's "Test of Time Award." Back in the past Beowulf won a 1997 Gordon Bell Prize.

Beowulf, and other experimental projects involving commodity parts thus ushered the cluster era.

The New HPC Business Model

Fischer, at his Beowulf talk, laments what he called the *"maze of architectures"* that existed in the first half of the '90s. He's referring to available HPC choices that included vector processors, MPPs, minisupers, and technical workstations. That was compounded by the uncertainty about the continuation of several architecture styles, and even about the existence of the companies producing them.

While the microprocessor revolution took over in the early '90s with the resulting reduction in specialized process technology, there was no immediate coalescing of processor architecture. There was the MIPS architecture from SGI, ALPHA from DEC, SPARC from Sun Microsystems, x86 from Intel etc. That, and the variety of system architectures, was about to change.

By the turn of the century the HPC landscape was much different than how it was twenty years prior. No more vector processors (with the exception of NEC in Japan). The MPP companies of the '90s were gone or switched to software and other products. Computer companies write compilers and libraries, but not operating systems. Communication protocols within and without the systems are the same across products.

We can talk about two trends that altered both the engineering and the business makeup of HPC. One is *standardization*, and the other is *commoditization*.

By *standards* I mean the growing commonality in the components of the software stack and high-level view of systems' architecture. In particular, the adoption of Linux variants for the operating systems. The architecture could almost universally be described as a collection of server nodes in similarly-looking racks connected by one of the two or three generally available network technologies. The servers themselves share a similar structure of two processors on a motherboard with memory DIMMs attached via memory channels and addressed by both CPUs. The CPUs are microprocessors adhering to the IEEE standard for floating-point operations. Over time there were fewer and fewer microprocessor implementations, and an overwhelming majority are x86-based since, driven by the PC market volume, it did better by the Moore's Law progression curve. Accelerators' return to HPC after early occurrences in the '80s (see Chapter 8) in the form of GPGPUs has very quickly become a standard feature of most of the HPC offerings.

There are exceptions to this over-simplified picture. For instance, Cray has a propriety interconnect solution. Some vendors offer *fat nodes* of 4 or 8 processors. But these represent a very small volume of the market.

Commoditization is the reliance on parts that are mass-produced and are also used in the consumer market. Or, at least, outside of the HPC space. It is true that for HPC the vendors apply the high-end versions of their product lines. These will include the larger and faster memory parts and the processors with more cores and faster clock. But, almost always, these are derivatives of the parts built for the consumer market, that is, the PC and workstation users; not the other way around. The shift to x86-based servers amplified both *standards* and *commodity* trends.

These engineering-inspired changes in how HPC systems are created produced a major shift in the HPC ecosystem. It has been a gradual process that started in the early '90s. By the turn of the century is was clear that gone are the days of in-house design, development and manufacturing of proprietary systems. The self-contained "silo" structure of the HPC business sector has morphed into a set of inter-dependent entities each with narrower function compared to the all-inclusive nature of the companies of the past. Existing companies adjusted and new companies appeared. A notable exception is IBM with its POWER architecture that is done entirely in-house. But even here the software options are a UNIX-based operating system or the open-source Linux. And IBM created an OpenPOWER alliance – a forum for shared development with its partners.

The use of common components of both hardware and software created a sector we can call *technology providers*. These include companies that design and sell the processor chips, whether manufactured in-house or elsewhere. Intel and AMD provide x86 chips. Nvidia became a major provider of the GPU-based accelerators. Memory cards were always outsourced to companies in the Far East and the U.S. Storage devices are largely produced by companies specialized in digitized storage. The same is true for the equipment and parts needed for power supply and cooling. Relatively late arrivals were companies that developed high-performance interconnect networks for HPC clusters. Mellanox stands out with its InfiniBand products during early 2000s. The technology providers have sales and marketing organizations that are aimed not at the HPC end-users, but at technology customers, who are the companies building the computer systems out of the various components they acquire.

These are the old HPC companies that survived past the '90s and became, essentially, *system houses.* The buyers of HPC systems, who use them in-house to support their institutional mission, acquire them from companies such as Dell, HPE, Cray, SGI, Penguin, IBM (which also remained a technology provider), Lenovo, Fujitsu and others.[2] In the emerging business model a system house can, and often does, choose components from multiple technology providers. The model allows products with Intel and AMD x86 processors, for example. Similarly, with the choice of processor interconnect, storage devices, and memory parts. The differentiation between system providers is about the combination of building blocks chosen for a given product line, their packaging, the engineering that goes into the server boards, the modules and racks that house the servers, and the cooling and power supply subsystems.

This clear separation of responsibilities between the technology providers and the system houses gets a little blurred when it comes to what is known as *chipsets.* These are the chips that enable the data flow between the CPU and the memory, storage (IO), and other systems (the network, for example). The CPU vendors, such as Intel and AMD, provided such chipsets to their customers, but the larger system houses such as HP, IBM, SGI, Cray etc. often developed and deployed their own chipsets. That was an added differentiation factor between the system houses. However, as process technology advanced and the gate density on the CPU chips increased some of the chipsets repertoire was integrated into the CPU chip. In particular, this is the case for memory and PCI (Peripheral Component Interconnect) controllers.

The software delivery structure is more of a mixed bag. Commercial applications have been mostly provided by ISVs (Independent Software Vendors) who confined themselves to a single category of application. Examples of such categories are Engineering Modeling, Drug Exploration, Seismic Oil and Gas Exploration, Computational Chemistry, Financial Currency Trading, and more. Some application types are produced by, and for the use of, a single organization. This is the case for weather models, for example. Scientific applications are not sold commercially but often shared among institutions. The HPC system houses have significant software teams, though. They produce their own middleware and system libraries, and several develop and maintain their own compilers and numerical libraries. Others use open-source compilers and math libs. The most common operating system is Linux with minor adaptations to cater for features of the specific system. An important ingredient that supports the emerging HPC environment is the adoption of standard open-source libraries for expressing parallelism for shared and distributed memory systems – OpenMP and MPI, respectively.

---

[2]SGI and Cray were acquired by HPE later on.

The interplay between the technology providers and the system houses leveled the playing fields. Small companies now have access to the same technologies as the larger ones. Easy access to different solutions of subsystems allows for greater flexibility and faster design and development. The competitive landscape changed from *'who owns that latest technology'* advantage, to *'who can assemble a better high-performance system from the available parts'*. The system houses can now focus their resources on the quality of packaging, assembly of parts, and maintainability.

The HPC user community benefitted well from this new business model. Similarity in the architecture and the use of common processors allow for greater compatibility and portability between systems from different vendors. End-users can more easily switch vendors between procurement cycles. The competition for winning the user customer is now about features that matter to the users and that multiple vendors can strive to provide.

On Solving the CISC vs. RISC Issue

In Chap 8 there is a brief discussion of the pros and cons of the Reduced and the Complex Instruction Set Computing (RISC and CISC) – the two approaches to designing the instruction set for a processor. In short, RISC is simpler to build and results in a more compact hardware. CISC results in a more compact code and offers a richer set of instructions.

High volume products drive bigger investments in development and reduced costs per unit produced. The PC market was the high-volume market. At Intel they concluded that its processor – the x86 CISC architecture chip, will be adapted to the server and the HPC market. For reasons of backwards compatibility, and after investment in applications for over a decade, its instruction set had to be kept. But it takes less silicon and a simpler design to manufacture a RISC processor. The Intel processor architects (and others) wished to get the best of both the RISC and the CISC worlds. They found a way:

Build a RISC processor that will execute CISC-generated binary code. The core of an x86 processor, since at least the mid '90s, executes a RISC-type instruction set. We need not know what it is since compilers for x86 generate its CISC-based instructions, allowing for continued use of 'legacy' codes and a smaller footprint of the binaries produced. The RISC details are not visible to the programmer. The magic that glues it all together is known as *microcode*. It is the layer that processes the CISC binary. The microcode, occupying a fast memory area on the chip, takes CISC instructions and outputs RISC instructions that are fed into the functional units of the processor. Thus, the x86 processor behaves as a CISC CPU to the outside viewer, and as a RISC processor internally.

There is an additional, and very important, benefit to the use of microcode. Being a code, a program, that is placed on the chip, it can, in principle, be modified if necessary. Intel, and later AMD too, implemented a mechanism by which corrected microcode can replace the code previously installed. This feature allows bug fixing without having to replace the hardware. It was first implemented by Intel after a rarely occurring bug, known as the FDIV bug (see [47]), resulted in wide recall of its x86 chips. Being able to *patch* the microcode served Intel well since.

It is interesting to note that the first phase of the ASCI Red system described earlier started with a RISC chip (the i860), and when the system was upgraded it was populated with the x86 Pentium Pro CISC architecture chips (see page 105) – a more powerful CPU.

On Accelerators

The discussion about standardization of HPC system architecture, CPUs, and software, combined with the note about the instruction sets, is a reminder of another element of the HPC environment: Accelerators.

A recurring theme of HPC, accelerators for numerical or AI workloads seem to be quite a common fixture in todays systems.[3] Clearly, having attached processors makes life more difficult for the user. But the economies of building high-performing chips without the silicon needed for managing the OS, storage, interconnect, and parallel execution are too attractive.

This is a case where hardware considerations overruled users' convenience.

---

[3]At the high-end, as given by the TOP500 list, accelerators are present in about 30% of the systems. Their performance share is close to 60% (2022 data).

CHAPTER 18

HPC at Intel

The Role of a Major Technology Provider

INTEL was an HPC player back in the '90s. Not only was it the major provider of the x86 microprocessors that enabled the "killer micro" revolution in HPC, it also took on the role of a systems house. As mentioned in Chapter 12 Intel designed and manufactured several massively parallel systems. A phase that culminated with ASCI Red at the Sandia National Lab.

Having achieved the Teraflops milestone Intel pulled back from its role as a system integrator and vendor. It dismantled its HPC division in 1995. The main rationale for this action was that being a systems house did not fit with the company's business model. Intel's strength was in the manufacturing of processor chips and other components. The fabs were, and are, the backbone of the company. Its customers were the companies that build HPC systems (and PCs too). It would complicate the relationships to have to compete with one's customers.

An unintended consequence of this move was that Intel personnel neglected, or did not see the need, to talk to HPC users and their organizations. That changed when the same lab where it had its greatest HPC achievement – Sandia National Lab, selected its successor HPC system to be based on processors from Intel's arch-competitor, AMD, in mid 2002 (installed in 2005). That system, named Red Storm, was built by Cray and its architecture conceived by a computational scientist from Sandia, Jim Tomkins, with Cray as the integrator and design partner. Cray, at that time, was using AMD CPUs. Red Storm had a similar high-level architecture to that of ASCI Red. The tightly-coupled systems were partitioned by function (compute, service, I/O, system for ASCI Red) or by workload (classified or not, in the case of Red Storm). Red Storm was Sandia's flagship system from 2005 to 2012. It ranked #6 on the TOP500 list when launched, and #2 the following year after an upgrade that nearly tripled its performance. A second upgrade, in 2008, doubled the system again, and yet, as a reminder of how fast technology moves, Red Storm ended up as #9 on that list.

DOI: 10.1201/9781003038054-18

It is noteworthy that by that time, early 2000s, top HPC systems costs were about an order of magnitude higher than they were merely 20–25 years before. Red Storm initial installation cost about $90M, compared to the $10M or less of the late '70s supercomputers (10 years later the largest Cray Y-MP would sell for over $30M). This is in spite of using commodity parts and the advances in process technology, and an indication of the even faster pace of growth in HPC systems size and complexity.

The Red Storm announcement was a painful wake-up call for some in Intel. Mainly for its Sales and Marketing organization (a datacenter group did not exist yet at Intel). *If only they would have kept in touch with their prized customer..*

As fate would have it, this event brought me back to direct involvement with HPC. I was reassigned to start what would become an HPC practice at Intel. My first task was to go and meet with Bill Camp who headed the high-end computing division at Sandia. I remember the lunch in Albuquerque in the summer of 2002, where the typically mild-manners Camp gave me an earful of where Intel has gone astray. Turned out Intel's troubles went deeper than just not communicating with the HPC end-users. He explained to me that Intel was lacking the right product to be considered for Sandia Lab's next capability system. That was all the more unsettling considering that the system to be replaced was the Intel based ASCI Red, the first teraflops system in the world.

Which brings us to the story of Itanium – a topic Camp raised as a misstep by Intel.

The Itanium Story

In 2001, Intel released the first processor, code named Merced, in a product line called Itanium (formerly known as IA-64). It was the culmination of about 12 years of research and development, a project that started at HP Labs and that Intel officially joined in 1996.

There are several important lessons that can be gleaned from the Itanium saga (more below), but the salient facts regarding the Sandia Red Storm decision are these: AMD's Opteron had 64-bit floating-point hardware. The Intel Xeon was still at 32-bit floating-point functional units. Intel's 64-bit offering was the Itanium. It underperformed HPC workloads even after several delays of its launch.

The Intel-HP team thought the transition to 64-bit microprocessors for the emerging Enterprise server market and for HPC is an opportunity to introduce a clean new architecture supporting large memory that will do away with the x86 CISC 'baggage'. The idea was to go beyond just RISC: Execute RISC instructions, but operate on multiple of them concurrently. The fancy name given to the architecture was *Explicitly Parallel Instruction Computing* (EPIC). In fact, they revived a concept that was tried before – the *Very Long Instruction Word (VLIW)*, in which the instruction format contained several machine instructions. The 128-bit

Itanium instruction bundle has space for three instructions. The Itanium's independent functional units with its intra-processor paths and registers allowed, in theory, up to six instructions to execute in parallel (of course, not all of the same operation type).

The smart engineers who worked on Itanium understood that Itanium cannot deliver on its promised performance without the compiler carefully scheduling instructions that can be executed in parallel by filling the VLIW. They bought off on the idea that opportunities for such parallelism exist in abundance, and that the compiler can be taught, within a short time, to detect them. That did not happen. Note that here we are not looking for data parallelism, where the same operation is applied to many array elements. This is a case of finding concurrency within a single instruction stream.

The wishful thinking regarding the compiler (and the applications) lasted longer than it should have because building the hardware turned out to be more difficult than expected. It took years before there was a prototype to test the architecture against real-life and real-size codes. So, the hype persisted.

To ease the issue of not having many applications run on Itanium and to help in transitioning to Itanium an x86-compatibility feature was added. It was a translator from x86 binaries to Itanium code. Almost needless to say, it did not produce quality code, and did not open the market up for Itanium.

The delay in finishing the CPU construction meant that when the first-generation Itanium, Merced, and launched in 2001, it was built with the 180nm process technology of the mid '90s. The competing CPUs were one or two process generations ahead. The Itanium line survived, longer than it should have, due to a combination of factors: A strong case of 'group think' within the Intel leadership,[1] the enormous power and influence of Intel's marketing, the reluctance of Intel's software group to admit its compiler cannot be taught to produce efficient Itanium code in a reasonable time-frame, and a tight contract with HP Intel did not wish to break. After HP (later HPE) realized Itanium cannot succeed in HPC, they still wanted to use it for enterprise applications, especially database apps, where it had some success.

So Intel continued to produce future generations of Itanium. Almost invariably, they arrived later and with less than the promised performance. Relative to Xeon, Itanium was behind in process technology, had slower clock than the x86 processors, and consumed significantly more power. Itanium remained a low-volume product, yet consuming engineering resources that could have been put to better use elsewhere. Even worse, Intel held back adding 64-bit arithmetic hardware to the Xeon (x86) servers, so it would not compete with Itanium. Of course, AMD did not hold back and some companies (notably Cray) did not use processors from Intel until

---

[1]In the spirit of full disclosure: I was an early Itanium skeptic in an era when such dissent was not looked on favorably.

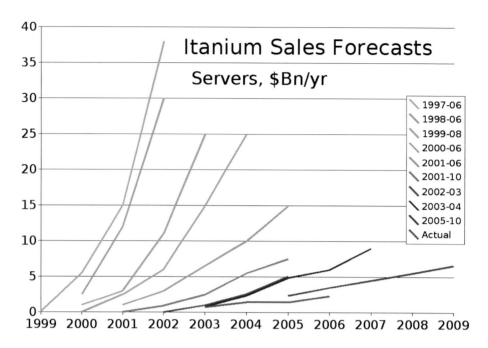

Figure 18.1: Itanium Sales Forecasts History Over the Years.
Source: Arch dude –
http://en.wikipedia.org/wiki/Image:Itanium_Sales_Forecasts_edit.png.

years later. Intel's share in HPC dropped. There is an interesting side story here about how the Xeon development team quietly prepared for 64-bit capability and were able to quickly spring Xeon x86-64 on the market when received the OK in 2004. More details about the history of Itanium can be found at [48].

To me, one of the more enduring lessons from the Itanium history is the power of Marketing in keeping a myth alive. There were many technical and performance claims that were misleading or flat out wrong, but the sales forecasts for Itanium outdid those easily in how wrong they were. One account (not independently verified), shown in Figure 18.1, was created in 2007. It shows that in 1999 Intel Sales forecasted that in 2002 the annual revenue from Itanium servers will reach $38 billions. In 2000 it was revised down to $30B, and later that year revised to annual revenue of $25B in 2004. In 2001 the forecast was that Itanium server sales would reach $15B in 2005, and later that year downgraded to about $7.5B. The 2002 forecast started even lower, but predicted close to $9B in 2007. Then there is the 2005 forecast that says it'll be 2009 before Itanium sales reach $7B. The actual numbers? – for years 2003-2006, low at $1B to a high of $2B, annually.

With all that said, there was a (short) period when Itanium showed a strong presence in HPC land. Even at HPC high-end space as represented by the TOP500 list. In June 2003, Itanium at its second generation, 17 out of the top 500 HPC systems listed were Itanium. Its peak presence on the list was the November 2004 list, with 83 systems (16.6%), including two at the top 10. It's downhill from here – eerily close to half-lifetime of radioactive decay: 44 systems in 2005, 22 a year later, 9 in 2007.

Itanium became less and less competitive for HPC workloads, and it is no coincidence that in mid 2004 Xeon acquired 64-bit native support. Indeed, while Itanium was fading from HPC, Intel began to do very well with Xeon.

For HPC the greatest successes for Itanium on the TOP500 list were two systems that were built and installed in 2004. A prior system deserves an honorable mention: An Itanium cluster installed at the DOE's Pacific Northwest National Lab (PNNL), reached in the summer of 2003, after an upgrade, the #5 spot on the list.

The first one, named Thunder, was a 1,024 nodes cluster with four second-generation Itanium processors per node. The 4-way Itanium SMP had a large memory (8GB/node) compared to the more common 2-way x86 servers. Thunder was supported by the first deployment in production of the Lustre Parallel File system worldwide. It was also the world's fastest Linux cluster when launched. Thunder was designed in close collaboration with scientists and system architects at Lawrence Livermore National Lab led by Mark Seager, at that time the Assistant Department Head for Advanced Technology in the Livermore Computing Center. The systems integrator was a small Silicon Valley startup – California Digital Corporation. Remarkably, it took just 5 months to build and deliver the system. Thunder was #2 on the June 2004 TOP500 list, behind only the NEC *Earth Simulator* (its last time on the #1 spot on the list), and at about one 20th of its cost.[2] Seager expresses Thunder's significance thus: "Thunder was hugely successful. It put institutional open computing on the map for NNSA[3] and opened the door for commodity microprocessor Linux clusters for the ASCI program."

The second large Itanium system followed soon after. This is one I was more closely involved with, as it got me working together with past colleagues at NASA Ames Research Center. Walt Brooks, the chief of the NASA Advanced System division at NASA Ames, invited Intel (the sales rep and me) to a meeting in Spring 2004. He outlined a proposal for a top HPC system designed and built with SGI's Itanium product, Altix. The proposal quickly turned into a high-pressure project. NASA, SGI, and Intel engineers designed a modular system made up of 20 'blocks' of 512 Itanium2 processors each. A unique feature for system that size was its shared address space through the SGI NUMAlink interconnect technology. The system was named Columbia, in honor of the same-name ill-fated space shuttle.

We believed that if put together by fall it will top the next TOP500 list. It almost happened. Columbia was built fast and in record time,[4] and performed as

---

[2]Seager discloses that Thunder cost \$13M while the Earth Simulator's price-tag is estimated at \$500M (a figure that included the building housing it).

[3]The Department of Energy's National Nuclear Security Administration

[4]the NASA-SGI-Intel team was recognized for this achievement with a NASA-sponsored award presented in Washington, DC a few months later. I was one of the Intel representatives there.

expected. Then rumors started of another system that would have even higher performance. IBM was working with Livermore Lab (and Seager) to stand up a large BlueGene/L system. It was to be much more powerful than Columbia and we did not think it could be built in time. We were partially right. Turned out that merely half the BlueGene's intended size would outperform Columbia. And the TOP500 list rules allow a system that has a customer to submit the benchmark while the system is still on the factory's floor. We found that it applied to a part of the intended system too. Half of the planned IBM/LLNL system was built and benchmarked for the Fall 2004 TOP500 list. The 'drama' ended with Columbia capturing the #2 slot. It stayed a top-10 system for two more years.

These high-end successes of Itanium did not translate to downstream acceptance by the broader market. Technically, they had unique system designs, but they were more 'special' from a business perspective. Seager talks about the Thunder deal:

"I convinced Intel to price 4-way itanium processors and chipsets for HPC different than for enterprise. Enterprise surcharge for 4-way was excessive. With special HPC part numbers (SKU – Stock-Keeping Unit) they were able to separate the markets and not undermine the enterprise margins. Part of the argument was that AMD didn't have a 4-way product, and they needed to grow the Itanium market share in a way that didn't inadvertently help AMD. That carried the day inside Intel."

In the case of NASA's Columbia, Intel provided SGI with a one-time price for the 10K Itanium chips that was much reduced from the standing favorable pricing SGI already enjoyed. Rumor, deems plausible by people in-the-know, has it that even with the O(10) discount relative to the list price, Intel more than recovered the cost of manufacturing the Itanium chips.

There was another benefit to the purchasers of these large systems that could not be scaled. That is the attention they received from Intel during and after the installation. Seager again: "The other thing that Intel did was that they had some of the compiler guys directly addressing the problems that our guys were having while we were fielding the machine and that also made all the difference in the world. And since these were enhancements of the existing compiler tool set, a lot of the changes that they had to make also improved the compilers for Xeon, which we liked a lot."

The NASA folks got similar close attention from Intel and SGI. This mode of vendor-user cooperation was successful, in no small measure, due to the users' own programming expertise. Such pools of sophisticated software and application teams is non-existent in most customer sites.

Columbia's architecture was more specialized than Thunder's with its blocks of processors connected by proprietary network and supporting global address space. But Thunder, as Seager emphasizes, was a demonstration of 4-way, large node memory, Linux cluster that could have been widely replicated, had there been a more forceful marketing effort to promote the concept. He points out that Thunder was within 10% of the Earth Simulator's performance (on Linpack) at about 20th of the cost.

The Itanium story can serve as a case-study of several important lessons:

■ It is hard to introduce a new architecture. Merely coming up with a clean and innovative design is not enough. Computer architecture, as a technology, has to be supported by a complex ecosystem that includes system software, committed vendors, assurances of future generations of the processor, and, above all, a way to migrate user applications to the new architecture. Though successful in some aspects, Intel failed to make it easy or worthwhile for general HPC workload to be ported over.

■ It is wrong to assume that software will 'eventually' resolve complexities that stand in the way of efficient use of hardware resources. Perhaps oversimplifying, the gist of the matter is this: Software engineers were able to hand-schedule VLIW instructions, mostly filling them up, for small kernels, and then generalized that a compiler can imitate their coding process on a full-scale application. They may have also assumed that the range of opportunities for filling VLIW exists throughout the code. In any event, the hardware development proceeded based on false expectations. The compiler technology was not advanced enough, and perhaps still isn't.[5]

■ Marketing hype will eventually succumb to reality. It is a testament to Intel's influence as a technology provider that Itanium lasted as long as it did. Hardware vendors were provided incentives to introduce Itanium products. Software companies were compensated for porting operating systems and utilities for Itanium. The large systems, including the two described above, were given to the customers at much discounted prices. The writing was on the wall, and all it took is revealing a 64-bit Xeon to diminish Itanium's relevance for HPC.

■ There are side effects and unintended consequences to a strategically misguided business decision. To Intel's credit, unleashing the 64-bit Xeon when it did allowed it to recover from the continued loss of market-share in the server and the HPC markets suffered due to the focus on Itanium. However, hundreds of engineers, both at Intel and at HP, could have been engaged all those years in more promising projects. Hopefully, some skills and techniques developed then were useful elsewhere. That said, careers were disrupted and some departures of people may well have been triggered by desire to avoid

---

[5]Intel appears to have repeated the same mistake just a few years later with the Xeon Phi. Many-core x86 design, first an attached processor, later with a stand alone version. It, too, was pushed hard onto the market by Intel and it, too, failed to deliver the expected performance. Here it was more of the challenge of managing the data flow that was necessary for sustaining the compute power on the chip.

working on what seemed to be a dead-end project. Of course, we accept many failures of startup ventures, but the Itanium project was on a much greater scale.

Getting on Track Again

In the early aughts it was the 'Field' (the sales organization including its technical support team) that was the driver for getting Intel re-engaged with the HPC user community. The people most eager to participate were involved with Intel's supercomputer division in the days of the Paragon and ASCI Red. With their help I started visiting HPC personnel in national labs, government agencies, universities, and industry. These meetings with individual organizations were very valuable and would continue, but after participating in a technical meeting that had the roundtable format, I thought this format would be useful for creating the dialog between Intel and the HPC user community.

A central lesson from the Red Storm affair was that Intel needs to engage with influential HPC end-users and reach leaders in the HPC community – mostly in the public sector. This target audience has to meet not with one Intel visitor, but to have a dialog with Intel executives as well as with system and software architects. It has to be a two-way dialog, so the number of participants has to be kept low, with talks from both sides – fairly short, with significant time allowed for open discussion. The goal was for Intel technologists and management to hear directly from some of HPC's biggest users. Intel is now a technology provider, its direct customers are not the end-users, so this kind of Intel-users dialog was not likely to happen in the normal course of business.

A task force was called to plan the first Intel HPC Roundtable. Objectives defined: Convey that Intel is engaged, involved, and listening. Provide technologies and products roadmap. Uncover gaps in Intel plans. Explore avenues to collaborative projects. Learn of users' HPC plans.

The practical implications were that is would be a by-invitation event, and the attendees will be asked to sign a non-disclosure agreement to enable disclosure of future technical plans. The meeting location to be Santa Clara, close to Intel headquarters, to get Intel's management attention and attendance. In Feb. 2003, we gathered for two days with 25 users attending (we invited over 80 people) and over 30 Intel people. We did have more Intel talks than users, but allowed for immediate feedback and a long discussion session. The majority of the user attendance came from the DOE national labs but there were several from other government agencies. Their critique was frank and blunt.

The feedback can be distilled into a few key messages to Intel: AMD's Opteron (with its 64-bit support) is a threat to Intel's presence in HPC. The Itanium roadmap is disappointing and not competitive. Concerns at the platform level of architectural direction and execution that had to do with insufficient memory bandwidth, performance, cost, and time-to-market. Gaps in interconnect solutions (for Itanium) and a request for opening the spec for I/O port access. Software lacks maturity (Itanium compiler, Linux tools, enabling of open-source tools).

It is interesting that at that first HPC roundtable the users did not reach the stage of giving up on Itanium for HPC. At least, they did not express such a sentiment. That would come later, in a matter of a couple of years. In 2003 they were 'merely' disappointed and asked for improvements.

On the plus side, the audience were encouraged by the apparent Intel's commitment to HPC and wished to see it acted on. They wanted to continue the dialog and the exchanges and requested that we repeat such events regularly. We ended up doing such an HPC Roundtable twice yearly for next five years. There were Spring and Fall meetings one on the West Coast, one on the East Coast. The later meetings had 50-60 users attendees, while we tried to keep the roundtable flavor alive. By 2008 the intimate roundtable format was transformed into a large-audience event, named "Intel Day" and attached to the annual supercomputing conference.

These HPC roundtable events were, of course, just a forum for communication and exchanges. There were always lively and informative. But, by themselves, they can only build confidence and trust between Intel and HPC planners at large. It helped that roadmap information was given out to the organizations that sent representatives. It was helpful for the whole HPC market that the much larger non-HPC Enterprise market started, due to the 'big data' revolution, to acquire and require large data centers. Architecturally, the enterprise clusters look like most HPC clusters (though the latter employ high-performance components). That meant that at Intel, as in other companies, the HPC activities were now seen as synergetic with those on behalf of enterprise workloads. HPC teams fit nicely as the high-performance segment of the Data Center Group.

Indeed, Intel grew its direct involvement in the HPC sector. The 'Field' got educated about HPC users needs and priorities. The processor and system architects sought input from internal and external HPC practitioners. The software organization was paying close attention to performance of codes and metrics relevant to HPC (and not just those synthetic SPEC benchmarks). I was the instigator for what can be called 'weather codes practice' that was a team of several application engineers optimizing weather models. The company's org charts included entries with HPC in them. The strategists and upper management saw HPC as a technology driver for the datacenters at large.

For HPC-world this fits with how Seager, from the Livermore Lab, describes the interaction between the large HPC government centers and companies such as Intel:

"The DOE Labs and the NSA have a high level of sophistication in terms of understanding the architectures, the applications, and how to build systems and so forth, that the rest of the HPC market doesn't have. And so we were able to convince Intel, begrudgingly, that going after the large systems is worthwhile because you take your lumps early, and then you're done. Everything else is a piece of cake, relatively, because they were all smaller scale then the early large systems. Each system has its own challenges, and each application will find new bugs in the compiler or the system's software. But it's minor compared to the level of problems that we dealt with. And we drove technology like the Lustre parallel file system. We were the guys that showed up and paid for Lustre, and we were the first people to put Lustre into a production environment. And that made a big difference.

"So we coined the term *lighthouse account*. We were the lighthouse account and showed the way. And for a while that argument went a long way."

Intel also acquired HPC talent from outside. Two from the DOE national labs that were mentioned before joined Intel for a short period after their retirement from their government jobs: Bill Camp from Sandia and Mark Seager from Livermore lab. From IBM Intel hired Al Gara who was the chief architect of BlueGene, and Robert Wisniewski who was the software chief architect for BlueGene.[6] There were a few others brought from outside, from government and the private sector, to fill management positions.

Were the changes made by Intel effective? – the answer is a qualified *yes*.

The revenue numbers from HPC over the years are not available. Even if they were, it is hard to estimate the downstream impact of HPC on the much-bigger enterprise server market. The HPC research analysts firms provide details on market share by system houses and not for technology providers. But we can look at historical data of the TOP500 lists. They represent only the largest HPC systems, but the distribution of processor choices extend to the whole market, more or less. The Itanium decline was mentioned above, and there was a contemporaneous uptick for Xeon. Here are a few data points from the TOP500 in terms of number of systems (and not of performance share):

In June 2003 19% of the 500 systems were populated with Xeon (or Intel x86; not counting x86 from AMD). At the end of 2004 its share was 47%. Two years later the Xeon share was about the same, with AMD's Opteron at over 22%. Intel's x86 processors' share continued to rise – 71% in 2008, 85% in 2014, peaking in 2018 at over 93%. It is fair to conclude that Intel was successful in HPC during those 15 years or so.

The TOP500 entries of 15-18 years ago included a broad mix of processor types: IBM's BlueGene, Itanium, IBM's PowerPC, HP's Alpha (formerly by DEC), IBM Power, x86 of several generations from both Intel and AMD. By 2022 some diversity remained, but over 96% of the systems have x86 processors (with about 30% also with accelerators, the vast majority of them made by Nvidia).

---

[6]Seager, Gara, and Wisniewski became Intel Fellows and all four left Intel after a few years (Camp left after only a couple of years).

But not all was smooth sailing. There was the matter of Intel's attempt at developing an attached processor, as a reaction to Nvidia's GPUs. First there was *Larrabee*, an attached processor chip with x86 cores targeting both graphics and general computing. It was abandoned in 2010. There were two other research projects of that nature whose names points to there intended positioning: The *Teraflops Research Chip* (2007) and the *Single-chip Cloud Computer* (2009). What followed, starting in 2010, was a series of products that came under the heading of *Many Integrated Core* (MIC) architecture. This product line was marketed and branded *Xeon Phi*. It had the advantage, compared to other accelerators, of having x86 cores, and, therefore, supporting existing x86 software tools and binaries ([49]). Though not a commercial success Xeon Phi appeared in several large HPC systems from 2012 onwards. As was the case with Itanium, the product was marketed aggressively and was accompanied by heavy investment of software engineering. The later generations of Xeon Phi had impressive performance potential, but the challenge was feeding the data sufficiently fast to make use of that capability. Its end-of-life was announced in 2020.

History repeats itself, and as Opteron caused a decline in Intel's x86 in HPC at the turn of the century, so does AMD's EPYC CPUs since their introduction in 2017. As always, competition is a driver for innovation, and the user customers will benefit from it.

Intel continues to be a power player in HPC not just in terms of volume, but also as a partner for the highest-end systems. It is the contractor, working with Cray, for the second exascale system in the U.S. – the Aurora system to be installed in Argonne National Lab in 2023, after delays due to the termination of Xeon Phi development and related difficulties Intel had getting to 7nm process technology (more on the march toward exascale in Chapter 23).

CHAPTER 19

High Productivity in HPC

Initiatives for Achieving Higher Productivity from HPC Systems

M UCH of the history, stories, and 'legends' surrounding HPC are about single-figure performance numbers, such as theoretical peak performance, or the fastest execution time for a given application. While this was going on, people began to look beyond the system as a closed stand-alone object. The computer system does not exist in isolation from its environment. There are people who maintain the system, develop its software, and creates a myriad of applications. Surely, there are other metrics to consider when evaluating the value of HPC.

From early days there was the tension and differentiation between *capability* computing – the ability to execute a large application that require all of the systems resources to itself, and *capacity* computing – where the system is to process a workload composed of smaller jobs, multiple of which in the system at the same time. This capability vs. capacity tension is more pronounced since clusters became the more common form of HPC systems.

Indeed, in the early 2000s there were some attempts to address the issue of defining *value* of HPC that will take into consideration the broader HPC environment and the diverse ways its systems are being applied. This led to coalescing around the notion of *productivity* as a way to assign value to HPC. This way of thinking is not just a methodology for ranking systems, perhaps not at all, but a mechanism for future planning. Some of the *productivity* debate and research would later be recast, at least in the U.S., as *competitiveness*. Indeed, the U.S. Council on Competitiveness established a High Performance Computing Initiative, now called Advanced Computing Roundtable ([50]). With its slogan *"To out-compete is to out-compute"* it became a forum for raising awareness of HPC to policy makers and funding sources.

DOI: 10.1201/9781003038054-19

HPCS

Agencies within the U.S. government, concerned with national security, 'big science', and space (in the stellar sense), have been quietly supporting HPC companies for years. The beneficiaries included Cray, SGI, MPP companies, DEC, and projects by large companies such as IBM and HP. The list is long. We saw that the '90s was a tumultuous period for HPC companies and architectures. Indeed, planners and strategists in these agencies became concerned about the prevailing trends. As a result an inter-agency long-term program was created in 2001 to address to address leadership computing in the U.S. It was led by the Department of Defense's DARPA, the Defense Advance Research Projects Agency, and augmented with participation of DOE national labs and HPC practitioners from several academic institutes. The program was titles *High Productivity Computing Systems* (HPCS). The HPC private sector was (selectively) invited to propose high-end systems. At the end IBM and Cray were selected for funded projects.

The HPCS program's focus was on *leadership computing* of HPC. In fact, their concern was that the emergence of clusters as a medium for HPC work might hurt innovation at the high-end. Their interest was more in capability computing than in capacity computing. In fact, the tangible goal of the HPCS was to realize several petaflops-class systems, mainly through projects they funded with Cray and IBM.

Still, the 'think tank' of advisors to the program grappled with the issue of defining productivity and how to measure it, and much of it applies to the broader HPC market. HPCS's activities span the period of 2002 to 2010. In 2008, the advisory team produced a comprehensive report ([51]) on productivity modeling, evaluation, and metrics.

At the outset, the report recognizes that *performance* as commonly understood – a measure of how fast an application or workload runs, should be looked at within the broader context of *productivity*. Much attention was given to *performance modeling*, and we return to it in Chapter 31. Where physical devices are concerned the report seems to equate *productivity* with what is often referred to as *efficiency*, that is the fraction of the theoretical computational performance that is realized for a given code or workload.

Using this logic we see that HPC's obsession with performance is really the same as seeking high productivity out of the underlying devices and the physical system. However, we cannot devise universal high productivity (or, high performance) systems because there is a wide diversity in computational profiles of applications.

The questions, then, become *what is the target application or workload profile?* and *how do we optimize the assembly of components to cater for the diversity of applications and its changing over time?* and *should we promote radically different architectures for capability and capacity computing, or even designs that specialize to a class of applications?* and *is it possible to have a set of building blocks from which to build a system customized to any given profile of workload?* etc.

In any event, in the context of productivity, more factors enter the equation. Physical devices by themselves will not solve the high productivity challenge. There is the system software, the applications' algorithms, the uptime (availability) of the system, and, of course, the human factor – the scientists and engineers, the developers and the programmers. The totality of the above makes the productivity universe.

The report cited here shows that the authors struggled with how to quantify productivity. It is understandably difficult, as it involves hardware, software, algorithms, and human factors. There is a segment in the report where productivity is quantified as *speedup gained relative to effort that was put in*, where *effort* is measured by the number of lines of code that resulted in said speedup. The authors recognized the inadequacy of this measure for effort. To quote them:

> *"We share the concerns of the entire HPCS community that the SLOC[1] metric does not fully capture the level of effort involved in porting and optimizing an algorithm on a new system; however, it does provide a quantitative metric to compare and contrast different implementations in a high-level language."*

The report quotes ratios of speedups to lines of code implemented for several applications, with the idea that a higher ratio indicates higher productivity. This is a metric that links human productivity to the system's performance. As expected, when different systems were compared, over several codes, their *relative productivity* varied greatly. This attempt at quantifying productivity did not provide much enlightenment. Worse than that, counting lines of code takes no account of effort put into algorithmic and numeric considerations. It seems that this methodology requires using the same person, or people with similar level of skills, for the measurements. A more relevant measurable would have been the *time it took*, and not the number of lines of code produced. At the end, of course, we would still get an answer that varies with the code tested. We are not closer to determining, in general, the better system, nor the better language (unless all that matters is a single application; which sometimes is true). Another flaw with the methodology used is that it measured incremental improvements. There was no accounting for how well a tested system performed on its base run; that is, the efficiency of the system before any changes were made. If a system happened to be very efficient initially, then its speedups would be more modest.

The HPCS program team went on to tackle *productivity* in its broader sense, beyond just software productivity, by searching for a productivity metric from the business and from the system perspectives.

The business perspective assigns *value* derived from the HPC system as a relevant measure. It is not sufficient to have high utilization of the system. Its output

---

[1]Source Lines of Code.

has to be of value to the enterprise in terms of advancing its *competitiveness* by providing ways to improve its products and innovate. This thinking leads to metrics that are expressed as ROI (Return On Investment) or Cost-Benefit ratios. When all is said and done, this is the one important metric for the commercial value of HPC (the other one – the value to science, will drive yet-unknown benefits to society). We return to this topic toward the end (in Chapter 35).

The system perspective of productivity is a measure of the output of the system relative to the costs associated with its operation. The formulation involve multiplying the system's *utility* by the *efficiencies* of the project, its administration, the job's, and continue multiplying by the *availability* and *resources*, all divided by the *cost*. The resulting formula is not particularly actionable, as some items are subjective (and there is a complication with the choice of units for the variables). But it provides a framework for thinking about what is meant by the system's productivity. The formula tells us what factors to consider and that for higher productivity we want more of what is in the numerator, less of the cost.

The HPCS program set out ambitious goals. It was well funded. However, taking a historical perspective, it can be graded as *mixed results*.

There were two finalists hardware projects, one each with Cray and IBM. Cray proposed the *Cascade* system. Its design was driven by the observations that, (1) Applications vary widely and no one parallelization model fits all. For higher productivity it is best to offer multipole processing technologies (commodity microprocessor, vector processing, multithreading), and (2) The architecture has to support programming productivity by addressing such features as compute/bandwidth balance, threads, synchronization, and latency tolerance. The resulting product line, with its rich programming environment and a high-performance and feature-rich interprocessor network, was very successful and a major presence in HPC since its introduction in 2012.

IBM's project was named PERCS (Productive, Easy-to-use, Reliable Copmuting System), built with the POWER processor plus enhancements. The contract for its first large system (NSF contract for the University of Illinois) was cancelled in 2011 due to complexity and cost considerations. Later, smaller configurations were delivered to some 20+ sites, notably to some of the large weather centers.

HPCS also sponsored studies of what they termed "emerging architectures" as potentially better suited for HPCS productivity goals than the homogeneous multicore processors. These were the STI Cell processor, GPUs, and Cray MTA. Commodity multicore CPUs and GPUs are with us today. The Cray MTA (Multi Thread Architecture) was an innovative system designed by the Burton Smith. Only a few systems were shipped to non-classified sites and the product was discontinued after its third generation. The Cell processor, built from IBM PowerPC cores contributed to the first petaflops system – the Roadrunner (see Chapter 16). A shining achievement, but not of a sustained benefit, since it was a one-off relatively short-lived system.

On the software front there was a big investment in languages for parallel programming that were extensions of existing languages. Collectively referred to as PGAS (Partitioned Global Address Space) languages, there was UPC (Unified Parallel C), Coarray Fortran (discussed at some length in Chapter 32), and Titanium for Java. Their use today is marginal at best. More on programming languages in a section below.

The Community Perspective

The *productivity* discussion swept the HPC community at-large and was subjected to much talk and serious research. Indeed, The International Journal of High Performance Computing Applications dedicated its whole Winter 2004 issue to HPC productivity. Its ten articles were contributed by luminaries of the HPC community, and covered such topic as software project management, framework, metrics, and models for productivity of supercomputers and programming languages, as well as associated metrics and models for performance.

An example of the work and thinking on productivity is the article by David Kuck – *"Productivity in High Performance Computing"* ([52]). Kuck is computer scientist who was one of the early pioneers of software for parallel computing when he was at the University of Illinois Urbana-Champaign. By this time Kuck was a Fellow at Intel, which acquired the company he founded, and that developed compilers and software tools for parallel processing.

Kuck frames his approach by pointing out that the outcome of enhanced productivity is (or, should be) the ability to produce better work, faster, and with greater ease. Of course, not all three – quality, speed, and ease – have to advance with every step toward a more productive environment. The context for Kuck is not just that of the high-end of supercomputing, but of a landscape trending toward commodity-based clusters.

And he proposes the ingredients necessary for the outcome of higher productivity: Better parallel architectures for individual jobs; better run-time support for control, performance, and reliability; and software engineering methods for easier development of applications. Kuck adds another item, a burden on the users (that is accomplished almost naturally): Use larger and more complex problems (this is the *weak scaling* argument – see also page 255). He concludes by suggesting indicators for assessing productivity improvements: Track performance of individual jobs; Monitor resources used for very large jobs; Broaden applications diversity; and, enhance quality-of-service in terms of system uptime and components failure.

For the most part, the HPC community sees the productivity issue limited to the human effort of developing 'efficient' codes. The focus is on programming languages, computational algorithms, optimizing compilers, parallelization methods and aids, and numerical libraries. These are the tools that would make the work of the scientist and programmer easier and faster.

This view is akin to the optimization of one component within a complex process. The process being the development and execution of an application on a computational facility. There are potential side-effects to the making of code development easy and fast. How do we account for the computational resources that are needed? The cost of the system? The performance of the code? Its portability between different systems? etc.

What seems to be missing from the 'productivity' debate during the HPCS years is bringing together, holistically, the hardware, software, and applications requirements and opportunities. The Cray and IBM hardware projects paid close attention to the software requirements and wants in their design, and did their best to assist. The software engineers develop their tools with the known architectures in mind. But the projects are disjointed. That said, the HPCS program seems to have taken the first steps toward the co-design approach. We get to it in Chapter 23.

On HPC Programming Languages

The HPCS report ([51]) contains a comment that seems to express some frustration:

> "Programmers *do* value productivity, but reserve the right to define it. Portability, performance, and incrementality seem to have mattered more in the recent past than elegance of language design, power of expression, or even ease of use, at least when it came to programming large scientific applications. Successful new sequential languages have been adopted in the past twenty-five years, but each has been a modest step beyond an already established language (from C to C++, from C++ to Java). While the differences between each successful language have been significant, both timing of the language introduction and judicious use of familiar syntax and semantics were important. New "productivity" languages have also emerged (Perl, Python, and Ruby); but some of their productivity comes from the interpreted nature, and they are neither high-performance nor particularly suited to parallelism."

Introducing new programming languages is hard. The situation is similar to that of introducing new hardware architectures, as we saw in the case of the Intel Itanium debacle and the Intel Xeon Phi. Even being a technically sound product is not sufficient. An ecosystem of dependent products is necessary for capturing a share in the market. The same applies to programming languages.

Consider the example of Ada. The U.S. Department of Defense funded the development of a language that will replace both Fortran and Cobol (as well as a number of other languages used at the time). In the '90s the DoD, with its considerable procurement power, mandated vendors to offer Ada and its developers to code in Ada. The attempt in unifying around a single language failed, the mandate removed (1997) and Ada exists today only in some legacy codes.

The HPCS Language Project accepted and funded three new languages for parallelism in HPC, starting in 2006. One of these, Fortress from Sun Microsystems, was dropped from HPCS within a year but was released later as open-source. Its development was discontinued in 2012.

Cray created the Chapel language to support multithreaded parallel programming. IBM's new language was X10 and it implemented a PGAS model. Neither language gained traction outside of a limited group of users.

It turns out that, at least for HPC, it is not likely that new languages will gain wide acceptance. There is too much invested in existing applications, and many new implementations are built upon prior codes. Vendors and user facilities are reluctant to support additional languages.

Languages that are based on existing languages, such as UPC and Coarray Fortran, stand a better chance of acceptance. Extending the language still allows its use on older codes (that do not make use of the new extensions). This approach also allows to migrate to a parallel version incrementally.

Nevertheless, even languages that added extensions for parallel programming are not used much. Two other mechanisms were found to be more effective. From earlier on there was the use of *directives* to express parallelism (and vectorization) in shared memory systems. This approach was standardized as OpenMP. For distributed memory systems the HPC community adopted the MPI library as the means to parallelization and synchronization. We return to this topic in the 'Fortran chapters' (starting in Chapter 32), and only mention here that a library can be called from different languages, such as Fortran, C, and C++ – the popular HPC languages. It is also easier to update compared to changes in a language syntax.

To conclude the notes about languages and productivity in HPC, here is a relevant anecdotal research summary:

Inspired by the HPCS program, Eugene Loh – a principal software engineer from Sun Microsystems (which was acquired by Oracle by the time the research was published in 2010), sought to find "The Ideal HPC Programming Language" – the title of his paper. The subtitle is "Maybe it's Fortran. Or maybe it just doesn't matter" ([53]). Loh did not really answer his quest, because there was no direct comparison of languages and the target codes were relatively small synthetic benchmarks. But there were other interesting observations.

First, a programmer was asked to rewrite the codes while not being restricted by any language rules or syntax, but so the 'code' is most expressive and clear. The result was a much more compact and readable code. It also was Fortran-like. Which, as the author admits, may have more to do with the original being mostly Fortran and the programmer's familiarity with the language. Next, they put together a team to rewrite the codes in modern Fortran with the human productivity in mind, as well as readability, verifiability, and maintainability. The empirical findings are interesting. For the five codes examined, the number of lines of code was reduced by 3.6 to 11 times, while the performance hit ranged from no change to degradation of mostly 2-3x and a case of 6x slowdown.

The paper contains a technical discussion of algorithmic and numerics causes for complexity and length of the codes, ways to simplify the code by eliminating costly tests at the expense of a few additional values, with little effect on performance, and where compilers can mitigate performance loss (for example, interprocedural analysis and inlining functions or routines).

Musings about Productivity and Performance

The goal of the productivity efforts is to find ways – with software tools that include languages, compilers, and libraries, to extract more performance from codes, with less effort. In HPC we cannot divorce productivity from performance. In the grand scheme of things, we can think of productive HPC as one where we maximize performance, at a shorter development cycle, with less resources, at a lower cost.

On the face of it, if a user spends less time working on the code its performance will not be a good as it can be. This is almost always true, but it is not necessarily a bad thing. Think of the time saved and the likelihood that the research results arrived sooner. The calculation is different for a code that is to run many times.

The two, productivity and performance, are not always in conflict. For example, a user may resort to library calls to perform numerical procedures. Library calls incur overhead. But if the task the library routine performs has been optimized by an expert coder and is not too small a task, then the library option performs better than in-place code. That's the case for many math routines when they process datasets is large enough. The same is true for calls to GPUs. Both productivity and performance benefit.

On the other hand, a quick-executing task that is being performed in many locations within the code, can be turned into a function to be called where needed. Productivity will be gained as the code can be developed faster, be more compact, and more easily maintained. Performance may be degraded. This may still be a better approach if the code is to be used once or just for short period. There is saving in development time and a quicker path to results. It can be argued that it is justified economically as people's time becomes more costly and the hardware cheaper.

The trade-offs between performance and productivity is a complex matter. The cost-benefits analysis includes factors such as the potential speedup, the time and human resources it takes to achieve that goal, the cost and availability of computing resources, how soon results are needed, the number of times the code will be run etc. See also Chapter 31.

Productivity remains a subjective term. It is clear that productivity can be measured by some *output* metric. It is also clear that just a measure of computational speed for a given level of effort (programming, tuning) is not an adequate measure. So we get into a mix of *quantity* and *quality*, the latter much harder to put a value on, though at least as important.

We return briefly to *productivity* in the context of Fortran (Chapter 34 and in particular in page 303) and it is embedded in the *performance* discussion (Chapter 31).

CHAPTER 20

Weather Models' Impact on Our Lives

Applications of WRF and Other Models

AFTER seeing the scope of the capabilities and the popularity of the WRF model among weather models researchers (Chapter 15), it is appropriate to ask if and how WRF and other models contribute to society outside of the research community. It turns out they do, and in a number of interesting ways.

Though much of the content here is derived from NCAR, the applications are by no means unique to it. In every continent and in most countries weather models are applied in similar ways. NCAR's WRF is a particularly convenient model to explore because it is used globally and information about it is easily accessible.

The home page of NCAR's Research Applications Lab (RAL)[54] lists the main aspects of life and the economy that benefit from its models (more on NCAR's models in Chapter 26):

- Agriculture and Food

- Air Quality

- Aviation

- Climate

- High Impact Weather

- Human Health

- National Security

- Renewable Energy

DOI: 10.1201/9781003038054-20

- Surface Transportation

- Testing and Evaluation

- Water

- Wildfire

The scope is breathtaking, and this chapter cannot do justice to it, but only give a broad-brush idea of the everyday usefulness a good weather model bestows. In 2020 RAL had staff of some 180 people. It has been growing at a rate of about 5% each year for the previous 25 years.

The obvious and direct impact of weather and climate models is in providing weather forecasts and the basis for reported climate change predictions. The much broader impact in areas listed above is less transparent. This is because the models serve as support tools, behind the scenes, in those sections of economic activity. RAL's 'products', the data and predictions it provides to its clients, is generated by a set of tools developed there. Some are models that are derived from WRF, some are add-ons to WRF, some are specialized forms of WRF.

Take agriculture and food production: Modern farming uses support tools for decision making regarding crops growing. WRF-based models of high resolution use observed vegetation data with predicted weather and near term climate to calculate a quantity called the *Leaf Area Index* which is green leaf area per unit of surface area. The index allows estimates of crops as it points to amount of light that falls on the leaves. This kind of simulations informs farmers of actions to take, and helps economists and planners prepare for the next season's food supply. The tools used include several variants of a high-resolution land surface model, WRF-Crop modeling system, and others.

WRF-CHEM, which adds chemical processes and aerosol simulations, is used for regional and local air quality forecasts. It is now possible to provide 14-day alerts of atmospheric conditions conducive to deterioration in air quality allowing authorities and communities to take steps that protect the vulnerable.

It is obvious that daily weather forecasts alert airlines of locations of storms and high winds, and inform them on where flying paths are safer. But WRF products do more to increase safety of flights. One of them provides an hourly assessment of potential conditions for ice to form on the aircraft. Another warns airplanes of lightning threats. In 2012 NCAR noted that for more than 15 years there was no downed airplane due to wind shear. The credit goes to a wind-shear and turbulence warning system, based on high-resolution WRF models around airports.

High Impact Weather includes events of water, fire, and wind. High fidelity local-regional runs of the model track progress and strength of developing hurricanes, including landfall timing and surge levels. When high levels of precipitation, rain or snow, is predicted, high resolution topography-aware variants produce warning of flood risk by location and risks to reservoirs and dams. In fact, the National Weather Service (NWS) adopted an implementation of WRF called WRF-Hydro as its national tool for assessing hydrologic risks in the U.S.

A public health-related application of weather models, that includes WRF-CHEM, is forecasting air quality and levels of pollution. The models are used for advanced alerts in various cities of days with poor air quality, and to study and warn about conditions for the spread of vector-borne diseases.

On the national security front, RAL is developing methods and tools that link the state of the atmosphere with the spread of hazardous gasses released by accident or intentionally. Agencies responsible for national security apply such modeling into their emergency response systems.

The utilization of the more common renewable energy sources – solar and wind, are helped by regional and local modeling of both wind strength and gusts, and the amount of direct solar radiation based on time and cloud cover. These help manage the resources and prepare for both excess and shortfall.

Timely and targeted weather forecast is critical for surface transportation. Safety and economy benefit from advance notice of hazardous road conditions and the timing for treating roads in the winter. Tools are being developed that incorporate weather predictions helped by connected-vehicle data.

One of RAL's functions is to evaluate the accuracy of NOAA's hurricane forecasts so its predictions are improved. A more accurate hurricane's time of arrival and its path saves lives and costs in the billions of dollars.

WRF-Hydro is instrumental in managing water resources and for understanding of the world's water cycle. It informs utilities and governments about reservoirs and underground water supply as well as river navigation. And it is used to warn of flooding, both timing and size.

High resolutions models, such as WRF, help fight wildfires. There is a physics module, called WRF-Fire, added to the main model's body, that uses weather, terrain, and organic fuel data to show how a fire will grow and evolve. Strong wind warnings are issued to help communities prepare and reduce preventable damage.

Needless to say, such applications of the base model, WRF, save lives and helps avoid potential economic damage that is sometimes hard to recover from.

To get an inside perspective on one area where WRF is applied I spoke to Roy Rasmussen, a senior scientist and lead of the Hydrometeorological Applications Program at RAL. The group has about 30 on staff, and is one of six at RAL. Rasmussen's focus is on the environment's water cycle. His story shows the sometimes circuitous road research and application takes:

"I arrived at NCAR some 20 years ago. Started to work on water cycle, but my real work was on weather. Then they put me in charge of the climate program, and I had to learn a lot about climate change. I learnt the CESM (Community Earth

System Model; see Chapter 26 "The NCAR Models") and discovered it doesn't describe the water cycle. So we needed to go to weather modeling scale. I realized that would be very 'expensive', so we did short test runs. And the results were so good we went for 10-20 years of simulation. And that's where we are now. We modeled the Rocky mountains using a regional model – a region of maybe a 1,000 by 2,000 km, with 2 km horizontal grid resolution. To my surprise the model did a very good job predicting the snowfall and snowpack. That started at about 2007, and a 2011 paper we wrote on that got a lot of attention. At the same time a scientist at ETH in Switzerland did similar studies on the Alps and the Mediterranean. He learnt, for example, that some areas in the region are going to be hotter and dryer. It was hard to learn that from a 100 km grid model. We both learnt that we can run high resolution (regional) model for long periods of time. We took the temperature and moisture values from the global climate models as boundary conditions, being the most important variables for our studies. We called this *pseudo climate model* since it didn't use the dynamical changes. But it is the temp and moisture that all the climate models get right. They differ in their dynamical signatures, so you'd need to run ensembles of different models. We couldn't do that, but we could do interesting studies based on temp and moisture.

"The dynamics does matter for studying climate. But I decided it was more important to me to capture the formation of thunder storms than to get the dynamics of climate change perfect. And I was clear about what I'm doing: Just adding temperature and moisture, and here are the results. It's not where there's going to be a thunder storm, but whether that there're going to be any.

"Let's look at hurricanes. They are not captured correctly even at 10-20 km grids in climate models today. Can you believe in their future predictions? – I would say it depends on the question you ask. What is 'climate'? – it's 'average weather'. So, if you can get 'weather' right you can get the climate. That's the approach I've taken. That's not to say that one is right and the other wrong. I need to get the snowpack right, and this approach does it."

Water resources cycle studies rely on predicting snowfall and snowpack. And that means having to run models over mountain ranges. Which, in turn, means the models have to employ a very high resolution grids to account for the local topography which the precipitation patterns. Rasmussen continues:

"The world's mountain ranges (Rockies, Alps, Andes, Himalayan) are very complex. In fact, the Himalayan are so complex the modelers have given up.

"We've created a hydrological WRF model. It is a nested version of WRF with climate signals added to it. We are now able to do 20 years of the continental U.S. In historical runs we use the ECMWF model results to refresh the boundary conditions. For future runs we use the average of a number of climate models. We don't use the ocean model that much. My philosophy is to learn how to use the processes in WRF (which does not include an ocean model). I believe the WRF knowledge can be transferred to MPAS (a model described in Chapter 26).

"How do we handle the Andes? how small does the timestep need to be? etc. Those are interesting questions for climate time scales. The Andes is the longest mountain range in the world (above sea level). Can WRF handle simulating its weather? Can MPAS? – we don't know. I'm organizing a team of about 30 scientists to do research on South America. That includes El Nino, La Nina, etc. My goal is to create the datasets that students and researchers can do their research on. Learn the processes and the model – what they do well, and what they don't do well; how to do simulations over south America. And that knowledge can be transferred to MPAS."

Applying WRF in other areas also require very high resolution, that is, very small grid size. To provide useful information for agriculture planners and farmers, the grid point intervals have to be under 4 km, according to Rasmussen.

Consider what RAL calls high impact weather, or sever weather. Extreme examples are hurricanes and tornadoes. For tornado prediction the grid requirements are considerably more demanding. It got to be sub-km to capture the fine features. To be useful it has to be a quick run over a region that can provide accurate prediction at least a few hours ahead of the event. This will have to wait for the next-generation computer systems for sever weather phenomena, say Rasmussen. At present, the state of the art is at the stage of alerts that above a certain area, over a period of time of some hours, there will be conditions that favor the formation of tornadoes there. Such an alert, expressed as *tornado watch*, can be issued more than a few hours in advance. It allows people to prepare for an event that may or may not come their way. *Tornado warning* is issued when an actual tornado is detected. At this time, the detection is by radar or human spotters; not computers. It gives people 10-20 minutes warning, typically. If people are alert and not outdoors or on the road, the time allows finding a shelter, but not to evacuate. Obviously, a reliable prediction of a tornado forming over a limited area, even one hour ahead of its occurrence, would save life, if not property.

WRF helps the renewable energy sector in the U.S. It is applied in operational wind farms. Rasmussen explains: "They are very sensitive to wind speed and above certain speed they shut them down. And then the supply switches to some power station. But that takes time, so they need accurate forecasts some 5-6 hours ahead." A very fine resolution wind speed forecast over the wind farm area, with topology details, saves turbines from breaking and allows smooth operation of the energy supply.

WRF is also being used commercially. The Weather Company, that owns weather.com and Weather Underground, relies on WRF for its forecasts. It created a propriety version for very high resolution version for its aviation industry customers.

There are lesser known, but no less interesting, commercial applications of WRF. Here is one such case:

I came across Milan Curcic in connection with the state of the Fortran language (see chapter 32), only to learn of his commercial enterprise based on WRF. Curcic is a meteorologist and an oceanographer, and a researcher at the University of Miami. He also founded a company, Cloudrun, [55] that offers WRF modeling Software-as-a-Service on Cloud computing (discussed in Chapter 25) platforms. Its customers can run WRF from their desktop browser. They can select any region in the world and setup a schedule for receiving forecasts.

Public Cloud providers have some HPC configurations available. However, "as of now", Curcic says, "what we use are not HPC instances. They are basically shared-memory machines, and we tailor the domain and the resolution so that it can efficiently run on that. We also have the capability to run in distributed memory, by spinning up multiple compute instances, and then do the MPI across them. But the problem is that the interconnect network between the instances is not optimized for performance. Once we try to run it in parallel, we get very high latency which is not conducive for HPC. With more HPC cloud providers around we are definitely looking in that direction. But for now, we can easily get pretty high-end 32-core shared memory instances that run WRF pretty efficiently."

It turned out the Cloud-based business model was not suitable for academic researchers. Curcic: "We had encouragement from academics, but once we launched, we realized that academic researchers were not ready to commit funds to this format, because of the way academic and research grants operate. They have slow feedback cycles. Funding agencies still don't like funds being used on external resources. They want people to run models on their own supercomputers at the large research centers, such as DOE's Oak Ridge, Livermore, or NSF's NCAR or wherever your partner institution is."

But small businesses started to show interest. Curcic described to me some of the ways people use WRF.

"There is a consultant-coach for high-end sailboat racers. He is a meteorologist who coaches sailboat racers on the weather on the day of the race and prior to the race, on which trajectory is best to take during the race. I learnt that modern sailboats that are used in races can take realtime feeds of the data to update the optimal trajectory of the boat. That is, the racers can get information during the race. The coach is running a very localized custom LES (Large Eddy Simulations) scale WRF simulations for coastal areas and small bays wherever he has races that are going on. And right now (mid 2020), we are preparing for the America's Cup in New Zealand, which is coming up next year as well as the Tokyo summer olympics.

"Another example is in the area of energy applications and energy predictions. Weather has a very high impact on energy load from consumers usage. When you have a heatwave, for both residential and commercial, buildings or structures will use considerably more AC which creates a load on the power grid. And because you can't easily store the energy that you produce, the optimal state of the power grid is to produce exactly as much energy so as to match the demand, and that can be sold to consumers. in Europe, for example, different countries trade energy between them. It's very much like stock trading except that they trade specifically with electric power and between different countries and markets."

We mentioned above that WRF is applied in support of wind farms. Curcic's experience also touches this aspect:

"We recently put a bid to a Dutch business that predicts loads on wind farms for calculating the output of the energy from the wind farms. They take weather forecasts as input.

"And another example, which is also a current customer is a research institute in Canada. It's a government Institute and they use WRF throughput service to downscale a climate model for the whole of the 21st century. From 2006 to 2100 for the metro area of Ottawa and Montreal, to look at the impact of climate change on the urban areas. So this is all WRF on 'under the hood'."

The last example of how WRF can be applied is interesting in that it shows how a regional atmospheric model can be incorporated into a long range climate study.

There are other companies that customize weather models to produce products other branches of commerce use in their operations. One such example is Tempo-Quest ([56]), based in Boulder, CO, that I learnt about from its CTO, Christian Tanasescu, who I knew from his many years at SGI. They offer GPU-accelerated high-resolution spatial and temporal regional forecasts, very fast regional WRF version, and real-time weather tracking and monitoring.

After detailing the areas impacted by weather (and climate), even if not a complete list, the question arises as to whether the impact can be quantified. As we might expect this is a hard question to answer. A first step is getting a handle of how weather variability affects the economic activity. For that there is an answer – restricted to the U.S. and based on 2008 data. A 2011 study published in the American Meteorological Society journal [57] concludes that variability in economic activity due to weather is estimated to be $485B, or 3.4% of the 2008 GDP. Another data point, and a jarring one, is that over 10 years hurricanes costs to the economy and to property amounted to about $350B ([54]).

An even harder question to answer is how much of the negative impact of weather variability is being mitigated by the tools and applications described above. But it is possible to appreciate that even a small percentage of reduction in damage is very significant, and that the return-on-investment in computing for weather and climate modeling is high.

NOAA's chief economist crafted a 2018 report titled "NOAA's Contribution to the Economy"[58]. NOAA's products are related to weather and climate. Its predictive and advisory tools are models that run on supercomputers. Its annual budget is $5-6B. Data collection and observation stations are a significant part of it. As is the human capital for research. All are essential for the agency's mission.

However, the investment in high performance computing – the systems and their ancillary equipment, is a small fraction of the budget, measured in the $10M's annually. The report lists areas where NOAA contributes: Emergency Management, Transportation and Warehousing, Fisheries, Insurance, Agriculture, Mining, Utilities, Construction, Manufacturing and Retail. The aggregate annual value of these enterprises runs in the trillions of dollars. Again, even a single digit percentage of increased efficiency or reduction in damage amounts to benefit far exceeding the investment. For example, NOAA estimates that just improving the forecast of El Nino impact can add, annually, some $300M to the U.S. economy.

Without supercomputers all these would not be possible.

CHAPTER 21

Computational Life Sciences

The Depth and Breadth of HPC Applications for Life Sciences

L IFE Sciences is a broad discipline with many sub-disciplines each of which deserving its own mention. Here we are concerned with these areas that are helped by, and require, computations – Computational Life Sciences, or CLS. It is a prime example of an interdisciplinary field combining biology and medicine with computer and computational sciences. At a finer granularity it brings to HPC applications in subjects such as bioinformatics, big data analysis and text mining, genomics, molecular dynamics, imaging, and a variety of medical sciences. Biological systems are very complex and it is the computational approach and methods that make it possible to progress from general descriptions to more quantitative statements.

A Scientific-Computational Perspective

Rick Stevens, is an associate lab director for Computing, Environment and Life Sciences at Argonne National Lab and a professor of computer science at the university of Chicago. I asked him to introduce the use of HPC for life sciences (LS) applications over time:

"One way to characterize LS applications is to look at space and time scales. If we go down to the molecular level, we think about modeling biological molecules, whether they are proteins or DNA or metabolites, and studying the interactions of those classes of objects. That's a whole area of life science computing. It involves micro-dynamics, quantum mechanics, molecular mechanics methods and so forth. This is the molecular space, where we can model interactions from a Newtonian standpoint, or model them quantum mechanically. There's been, of course, enormous progress over the last 20-30 years in those kinds of problems, largely coming from a little bit of improvement in methods and employing large-scale ensemble techniques. With the latter, instead of simulating one system, you do hundreds of thousands or tens of thousands of systems and then you aggregate the statistics

DOI: 10.1201/9781003038054-21

together. There's been huge progress at this molecular scale. We were able to predict quite accurately how this kind of brute force approach to that class of problems would improve as we moved from terascale to petascale to exascale. The results were aligned pretty much to our expectations."

We later take a look at how data analytics and artificial intelligence (AI) research is being incorporated to, and changing the face of HPC. With regard to LS Stevens notes: "What has changed recently is the idea that we can use AI methods, machine learning (ML) methods, to further enhance or accelerate molecular simulations. We can learn approximations of the force fields with ML, and then we can apply that. When we do that, we get factors of a thousand or so acceleration on top of what we got from high-performance computing hardware. We can also use AI to aggregate the ensembles. That gives us factors of a hundred in speedup. AI can be used to steer molecular mechanics simulations by understanding whether or not the molecular state that we are in is a new state or a state we have previously seen. And if it is one that we have already seen, we skip over it. If it is a new state, we study it. So, we actually can use machine learning to sample the configuration space in a much more efficient way than with classical methods."

Moving up from the molecular scale, Stevens continues, while highlighting what AI/ML methods made possible:

"The next level up are subjects such as protein interactions and drug interactions with proteins, or protein folding. We heard a lot about that in the last couple of years, because machine learning methods have really more or less solved it; at least solved many instances. And though we thought about it, in the abstract, some 20 years ago, there was really no sense that it was going to be solvable in our lifetime. Two things have changed: One, the ability for machine learning methods to be applied in spaces that we didn't understand before. And applying ML on many thousands of configurations meant that the protein folding configuration will be encountered at some point. Two, We now can produce the 3D coordinates of the folded protein. The initial approach to using machine learning to address that problem combined classical bioinformatics approaches where you would take proteins that are similar from a sequence standpoint, you compute them all to a sequence alignment which gave you a 2D object that showed you where the substitutions were happening between different instances of the proteins. This is used to build an evolutionary tree. But if you take that 2D object as input to the machine learning, it could actually go to the next level, which was predicting which parts of the protein are going to be in close proximity to each other. Taking that as input to another machine learning program, it could predict the 3D coordinates.[1]"

Recent (2022) developments of ML methods now allow all-AI (in the sense of ML and DL) protein studies to replace the classical bioinformatics combination with AI followup.

---

[1] AlphaFold, a deep-learning program from DeepMind, an Alphabet/Google company, is such an application.

Above the level of proteins we get to examining, computationally, a whole biological mechanism. For most of the cases at this level there is not enough knowledge or data to be able to invoke AI methods. The interesting cases of interactions between proteins or between proteins and DNA are still done by first-principles, that is, numeric, simulations.

This brings us to the important field of drug discovery. A proposed drug will only work if it binds to proteins in our body, and much search time is saved when finding the molecules that bind can be simulated. Here HPC is crucial for discovery of various cures. As Stevens puts it: "Drug target-identification has been hugely accelerated in the last five years by high-performance computing, in that we have machines now that can easily classically screen a billion molecules against the target, in a day or two."

This step is helped by AI techniques, which themselves require intensive computations of matrix manipulations in the learning phase. This is a case where it is hard to separate the improved hardware contribution from that of the new AI algorithms in accelerating the solution time of identifying a promising drug. This is how HPC helps identify promising drugs, that can be brought out for trial much sooner than in the past, and with higher probability of success.

That is not where computer simulations end for LS. Going beyond proteins we get to biological systems in what is referred to as *pathways*. Quoting NIH fact sheet ([59]): "A biological pathway is a series of actions among molecules in a cell that leads to a certain product or a change in the cell." Computationally, Stevens explains, it can be presented as a mathematical graph. For a bacterial organism it would be of order of a thousand nodes and a few thousands edges (a bacteria has about 4,000 genes). Progress has been made, but much is yet to learn before we get to the level of human genome and its pathways.

Stevens continues: "At the cellular level, there have been various efforts to simulate entire cells. Not at the atomic level, but at the abstraction that is more like pathways and biological interactions. Those models are more similar discrete event simulations, or games theory kind of models. The network they simulate is similar to that of agent-based models. And while there's been progress there, we haven't seen a major breakthrough on how to get some acceleration because of exascale. We cannot run individual models much faster, but we can run large ensembles, or whole population, of them. We can do certain kinds of population biology experiments on exascale platforms. For example, we can do simulations where we have, say, millions of different cell simulations, and when we impose variations on them we can perform evolutionary studies."

From cell-based evolutionary studies, scaling up, there are simulations of tissues and organs. Stevens points to the Physiome[2] Project. The project can be called

---

[2] "The physiome of an individual's or species' physiological state is the description of its functional behavior. The physiome describes the physiological dynamics of the normal intact organism and is built upon information and structure" (from 'Physiome' in Wikipedia).

"The Human Physiome" as it aims to explain the working of every component in the body, from molecules up, as a part of the *"integrated whole"* ([60]). The project is run under the auspices of the International Union of Physiological Sciences (IUPS).

Stevens on the Physiome Project: "The idea is to build a *digital twin* of a human. For that they need to model tissues, find out some kind of structural model, as well as materials, and a functionality model. To that they add a model of the circulatory system, and model some version of the nervous system. It is a 3D model with about 200 different types of tissues that humans have. The tissues make up organs, and so they construct a hierarchy of models for tissues, organs, and then subunits like the entire circulatory system etc. With such models they can do very interesting things. They can model, for example, a heart attack and watch what happens. Or heart arrhythmias or stroke. One can model what happens in an accident or maybe how metastasis happens in cancer, how a tumor moves around and colonizes other tissues."

We return to the *digital twin* theme in earth system modeling and engineering.

We see that these LS simulations are of great benefit for understanding biological processes, and in providing non-invasive means of studying cures and remedies. Stevens reflects on the computational aspects:

"There is much progress with these large-scale full-organism models because they're highly parallel. They can be parallelized in the spatial dimensions, of course, but also by, say, running the fluid dynamics of the circulatory system separately from the electrical components in the nervous system, and then aggregate them. So you could think of having both functional decomposition and domain decomposition on parallel machines. The progress is driven by computing at scale."

Another area of life sciences aided by HPC is genomics, discussed at greater length in Chapter 22. Stevens highlights DNA analysis as a computational tool that is used to find, for example, the genetic contribution to cancer. Again, progress there is being achieved due to the combination of greater compute capability that produces much more data that, in turn, makes machine learning methods effective. In particular, today's compute power enable the computationally intensive process of *imputation*. It is the process of estimation, or a structured interpolation, to fill in the missing pieces of the often incomplete genome sequence of a population.

Applying CLS

Here is a partial look at a small sample of institutes, out of many such centers and labs, that apply CLS on HPC systems for projects that can result in products and knowledge of near-term benefits to society.

For example, Riken, the renown research institute in Japan for a variety of disciplines (and home for the Fugaku exascale system), has a strategic program called *Supercomputational Life Science* (SCLS). Its mission statement centers on the use of HPC for efficient analysis of large amounts of data that is generated with the aid of modern measurement instruments to gain understanding of life systems. It states that such a capability is necessary for reliable predictions of the multilayer systems they study ([61]).

Closer to actual patients, the Cleveland Clinic, a nonprofit medical center in Ohio, uses HPC via cloud computing for its Center for Computational Life Sciences. It also employs AI techniques and even quantum computing for their *"Discovery Accelerator"* ([62]). CLS is applied to the discovery of material and treatments in healthcare.

Then there is the Mayo Clinic. I met Yuan-Ping Pang around 2003. Pang is an emeritus professor of biophysics and pharmacology at Mayo Clinic College of Medicine and Science in Rochester, Minnesota. Before his retirement in 2021 he directed a *Computer-Aided Molecular Design Laboratory* at the Mayo Clinic, and that is where HPC comes in. At that time, the early 2000s, Pang wanted to use computers for drug discovery. Inspired by the Beowulf concept (see page 136) and with limited budget Pang went about building his own cluster.

He secured a medium-size room and brought in 4-levels frame carts of the type used in bakeries and stores to pile up loaves of bread. They were lined up in several rows. A tower-style workstation would fit in one level of the cart. The setup was populated with about 470 Xeon-based desk-side workstations. The cluster's nodes were connected by Gigabit Ethernet cables and ran RedHat Linux. Its 940 processors produced enough heat (40 KW) to require a cooling unit. With a peak performance of 1.1 teraflops, it was not, of course, one of the fastest systems of its time, but built at a cost of $0.4M it had a performance-to-cost ratio that was 27 times better than that of the Earth Simulator, and about 40% better than the 10 teraflops innovative Apple Mac-based cluster at Virginia Tech at that time.

Figure 21.1 shows the cluster, named Kibbutz100, in 2002. We can see the wheels of the carts on the floor – a more 'mobile' version of an early 'server rack'. On the left, a glimpse of the raised floor for critical ventilation and cooling using a Liebert DS system.

Anecdotally, "Kibbutz" does refer to the Israeli communal village style of settlement. While visiting the Weizmann Institute of Science in Israel, Pang visited a Kibbutz, and found the idea of a community sharing the burden of work appropriate to the concept of a cluster.

This low-cost home-made Beowulf-style cluster affords Pang running continuously on a teraflops system for months on end. It had reduced an in-silico screening run from close to 2,000 hours to 5 hours on the Kibbutz100 cluster. A noteworthy achievement was the successful prediction of a substrate-bound of a SARS-CoV-1 proteinase after only 20 days from the release of the SARS-CoV-1 genome [63], which offers insights into how a water molecule regulates the proteinase activity that is relevant to other proteases, including that of the SARS-CoV-2 (we return

Figure 21.1: Kibbutz100: A 1.1 Teraflops Computer Designed for in Silico Screening and Multiple Molecular Dynamics Simulations. Source: Clinical Pharmacology & Therapeutics. Used with permission of John Wiley and Sons.

to this topic in the chapter about COVID (Chapter 30). For more on the science done with the original cluster see [64].

Pang then developed two additional clusters with newer Intel processors. His third generation of in-house cluster comprised of 100 Mac Pros with Intel's 32nm chip technology. Interestingly, the same network switch (Hewlett Packard ProCurve Switch 4000M) used in Kibbutz100 – one recommended by Don Backer and that was used in the original Beowulf cluster at NASA, was still in use in the third cluster until Pang's retirement in 2021.

The third cluster enabled Pang to determine protein folding rates from simulation and overlapping with experimental data [65]. As Pang puts it: "It opens new prospects of combining simulation with experiment to develop computer algorithms that can predict ensembles of conformations and their interconversion rates of a protein from its sequence in order to understand how and when a protein acts as a receiver, switch, or relay to facilitate various subcellular-to-tissue communications. With this understanding the genetic information that encodes proteins can be read in the context of biological functions."

The crowdsourcing project Folding@home Consortium (see also Chapter 25) offers a simple and convincing testament to the value of simulating protein folding: "There are many experimental methods for determining protein structures. While extremely powerful, they only reveal a single snapshot of a protein's usual shape. But proteins have lots of moving parts, so we really want to see the protein in action. The structures we can't see experimentally may be the key to discovering a new therapeutic. Using football as an analogy for the experimental situation, it's as if you could only see the players lined up for the snap (the single arrangement the players spend the most time in) and were blind to the rest of the game." ([66])

The field of computational life sciences provides a powerful tool for research and is of direct benefit to us in our everyday life. It is applied to modeling of biological systems that is helpful for diagnosis and therapy based on cell-level processes. Multi-level simulations contribute to what is known as *predictive medicine* where individual's health risks can be identified and thus addressed, perhaps preventing future issues. This is a start to personalized medicine. CLS is used for drug discovery and pharmacology, and to study the biology of cancer. Genomics, proteomics, and biotechnology cannot be done without computers. Then there are the research areas with longer term impact potential such as neuroscience, evolutionary biology, and large-scale studies of living organisms' data.

CHAPTER 22

Genomics and Beyond

Computations, Data, and its Impact

G ENOMICS is a hot topic in HPC for the last 30 years at least (The Human Genome Project kicked-off in 1990). A quick recap: There exists a special-purpose device that determines the order of the bases in a strand of a DNA sample. It is called *DNA Sequencer*, naturally. However, the human DNA, among others, is too complex (long) for the sequencer to handle all at once. In fact, the DNA strand being sequenced would contained several thousands '*letters*' of the DNA. The complete DNA would be made up of many millions such strands. That's where computers come in. The DNA strands are constructed such that there is a little overlap with their neighbors in the full DNA sequence. The *assembly* phase of putting all the strands together in the right order is a big pattern recognition problem. The NIH (National Institutes of Health) put together a short introduction to genomics ([67]).

Applying methods and software tools to very large and complex datasets that originated in biological systems gave birth to what we commonly refer to these days as Bioinformatics. It is most often associated with DNA and amino acid sequences, that is, with genomics.

There are many institutes dedicated to genomics. In this chapter we look at two programs under the auspices of the U.S. DOE Office of Science: The Joint Genome Institute ([68]) and the ExaBiome Project ([69]). Scientists in these programs collaborate since there is an overlap in the research areas of the two. Each involves several organizations, with the principals in the DOE's Berkeley Lab.

The Joint Genome Institute

To find out more about the computational aspects of genomics and why the information we gain can revolutionize medical treatments, I talked to Kjiersten Fagnan. Fagnan is the Chief Informatics Officer of the Joint Genome Institute (JGI). The

DOI: 10.1201/9781003038054-22

institute is a user facility located at the Lawrence Berkeley National Laboratory (aka Berkeley Lab and LBL). Fagnan is an applied mathematician, a fact that highlights the multi-disciplinary nature of modern computational scientific research.

When Fagnan started to model biological systems she had to supplement math with courses in anatomy that focus on material properties and structures of biological organisms. As she puts it:

"One of the hardest things in biology is moving from the toy problems to reality. You can come up with very good models in a very straightforward settings. You can model how a cell moves if you assume it only moves in two-dimensional space. But as soon as you want to consider movements in three-dimensional space, the complexity of what a cell is able to do and the chemicals and materials it is able to sense as well as the structures that the cell builds to move in 3D are incredibly complex and hard to capture with the types of methods we typically look at."

The fact that the interesting biological patterns only show up in the much more complex 3D models is what brought Fagnan to HPC and Berkeley.

"When you're sticking to simplified circumstances, then there's not so much the need for high-performance computing. But one of the reasons I wound up at NERSC[1] was because of the modeling we were doing of the sound waves propagating through the body. They turn into shocks, which made the problem non-linear. I had to use multiple computer nodes, and that was how I started with parallel computing."

It turns out that just sequencing a genome, as big a milestone as it is, is only a start. Fagnan:

"The genome is involved in regulating the body and determining what proteins are present or absent in a particular human. This is of massive combinatorial complexity, and you have to learn far more about that complexity before you could ever start to think about modeling it."

The purpose of looking at how human genomes are different from each other and finding markers for certain diseases is only one facet of this endeavor. A more challenging quest, Fagnan says, is:

"To understand how the genome serves as a blueprint for the activities in the body, like what drives the creation of certain proteins and the creation of certain metabolites that you might find in the process that is healthy in one person and perhaps diseased in another person. This is an incredibly complex process. And then when you consider the fact that you have microbes that have been evolving on the planet for something like 400 million years, and the diversity in microbial life that exists on the planet because of all of the different niches that they can find on the planet and different sort of mechanisms they've found for survival. It

---

[1]The National Energy Research Scientific Computing Center, located at LBL, is a primary scientific computing facility for the DOE's Office of Science.

is fascinating. Within your body, you have roughly five pounds of bacteria just hanging out and helping you stay alive."

Turns out, according to Fagnan, that what can be learned with the help of powerful computers is limited by the completeness of the data and what she calls the "huge bias" of the instruments collecting the data. It is the reality that sometimes low-abundance organisms have large effect, but are harder to detect experimentally.

Fagnan on what this means to the data analytics part of the challenge: "Trying to put those strings of information back together again into a blueprint for all these different organisms is error prone. There is noise in the data. There are a lot of repetitive regions, so you can't necessarily resolve one coherent long piece of a genome. You actually end up with breaks and sort of a lot of uncertainty in it. You take that information and you try to identify the areas where there are genes. Then you try to think about whether or not those genes are actually turned on or not in that particular organism from that particular environment. So, there are just very many, many layers of information that are still needed."

Yes, bioinformatics shows us how overcoming one major milestone – the sequencing of the human genome, only points to an even more complex and demanding challenge: One of understanding the workings of the totality of organisms, including the human body.

Fagnan had an interesting observation about the role of HPC in genomics research:

"In a lot of cases, you'll take a sample from the environment and you can only cultivate a small fraction of the microbial life that's actually in that sample. Much dies on transit or dies from having its environment disturbed. Therefore, there are huge chunks of this puzzle that are kind of missing still. The genomics field tends to be focused on cataloging, understanding, sequencing, and looking for new things that we haven't seen before, and thinking about new laboratory techniques to improve the quality of the data so that we can ensure a high quality in the reference databases. When finding something that's unknown, the researchers will look up in the databases and see what it might be or what it might be related to. We are still in this information-gathering phase. But there is enough data from the work that's been done for a few decades where you need high-performance computing to be able to analyze and put all of those data together. And so that's really what's driving the use of high-performance computing and genomics now.

"However, many of my colleagues, they will say they don't need high-performance computing. There's a significant camp that thinks that it's actually just incredibly painful to use HPC, and that's in part because those systems and those architectures weren't built with data workloads in mind. They were built for large-scale simulations. For PDE solvers. Additionally, the genomic data analysis problem involves comparing data that was collected from an organism to some reference database, trying to figure out what it is. The result can come back in less than a minute or it could take hours. You don't know how long it's going to take ahead of time. This is the high-performance computing environment we are in. We are terrible users because we can't really predict our resource usage ahead of time."

The problem these scientists have seem to be, at least partially, an administrative one. In practice, the researcher using an HPC system has to budget machine time which is difficult when they can't estimate the execution time.

Rick Stevens, from Argonne Lab, described in the Computational Life Sciences chapter the necessity and value of large HPC systems for the imputation process that is so critical for genomics studies (see page 174). He goes on to say:

"Now that we have, literally, millions of genomes that are available, you can calculate statistically across each gene what the very fine-grain distribution of variations are and use this pattern in the genome to predict quite accurately how it would show in new genomes. Imputation fills in all the missing parts, and then you can take genomes that have been imputed and run them through pipelines to predict things like incidence of cancer or predict various phenotypes, diseases, and so forth. This results from a combination of improvements in the scale of computing, and, to some degree, of algorithms. But it's also very much piggybacking on the increase in data that we have access to that has occurred over the last decade, enabled by the reduction of sequencing costs."

We see that the genome application of HPC is an example of data driven problem. As Fagnan explains:

"We have a couple of memory bound problems. One of them is genome assembly or metagenome assembly.[2] Assembling metagenomic samples individually can result in losing out on the low abundance organisms. But combining them together for a joint assembly generates enough of a signal of those low abundance organisms and makes them more discoverable. And this is only possible with a machine that has several terabytes of memory, or when the algorithm is adapted to make use of clusters. And for that we need a high-speed network because there is a huge amount of message passing between all of the nodes while generating the graph that is being constructed in order to build the assembly."

All this research has real life applications. Here are two important examples that tie together compute-based discovery and experimentation to addressing climate change:

"The JGI was heavily involved in creating high quality reference plant genomes. The goal is to make plants more resilient to changes in the environment. Many are feedstock plants such as soybean and corn. Other crops are good for potential biofuel production. All of these genomes are studied. There have been different experiments where people grow these plants in different environments and look at what genes are expressed. The phenotypic (observable characteristics) results are

---

[2]Metagenomics refers to the analysis of the gene content of a community of organisms; microbes, for example.

used to identify a gene that is being expressed that made this plant more able to survive in a drought-tolerant environment. The researchers then look at ways to increase that gene expression in other plants or potentially breed for it, or it just gives them an understanding of what's actually going on in the plants to make them more resilient to climate change.

"There are similar efforts to understand how you can actually do a better job breaking down lignin (organic polymers important in the formation of cell walls) in plants because that's one of the barriers to creating cost-effective biofuels. Biofuels are a little hard to produce because it's hard to break the plant material down into the sugars that are needed to actually scale up into biofuel production. Solutions are sought by studying fungi genes that cause generation of enzymes that are good at breaking down lignin. They are then transferred into other organisms that can go into the plant or into the biofuel mixture to actually help break down the plants."

Drought-tolerant plants and biofuel generation – helped by data analyzed on HPC systems.

Addressing the biology of humans as a whole system is much more complex. One of the important ingredients is our microbiome (the aggregate of microorganism in a specific habitat). Studying it starts with animals. Fagnan again:

"There is a project by researchers from UC Davis, UC San Francisco, and the Berkeley lab to look at microbiome analysis across different types of animals because we can't do human yet. But it turns out that the same bacteria that are in our guts are present in other organisms and other animals. JGI has generated and analyzed these microbiome data for a decade. We produce really high quality microbial references and also metagenomic data. Other researchers want to use the JGI data to augment data that they generate themselves to give themselves enough statistical power to be able to say something about the microbes that are present. Right now it's all about being able to aggregate enough data to be able to do statistical learning and to be able to leverage some of what's been developed in computatinal sciences and in mathematics and statistics to develop a better understanding of what's going on in these microbial communities."

We see that decoding the genome is a stepping stone and a foundation for understanding the much more complex total living systems. It also serves, through its role in driving the structure of amino acids and protein chains, to advance discovery and development of commercial products. Biofuels, for example. Here we cannot avoid the intersection of the public academic and government research findings and work done by private enterprises. Of course, the latter have a great incentive in keeping their data to themselves. It is a tricky balance to navigate. On the one hand, combining datasets and human resources will speed up discovery and share the costs of uncertain outcome. On the other hand, the rewards from the discovery may have to be shared.

Advancing the broad field of biology today relies, to a large extent, on vast amounts of data. The more basic research is public funded. Much data is collected by commercial interests such as crop producers, pharmaceutical companies, and others. For example, Genome sequencing and how genes' expression manifests itself in plants is central to the relatively new field of bioenergy. The U.S. Department of Energy (DOE) established the Bioenergy Technologies Office (BETO) which works in partnership with the private sector to develop sustainable energy sources.

Genomics also contributes to drug discovery (Chapter 21). Think of antibiotics. There is a complex process that involves use of both computers and lab work. It proceeds from the genome to transcriptome (the set of all RNA strands resulting when genes are transcribed) to proteins that can pick up secondary metabolites. The latter provide insights into natural products where antibiotics can be found. The discovery process is being helped, or driven, by modeling on computers. Refining the models require large amount of data. If and when large enough datasets are available, their size would require running on a large system. Unfortunately, the best software tools for the models were designed for laptop class of machines. However, the success of this line of discovery – from the genome to new antibiotics – requires enhancing the modeling tools to run on distributed memory parallel systems. Such work is done in JGI and by other research teams.

The way the study of genomes results in tangible benefits to our lives involves multiple aspects of interactions between theory, analytics, prediction, and experimental validation. The use of HPC-class systems is almost a constant. And certainly indispensable. The discovery process Fagnan described to me involves supercomputers for the large genomic datasets described above, and later on for investigating very large images data. It goes like this:

"Molecular dynamics calculations still take up a chunk of resources in NERSC, including for validation of genomic-derived predictions. The genome is the blueprint, and the researcher might think that because you found what looks like a gene coding region, that it is going to produce a particular protein. But you don't know for sure until you experimentally validate that. Until you actually see that and measure that this protein has been produced. There is prediction software out there that tells you what the protein might be. Now enters cryogenic electron microscopy (cryo-EM). Protein crystallography only works for the proteins that you could actually freeze. There are cryo-EM imaging techniques that are letting folks actually look at a whole host of new protein structures. Going from this predicted structure to actually experimentally validating that this structure is what you see is where a huge amount of work happening right now.

"There are national cryo-EM centers that generate a huge amount of image data to be processed. And that is one of these linchpins that makes it possible to validate the functional prediction that you get from the genome. You use the genome to make these predictions, helped by software to structure the predictions. But to validate it you need images. At JGI, we provide some of that first step genomic information and then folks can take that and validate that a predicted function is real."

Today there are genome sequencers that can fit on a desktop. The length of the sequences read allow a whole genome to be processed on a laptop. But to deal with errors that occur, the researcher might end up with many copies of the genome in question. It becomes a big data problem. And this is only the beginning. Fagnan explains:

"Let's fast-forward 10 years to a world where all of the genomic data comes out and it is perfect. We get perfect genomes, perfect metagenomes, we know exactly what organisms are present in different locations. The problem is far from solved. Just sequencing microbial organisms and all of the potential variations there, would probably take a yottabytes of data."

Where we are going to see the real benefits from genomic studies is not from individual genomes, but from collections of genomes from members of an organism, and even across organisms. There is a lot more to the subjects of matching traits to genes and the proteins they enable, how evolutionary features came about from dependencies between species, identification of causation from gene to protein to health condition, and more. All this is beyond the scope of this book. The point here is, the field requires HPC systems due to the size of the datasets involved. The quantity and complexity of the relationships calls for machine learning techniques for the discovery process.

The ExaBiome Project

The JGI is one of the collaborating organizations, with teams from the DOE labs in Berkeley and Los Alamos, that participate in the *ExaBiome Project*. The project is also under the umbrella of the Exascale Computing Project (ECP) as one of the projects preparing applications for exascale computing (see Chapter 23). It deals with the use of computers to study microorganisms, communities of microbes that sustain life through complex interconnections. This is the realm of *metagenomics*.

Kjiersten Fagnan made the case for approaching the ExaBiome's principals highlighting their '*performance assembler*' – not the programming language, but a genome assembler. It allowed some plant genomes and metagenomes, that were beyond JGI's capability, to be assembled. It made possible to take the raw data, perform quality control, then annotate it. This is the data representation that is useful to biologists and other researchers.

Katherine Yelick, Vice Chancellor for Research and Distinguished Professor of EECS at UC Berkeley, is the ExaBiome's Principal Investigator. Lenny Oliker from LBL is its Executive Director. They shared with me aspects of the work done at the project.

ExaBiome is about the application of computing, HPC systems in particular, to the study of microorganisms ([70]). As pointed above, much of the software for genomics was written for a shared-memory systems and that limited the size of data it could handle. At the same time the problems biologists want to work on – in metagenomics and microbiomes, for example, require analysis of very large data sets. Data sets that experimental instruments are providing. Yelick explains:

"Our goal in the ExaBiome Project is to take these problems that were really limited to a terabyte or so that you could handle on a shared memory machine, and turn them into something where you can handle multi-terabyte structures and data sets. We have demonstrated with the HPC versions of these codes that we can now run multi-terabyte data sets. Recently we ran a 30 terabytes data set that was part of a data set of 84 terabytes collected across all the oceans in the world. That's our exascale science problem that we will be working on: assemble that entire data set." By some time in 2023.

Yelick emphasizes that the challenge of parallelizing the genomics codes is quite different than that of numerical simulations of physical phenomena. Aside from the difference in the type of operations, and a more challenging difference, is that genomics data structure is irregular. We have no knowledge of relative locality of the data elements: "For one thing, these codes that we use for assembly don't have floating-point operations in them. They have very irregular patterns of data access. We build hash tables and filter data structures for counting things. There is no mesh that you put down and then chop it up into pieces and give each processor a piece. If we knew what the locality is in the genome we would know what the genome looks like. But since we don't know what the genome looks like we don't know how to divide it up in advance. This makes it interesting from a computer science standpoint. The algorithms are really quite different than a lot of numerical simulations."

Oliker adds: "There are several reasons bioinformatics and genomic analysis are late comers to HPC: data structure irregularities, low data reuse, high and complex communication patterns, and scarcity of floating-point operations. That, and the inherent lack of locality where you really have to go across the entire data set. For the ocean data mentioned before, when you actually do the calculation, the memory requirements could be a factor of 10 bigger. So we are starting to approach a petabyte of actual data at the highest memory use of the computation. And this obviously necessitates distributed memory computing. Then the question is how to orchestrate this in an efficient way not only for computation on regular CPUs, but also leveraging GPUs, since all the exascale and pre-exascale systems make heavy use of these accelerators, which are inherently better suited for the more traditional HPC applications. It all made it a really interesting project, trying to figure out how to connect the dots and be able to leverage these large systems."

Given the nature of the data structure and the processing flow for genomics it is clear that current HPC architectures are certainly not optimized for it. The one aspect that would be most helpful to improve is, according to Yelick, the communication overhead. Latency will always be there, but there is software overhead and there are hardware protocols that be better tailored to the genomics type of workload. The data characteristics listed for genomics are present for other data analysis applications – social networks data, for example. Meanwhile, the ExaBiome team needs to find ways to mitigate the communication overhead. Yelick explains:

"Existing hardware protocols require a bunch of handoffs that make it very inefficient to have fine-grained communication. So we try to coarsen the granularity of communication by doing dynamic aggregation. But unlike a typical simulation codes where one does domain decomposition to get chunks of data and exchanging boundary values, when you are building a hash table, this is not possible. Instead, we put items into buckets and say '*This bucket is destined for that processor, and this bucket is destined for that processor.*' And when the bucket gets full, we ship it off. This is a dynamic and adaptive way of trying to aggregate things. On the other hand, if we could lower the overhead of communication, even if latency is long, we will be able to overlap transfers. That is, to have multiple things in flight at once. As it is, having high overhead makes it harder to run these codes efficiently."

We also encounter here the recurring theme of the somewhat unexpectedly fast adoption of GPUs as the main processing units in recent large systems. According to Yelick: "When exascale took a turn very clearly toward GPUs, we had to do quite a bit of work to rethink some of the algorithms to figure out how to take advantage of GPUs and I've been surprised at the substantial speed-ups we get from GPUs."

In Chapter 19, on the subject of high-productivity in HPC, we mentioned several languages that started their development in that period (late '90s). Yelick was one of the principals, perhaps 'the' driving force, behind UPC (Unified Parallel C). It was later augmented by object-oriented library to become UPC++. It was fortuitous that both the ExaBiome project and UPC were led by Yelick, as it tuned out that UPC++ made coding for ExaBiome easier and allowed for faster development. Oliker states: "The global address-based methodology has made the task of implementing these complex irregular algorithms much simpler. Our success in successfully porting these codes onto different architectures has been due to the ability to leverage UPC++."

In addition to microbiomes, the project includes analysis and computations of protein clustering, and the software for this is written with the MPI programming model (and not UPC++). The team hopes to draw conclusions about which approach is more productive and how they complement each other.

One example of work done through the ExaBiome Project's access to large HPC systems is the analysis performed on vast amounts of microbiome data collected from oceans by the French Tara Ocean Foundation. A schooner named Tara has been roaming the oceans since 2003, collecting microbiome data with the objective of studying the impact of climate change ([71]) – after all, as the foundation's tag line says, *"Our future depends on the good health of the Ocean."* Tens of thousands of samples were collected, amounting to about 100 terabytes, which requires close to petabytes to process and analyze ([70]).

The ExaBiome's contribution is by enabling *co-assembly* (rather than multi-assembly). Yelick uses an analogy to explain the difference: "If I'm only shredding one book, it's going to be hard to put the book back together because I don't know where the errors are. But if I shred 50 copies of the same book, it'll be easier to piece the book back together because I'll have 50 copies to compare it to, and it's unlikely that errors will occur in the same place."

Oliker adds:"The TaraOcean teams came home with all this data, presumably not having an idea that somebody was concurrently spending years coming up with the computational capability of sequencing all those metagenomes in one comprehensive computation, which is the best way to get the highest accuracy and the highest quality for extracting the most number of genomes and phenotypes. In other words, instead of chopping up the samples into small pieces that you could fit individually onto a shared memory node, doing individual assemblies, and then merging them together (multi-assembly), we are able to do one large so-called co-assembly. This gives a much higher quality of the solution. It is quite fortuitous for us because there are not that many data sets that are large enough to be of interest to running these types of simulations on an exascale platform, partly because it didn't occur to teams that they could actually have the computational capabilities to do such a large-scale data analysis. And now, as more teams are learning about this capability, we are already seeing larger and larger data sets that are being collected."

There is another level of complexity to studying microbiome, better understood with another analogy, as explained by Oliker: "When doing a genome assembly of a single individual the sequencing technology breaks up the genome into little chunks called *reads*, and the assembly process is one of stitching them back together. It's like putting a jigsaw puzzle together, but without a picture on the box. In the case of microbiome there are tens of thousands of different microbes at different concentrations. Some are very similar because they're near species. Now imagine taking ten thousands of jigsaw puzzles at different concentrations, throwing them on the floor, and now you have to rebuild the individual pictures or the genome. This is the large-scale challenge of metagenome assembly."

The treatment of genomics codes is one example of the changing face of HPC away from just numerical simulations. And it is not just the data analysis. Some of the ExaBiome projects deal with understanding what proteins are coded in the genomes, and what are their functions. And here the other new comer to HPC, machine learning, comes in handy, by using a set of known proteins for a training set, then applying ML to further the investigations. We return to this theme in Chapter 24.

What is there to be gained from studying microbiomes? – One area is that of mitigation and remediation of climate change effects. It starts with identifying microbes that capture or release carbon. The ExaBiome's assembler and tools were used to compare this aspect on wetlands stretch that was overrun with salt water and later restored with fresh water. There is research into how wildfires change microbes and how it affects forests' recovery after fire. And, not surprisingly, in the realm of agriculture, researchers are trying to understand the relationships between microbiome environments and soil resilience to drought and what microbes are helpful for high yield of produce. It is expected that this may be applied to bioengineering of plants.

There is a huge amount of data from microbes in the human body. Assembling it all together – in a single assembly procedure, is a truly exascale task, in terms of both data and computations.

This work would not have been possible without the software tools that were adapted to distributed memory architectures.

The ECP, and with it the ExaBiome project, is funded until 2024. Yelick and Oliker see the large-scale assemblies mentioned above as the culmination and endpoints of the project, with the understanding that it is only the beginning of this scale of assembly and analysis, and the hope that subsequent projects will be funded. In particular, the continued incorporation of machine learning techniques to metagenomics.

V

The Epoch of Accelerators and Cloud

From Simulations to Analytics and AI

CHAPTER 23

Codesign

Multidisciplinary Teams Prepare for Exascale

FOR the most part application implementations were adapted to the processor and system architecture as they evolved over time. Programming models changed from serial processing on mainframe-type computers, to vector processors, to multi-processor shared-memory, to highly-parallel distributed memory systems, and the addition of accelerators.

It is not that computer architects were not listening to users and software developers. Some privileged users (due to their considerable buying power), such as some government agencies, would get briefings prior to product launches, but mostly the computer vendors would take external input and go back to their closed environment and do the best they can. There are some notable instances of direct and specific response to user demands.

The example that is consequential for general numerical HPC code is the introduction, first implemented by Seymour Cray, of the *fused multiply-add* (FMA) instruction. The operation is present at any matrix operation, and very common in scientific codes.

The intelligence community had enough influence in the '60s and '70s to get the supercomputer companies (CDC, Cray Research) to implement an instruction only they needed at that time. It is a rather strange instruction, called *popcount* (for population count) that returns the number of bits set to 1 in a word. It is a useful operation for decryption algorithms. The same agencies were also interested in more complex instructions for bit-wise operations, such as bit-matrix-multiply and bit-matrix-transpose (and got some of those implemented at times).

The popcount instruction story is an example of a feature that was done for a narrow and specific application, but, because it was there, found a much broader use. Today we find the popcount instruction in several microarchitectures where it is used in a variety of ways: machine error correction, neural networks, molecular 'fingerprinting', and even in chess programs and compiler optimization techniques.

DOI: 10.1201/9781003038054-23

The examples above notwithstanding, the more common occurrence is that software and application people make requests and express wishes that are only partially fulfilled; or not at all. This is true even for 'insiders'. Ken Miura, with his illustrious past at Fujitsu, says: "It is very difficult to get hardware designers to add instructions. I tried several times and they said 'no'."

It has always been clear to HPC practitioners that in an ideal world there would be an ongoing exchange between computer architects, system software developers, and application programmers. Of course, there have always been such conversations to gather feedback, to ask questions, to express wishes and requirements. But it has never been an all-party joined project from start to finish. First steps toward such a dialog were taken by the organizers of the Petaflops Technologies 1994 workshop and its follow-up meetings (Chapter 16), and later by the HPCS program (described in Chapter 19). The latter, with its limited success and market scope, did not include direct participation of technology provider companies.

At the close of the first decade of the 21st century, with the petascale era starting, the time has come to set out for the next monumental milestone – the next 1,000-fold growth in compute power that will usher in the exascale era. The challenges seemed to be even more formidable than in the mid 90s, and people began to talk more explicitly of the need for *Co-Design*, planning ahead cooperatively by multi-disciplinary teams of technologists, system architects, software engineers, and application developers.

A Technology Challenges Study

Repeating the idea from 1994 of studying the technologies that can be expected for petascale computing, a study was conducted in 2007 with exascale in mind. It was sponsored by DOD's DARPA but was composed of people from academia, national researchers, and engineers-scientists from Industry. Their work, over several months, is documented in a book-length report titled "ExaScale Computing Study: Technology Challenges in Achieving Exascale Systems" published in 2008 ([72]). Among the participants (listed in the citing of the reference) I single out Peter Kogge, from the university of Notre Dame, who was the study lead, and Bill Harrod, who was the program manager.

It should be noted that the study group's directive was to determine if 'mainstream' technologies will get us to exascale by around 2015. We now know it has taken 5-7 more years (depending how we count 'pre-exascale' systems).

The working group stipulated that technology advances alone will deliver the 1,000-fold increase over petascale within the same physical boundaries: An exascale would be "datacenter-sized, present-day departmental system would be petascale, and a few chips will have the capability of a terascale rack. That's not to say that the component count and power consumption will remain constant. Indeed, they identified the major technology challenges as:

- **Power**. The most challenging of the list. The more difficult part to solve is the energy used to move data; not the compute part.

- **Memory and Storage**. Lack of technology for both capacity and transfer rates desired by applications.

- **Concurrency and Locality**. Because clock rates and single thread performance are reaching the end of their upwards trend, there will be orders of magnitude more components and the burden of increasing system performance will fall to successful explicit parallelism at-scale.

- **Resiliency**. This refers to frequency of interruptions or failures of the systems. The concern arises from the explosion of component count and use of advanced technology at lower voltage levels. The count adds to the statistical possibility, and the latter increases the sensitivity of the devices to their operating environment.

The body of the study is rich in technical details, parameters, metrics, and projections. What was significant to me was that when it came to recommendations the common theme there was that of interdisciplinary cooperation.

The component devices are to be co-developed with system architects for optimum outcome.

The hardware architects are to cooperate with the software engineers defining future programming models.

There has to be co-development of algorithms and application with that of tools and run-time modules.

The programming models and application developers have to work with, and understand, the hardware resiliency pitfalls, so as to be able to recover and continue in the face of some component failures.

The value of the study is not so much in generating new information as it is in gathering disjoint aspects of what enables HPC, present a coherent picture of the status, and make projections. The scope of technologies examined shows the intent of looking at HPC holistically. Specific sections of the technology roadmap are assigned to: Logic, Memory, Storage, Interconnect, Packaging, Resiliency, Software Operating Environment, and Extracting Parallelism. More attention was given to hardware components: Of the group's nine meetings, one 'special-topic' meeting was dedicated to programming environments (with 'architectures') and one other special-topic on applications (with 'storage and I/O'). We see in the next section the additional actions taken by the software community.

The exascale study group highlighted technology trends, well-established or starting to manifest themselves, that would guide directions and possibilities of future developments: Device feature shrinking a la Moore's Law will continue, but without a proportional increase in performance (clock rates not increasing), though challenged by limits set by lithography's capability. The limit has been reached for extracting parallelism automatically out of a serial program. The same is true for mechanisms and methods for hiding memory latency.

The study group warned that the challenges cannot be addressed unless the hardware and the software communities work in concert.

It laments the decline in investment in research of alternatives that will mitigate or eliminate the challenges that are exacerbated by the trends above.

Addressing the challenge of continuing with past rate of increase in performance of computer systems means added complexity by the huge increase in device count combined with figuring out how to extract and express parallelism. That meant microprocessor design has become a very costly endeavor, requiring teams of hundreds of engineers and thousands of hours of computer simulations, that only a small number of large corporations can afford.

Contrast this with the vector processors era, when Seymour Cray single-handedly design his supercomputers using a pencil and paper.[1]

The report present several important observations, some even alarming: Critical applications demand performance levels well above petascale, that don't just use larger datasets but are more complex computationally. From the report: "Technology is hitting walls for which there is no visible viable solutions." There is little to no research into architectures that can provide further "explosive growth in performance." And apart from a few 'heroic' codes, they conclude, "our ability to scale up applications to the millions of processors, or even port conventional personal codes to a few dozen cores is almost non-existent."

In hindsight, we now know that exascale was achieved without resorting to major architectural innovations, and we did manage to continue and scale up applications, while staying pretty close to the trajectory of past rates of performance increases.

The International Exascale Software Project

While the high-productivity (HPCS) and the exascale technologies study were driven by the Department of Defense (in the U.S.), the next step, centered on software, was led by the U.S. Department of Energy (DOE) and included international participation. The International Exascale Software Project (IESP) received support not only from the U.S. DOE and the National Science Foundation, but also from agencies and universities in France, the U.K., and Japan, and from corporations – Cray, Dell, Fujitsu, Hitachi, IBM, Intel, and Nvidia. The list of sponsors has 30 entries.

---

[1]HPC lore includes the story of Seymour Cray being notified that Apple purchased a Cray Research supercomputer in the mid 80s to support their circuit design simulation (this is true, they bought a Cray X-MP – designed by Steve Chen, but a descendent of the Cray-1). To which he replies that he is using an Apple desktop for his work (possibly for drawings related to the Cray-3).

I was fortunate to be asked to represent Intel at the IESP meetings, and participated from its start at 2009 until my retirement from Intel at the end of 2010.

The initial phase of the IESP work is summarized in a report, the IESP Roadmap, that has been referenced extensively in the last decade (see [73]). It lists 65 names of participants and contributors from national labs, universities, research agencies, and the private sector. They come from the U.S., several countries in Europe, and Japan from Asia (later meetings had participants from China, too). The working group met in a series of workshop-type meetings in all three continents.

The central idea was to build on the concept of the open-source software for HPC, and add coordination and planning to what has been a disjointed collection of components. The underlying assumption was that without such cooperation the challenges of software for exascale computing will not be met. The IESP's stated mission was:

> "The guiding purpose of the IESP is to empower ultra-high resolution and data-intensive science and engineering research through the year 2020 by developing a plan for (1) a common, high-quality computational environment for petascale/exascale systems and (2) catalyzing, coordinating, and sustaining the effort of the international open source software community to create that environment as quickly as possible."

The thinking at the time was that the challenges for the software infrastructure that is needed to make exascale systems useful are so daunting that it requires, in the words of the IESP report, "the wholesale redesign and replacement of the operating systems, programming models, libraries, and tools on which high-end computing necessarily depends." And key to success of such an ambitious project is coordination and collaboration among all the parties involved, internationally.

The end-product of this open-source and common software infrastructure was called *extreme-scale/exascale software stack*, abbreviated as *X-stack*. It was to be able to support running on the whole of the largest systems, as well as on scaled-down systems. It was to be modular, to allow for alternative modules and participation of multiple developer teams. And there was to be an open-source option for all components.

Going from petascale to exascale means an increase of up to 1,000-fold in several aspects of the system, which defines challenges to software in terms of concurrency – expected 100s of millions of arithmetic units and up to 10 billion threads to deal with hiding latencies, and, consequently, in terms of resilience – recovery and continuation of execution in the face of errors or components failure.

The list of X-stack components, rated on scales of exascale *uniqueness* and *criticality* at the level of sub-components, is comprehensive: frameworks, numerical libraries, algorithms, debugging, I/O, scientific data management, programming models, compilers, operating systems, performance, power, programmability, resilience, and runtime systems.

A central tool for driving the IESP activities is what was denoted *Co-Design Vehicles* (CDVs). The idea was that applications to be run at the exascale level are supposed to inform the development of the IESP roadmap. But, at that time (circa 2010), only few existing applications were able to utilize such a scale of computing. Several elements were identified that were critical to exascale jobs: Programming models for many-core and heterogeneous nodes; handling the much larger datasets to be processed at that scale; making applications resilient and fault tolerant. They also wisely pointed out that the then-common insistence on bit-level reproducibility is not practical at the level of concurrency required for exascale.

The team then went on to look for applications that have a demonstrated need of exascale performance associated with science-based goals, with definable set of intermediary steps over some 10 years. It was stipulated that the development team has to be interdisciplinary and include expertise from hardware to software to algorithms; the code modular (for ease of modifications, additions, and simulations); and that it stresses the various dimensions of the X-stack.

When the IESP Roadmap was written there were two CDVs identified (Lattice QCD and Fusion Energy), with hopes expressed for quite a few more. Candidates were sought from a list of disciplines where it was understood exascale computing is needed: Materials, Energy, Chemistry, Earth Systems, Astrophysics, Biology, Health, High-Energy Physics, and Fluid Dynamics.

Rick Stevens, the associate lab director at Argonne National Lab (who we met already in Chapter 21), was also one of the leaders of the IESP. He describes the investment in preparing applications and the software stack for the exascale era:

"We started planning for the Exascale Computing Project back in 2007, but it finally became resourced at scale between 2014 and 2016. And at that point, we knew we had to fund the applications and the software. We had about 80 or so software efforts ranging from libraries to compiler tools to I/O systems, debugging and visualization and so forth. And we funded about 25 application projects ranging from climate to wind energy and earthquakes, cancer to material science, and more.

"We funded about a thousand people for the last five, six years. It's been a standing army of people who are working on this. And the applications are really interesting, because many groups inside the application space, once they knew what the hardware was going to look like, had a pretty good idea what to do. They had to re-optimize their codes for the hardware and for the hierarchical programming model. And in some cases, they made the decision to not maintain dependencies on libraries and do the coding themselves. In other cases, they decided to enforce the use of existing libraries, and then they partner with the library people. it has been a very interesting ecosystem."

The last chapters of the IESP Roadmap deals with practical matters of execution the plan. The relationships with vendors (systems and software) is spelled out. They basically reiterate ongoing understanding and practices within the HPC community: Products closer to the hardware, such as the operating system, compilers, and runtime utilities, will be provided and mostly supported by vendors. The user community may support some tools (debugging, compilers) and numerical libraries. The community will, of course, provide and support application-level components from algorithms to visualization, and, most important, programming models.

What is perhaps new, or just stated in stronger fashion than in the past, is the emphasis on open source across the board and the insistence that what is developed with public funds will be unconditionally available to the community.

In conjunction with the Roadmap report, the IESP's leadership team of about 10 people, from all the three continents, published a "call to action" report ([74]) following the first three workshop meetings (one in each continent – U.S., France, Japan). It described the expected requirements of applications in several areas and the challenges of applying them at the exascale level. From these observations of complexity, resolution, timescales, and data analysis, they make the case that the IESP is necessary to address three HPC existential factors: There is a compelling science case for exascale computing. The current software infrastructure is inadequate and replacing it is a formidable task. Lastly, there is no global coordinated activity of the HPC open source community to confront the challenge.

Formal meetings of the international working group continued until 2012. The organizing team created a website with a trove of information. It contains all the presentations given at the ten meetings that were held and serves as an historical record of the thinking and the process that took us from petascale to exascale, with their insights, foresights, and misses. The reader will find it fascinating to browse through [75].

How Did the Forecasts and Predictions Turn Out?

Now that the first exascale systems are operational, we can look back and compare reality to assumptions and predictions made over a decade ago.

The IESP Roadmap stipulated a timeline with several stages on the way to exascale. The 2010-11 thinking called for readiness of the software stack components in 2014-15; 2018-20 initial delivery of an exascale system with integrated software stack able to handle billion-way concurrency; exascale system in production in 2020.

Fugaku, at the Riken (the scientific research institute in Japan) facility in Kobe, was the #1 supercomputer in 2020. By the customary measure of the 64-bit HPL benchmark it rated at over 400 petaflops, quite a way from exaflops level. It did, however, pass the exaflops threshold (1.4 exaflops) in a modified HPL test of mixed-precision arithmetic suitable for AI computations. Fugaku was heralded as having ushered the exascale era. It is an outlier system in some ways. The CPU is an

ARM architecture (with vector extensions) and it uses no accelerators. Consider concurrency: With close to 160K CPUs, each with 48 cores, it can support 7.6M threads. In addition, its vectors can be up to 32 64-bit words long, so in theory, it is designed to support a concurrency level of 244M. Not quite a billion, but within reach. Notably, Fugaku is likely the first system to have cost $1B (from concept to research and design to construction of the system).

The first exaflops system, in the sense understood in 2010 – performing 64-bit HPL at over one exaflops, came officially online in the summer of 2022 following partial and limited operations in 2021. It missed the forecasted date by one to two years, but ± 2 years difference in a forecast made a decade earlier, at a period of major shifts in technology and architecture, is reasonable and expected.

The system, named Frontier (Fig. 23.1), and funded by DOE, was delivered by HPE employing Cray system design (having recently acquired Cray) with AMD CPU and GPU technologies, and installed at Oak Ridge National Lab (ORNL).

Frontier is constructed out of what we may consider a new kind *fat nodes*, made of one general-purpose CPU (AMD EPYC) and four GPUs. Similarly, the node of the second U.S. exascale system, named Aurora and to be installed at Argonne National Lab, will have two Xeon processors and six GPUs (Intel's Ponte Vecchio).

When we compare past assumptions with later reality, one of the glaring contrasts is related to GPUs. The Exascale Computing Study of technologies did not foresee the extent GPUs will play as the overwhelming source of compute power in getting to exascale, though there is recognition of the need to support

Figure 23.1: The Exascale-class HPE Cray EX Supercomputer at Oak Ridge National Laboratory. Source: OLCF at ORNL.

heterogeneous hardware environment for increasingly more applications. Mark Seager, who we met before (Chapter 18), known for his sometimes colorful language, summed it this way:

"The major thing that changed from the viewpoint that those guys had for the IESP discussions and where the industry actually went was GPUs. I don't think that they envisioned GPUs, (a) being as powerful as they ended up being for a broad class of applications; and (b) that the industry would, kicking and screaming, go there, which is what happened."

Rick Stevens provides deeper context to the underestimation of GPUs' role in the future:

"In the early planning we had these design points. We called them swim lanes. And one swim lane was loosely based on something like BlueGene, where you had a lot of fine-grain, homogeneous processors, and you just keep enlarging more and more. So, for 10^{18} flops, with about a gigahertz clock, you would have to have something like 10^9 processors. That gets you to one exaflops. And that was one of the conceptual swim lanes. We also had swim lanes that were much more on these fat node concepts. A fat node might be some big SMP or a vector processor. And you can think of GPUs as basically vector processors at some level. Not exactly, but it's one way to think about them.

"At that time, in 2010, classical vector machines had fallen out of favor. It was microprocessors, and everybody was just arguing about what the scale of the microprocessors is going to be. Whether GPUs became established, I would say it's still not completely resolved. In the U.S., we've got GPUs, because we have three big vendors making them. Outside the U.S., that isn't the case. In China, they've got a couple of accelerators, but they don't have really a U.S.-style GPUs at scale that are dominant. The same thing in Europe. They use U.S.-made parts, but RISC-V and other approaches there are more like what they were thinking about in the 2010s. There were two things going on. One is that the GPUs became useful enough and broke out of the scene and people were tinkering with them for a long time. And secondly, GPU use grew because of the vendors' investment."

Stevens demonstrated the uncertainty during the last decade with ANL's own exascale system, Aurora. In just a couple of years it went through three design concepts. First, it was Intel's Xeon Phi, then a dataflow engine, only to end up with a GPU-based solution.

The implications of this design decisions will, no doubt, be studied extensively. Some consequences are obvious:

With GPUs entrenched in HPC systems the norm now will be the added layer of parallelism of the many functional units in GPUs. This is another layer to the existing ones of vector instructions in each core of general-purpose CPUs, multiple cores in the CPU chip, the shared-memory parallelism on the multi-chip node, and, finally, across nodes within a cluster. Though not a new phenomenon, a significantly greater role is given to the GPU component.

Having dramatically increased the compute power of the node allowed an exaflops machine to be constructed with fewer than 10K nodes, the same level of node count we saw at the petascale era. The necessary increase in concurrency was achieved with the simpler functional units on the GPUs. System complexity is reduced (but not necessarily for the application programmer), as is power consumption, while the system's resilience increases.

More on the case for fat nodes in the chapter about *performance* (page 270).

In mid 2021, while Frontier's design was finalized and it was being built-up, Al Geist, a long-time scientist at ORNL and the CTO of the lab's *Leadership Computing Facility* (OLCF), presented a webinar talk about the exascale project ([76]). Geist recalled the trepidation they felt at the possibility of failing to built a workable exascale system within manageable power constraints, and even if they did, that it would be impossible to run applications at scale. Yet, they were successful. Not so much due to heroics of overcoming insurmountable obstacles, as for the fact that the hardware architecture turned out to less complex and less power-hungry, and the software environment did not have to be rewritten from scratch. Geist showed a 'scorecard' summary of how the 2009 predictions turned out in the actual system (Frontier) in 2022.

The most striking aspect, as highlighted by Geist, is that the number of nodes was predicted to be around 100K or 1B (there were two estimates). It turned out to be under 10,000, consistent with the unpredicted role of GPUs. It is important to note that Frontier's level of concurrency is, indeed, $O(1B)$, as predicted. Only that much more than envioned, by orders of magnitude, is packed into the node.

Let's pause here to consider that the peak performance of a single Frontier node is over 150 teraflops. Compare that with the 76 compute *cabinets* of the mid '90s ASCI Red that peaked at *one* teraflops. When comparing an ASCI Red node to one of Frontier's, we come up with around 700,000-times performance increase over a 25-year period!

Another gratifying achievement was that of the power consumption of Frontier. It was 'dictated' to be 20MW for 1 exaflops system, based on what was deemed possible and affordable. In circa 2010 some feared an exaflops system would require over 100MW. It turned out to be 29MW for the 1.5 exaflops Frontier – within the required power envelope.

There are several other elements Geist enumerated:

Frontier's memory size, at 45PB, is about 2/3 of what was predicted. The interconnect bandwidth between nodes, at 100 GB/s, is no better than at the pre-exascale times. Storage size fell within the predicted range (716 PB), and the system's I/O bandwidth came out better than predicted by about 25% (75 TB/s).

Finally, resilience was recognized as a significant challenge for an exascale system. It was predicted that such a system will incur an average of about 10 interrupts a day (it does not mean the whole system is down, but that there is an issue to be addressed). In practice, Frontier performs about 3 times better than expected.

One of the bold statements in the IESP Roadmap report was the need for a complete overhaul and redesign of the software stack – from the operating system to libraries, tools and programming models. We now know this was not necessary. Software components were hardened to support the higher scale of computing, and libraries were modernized to support heterogeneous computing. Yet, the OS is Linux and the programming model is a combination of MPI, OpenMP, with CUDA or OpenACC for GPUs. I asked Rick Stevens for his thoughts on the software issue, and it turned out the 'alarmist' perspective was not shared by all:

"If you look at the software side, there was a tendency, and there even still is a tendency in the system and software community to put their favorite research project between where we are now and where we're trying to go, and say, 'Oh, you can't get there unless you solve this problem', such as a new operating system or a new language or a new fault-tolerant scheme etc. And this happens over and over again. And it was happening then. The people saying we need a new OS were the OS people, not everybody else. The people saying we need a new language are language people. It is not that people outside are saying 'Oh, the current stuff won't work'. It's the people that have vested interest in getting R&D money for those things that were pushing that."

In fact, Stevens points out that for the first two exascale systems of the DOE labs, Frontier and Aurora, they opted for the more conservative choice of Cray's operating system over a newer version from HPE that involved micro-services suitable for cloud computing. The idea was to minimize the risk of less-tested innovations where possible, given that risks involving the hardware, applications, and libraries were unavoidable.

Stevens explains that the OS has evolved to better manage $jitter^2$ and $noise$. He emphasizes that it is hard to dislodge software parts that enjoy wide mainstream acceptance. We end up with what Stevens calls "MPI+", where the "+" was initially unclear – OpenMP or OpenGL and/or CUDA or oneAPI. He concludes: "I would say that we were pretty good at getting the programming model right."

And how did the X-stack turn out (or, what it actually is)?

The term *X-stack* was initially a funded R&D program, contemporary with IESP, that built upon some ongoing software projects. The IESP and the Exascale Computing Project (ECP), the DOE project to stand an exaflops system, each had a conceptual software stack. The work that spanned over 10 years culminated with an actual stack for Frontier. That said, Stevens clarifies:

---

[2]Jitter is the phenomenon of interference that applications can experience due to background processes initiated by the operating system.

"It wasn't something that was a concrete, *'Let's design once and make one thing'*. It was designed to have lots of components that could be replaced, and so it was more like a conceptual organization within which you would have multiple modules."

The actual stack on Frontier is a software package that gets regular update releases. The ANL's Aurora will have a compatible stack and applications will be pretty much portable between the systems; up to the use of interface libraries. This is an example of the "+" in the "MPI+": Frontier users use the AMD's programming programming environment HIP (Heterogeneous Interface for Portability) when using its AMD GPUs, whereas Aurora users will apply Intel's oneAPI to access its GPUs.

Stevens makes an interesting observation regarding the impact of availability of 'big data' in the context of genomics studies: "We did not understand at the time (2007-8) the impact that data was going to have. We were pretty good at estimating where computing alone was going to take us, but we were not very good at anticipating that once there was a million genomes' data, or some other huge amount of data, that our actual approach to the problem would change from a pure computational approach to a combination of data-driven and computation."

One example of a code revamped for exascale computing is the old chemistry code NWChem that was developed at the DOE's Pacific Northwest National Laboratory. It is an ab initio computational chemistry code designed to work on a range of systems from workstations to the supercomputers of an era in the past. One of its many applications is in biofuel research, and a team sponsored by Argonne National Lab that included members from several other labs was created to prepare it for chemical systems $1,000$ times larger than it was used for, to run on ANL's future exascale system Aurora. The new version, NWChemEx, is a complete re-write of the old code. As expected, a major thrust of the project was to design the application for effective use of GPUs. The core Fortran routines are complemented and managed by object-oriented C++ and python wrappers. The new modularization of the design makes it easier to adjust the distribution of work to reflect evolving hardware. This is supported by adapting to the oneAPI programming model developed by Intel. The new NWChem on the new exascale systems will be able to simulate a more realistic quantum chemical systems with comparable fidelity to that of experiments. This will accelerate finding solutions to sources of clean energy and more.

Lessons Learnt

What have we learnt, looking back on this multi-year experiment?

The IESP is explicitly a collaborative *software* project. It also touts "codesign" as a driving principle. And when it comes to HPC we cannot forget the hardware side of the equation. Especially when representatives of the systems vendors participated throughout the project. Were they there for the software only? – I asked Rick Stevens about the hardware-software interplay:

"Hardware co-design is harder largely because the ability to actually test an idea in a simulator is quite limited. Whereas in software, you can usually prototype it and try something, in the hardware co-design one has to extract from existing applications on existing hardware in order to understand such features as, for example, ratios or functional requirements and dependencies, and then relate that to the new hardware that was being discussed. One might create micro benchmarks that could be run on a hardware simulator. But often we try to understand a proposed hardware design point in terms of its ratios. For example, what is the ratio between some numerical operation and some logical operation, or how many outstanding memory references I need, etc."

But did the software community influence or drive hardware design decision? – Stevens explains that when the hardware architect asks a software engineer a "trade-off" question, such as doing something about outstanding memory references or a functional unit feature, the software person lacks the tools to provide an answer. Still, he says:

"We tried to do what can be called 'normal co-design', which means we tried to work out trade-offs between functionalities that would be in the hardware and those that we'd make in software. Now, in a codesign scenario of an embedded system where there is one app and relatively simpler architecture, where it is possible to instantiate with different options, one can figure out the optimal trade-off between hardware and software. But for the large systems we consider it is very difficult. It's not that we didn't try to do it. We tried to do it all over the place. It's just very hard."

This kind of trade-offs experimentation that involves the hardware comes to an end once the hardware simulator phase is over. Within the software envelope the process can continue indefinitely, in principle. Stevens concludes: "All things considered, a goal of the IESP was to create a conversation that allowed us to think big enough about the software to not make mistakes."

The IESP took the format of the planning activities of the mid 90s toward petascale computing (Chapter 16) and greatly expanded its scope and reach. The *codesign* aspect applied along several vectors: The project was *international* as well as multi-disciplinary. It went further into development, creating a complete software stack. And it spun dozens of application development projects. All that resulted in, arguably, unprecedented readiness of the system software, libraries, and applications when the hardware became available.

Interestingly, perhaps the result of the difficulty of hardware codesign, each geography has taken a different system approach for its initial exascale systems. The U.S. systems are scaled-up versions of pre-exascale system with fatter nodes of multiple GPUs. Japan's Fugaku is ARM-based with no GPUs. China (not an active participant in IESP) is doing its own thing hardware-wise using local device parts. And Europe has its European Processor Initiative for its own CPUs for its exascale project.

Collaboration has not stifled innovation.

CHAPTER 24

The Changing Face of HPC

How Data Focus and AI are Changing What We Think of as HPC

To be sure, the *face of HPC* has hardly been static for more than a few years at a time. We saw the transitions from mainframes to vector processors, to multi-processors, to MPPs, and to clusters. These were all hardware platform transitions, and throughout the years, until after 2000, HPC was still almost[1] only about scientific and engineering numerical simulations.

The changes in HPC over the last decade or so are different. They are about the *content* of HPC.

The last decade saw a broadening of the applications space that falls under the umbrella of HPC on a scale not seen before. This expansion has dramatically reshaped HPC's scope from that of numerical scientific computing in nature to an enterprise much broader and more complex. It entails major implications for the hardware requirements of the systems, the software environments, programming languages, and skillset demands of its users.

Not only the content of HPC is undergoing much change, but so do its platforms. The GPU, or the general-purpose GPU (GPGPU), as it is called more accurately to reflect its application beyond graphics, has emerged as the new form of accelerator, and is taking center stage by providing by-far the most flops for high-end systems. One indicator of the popularization of GPUs comes from the historical data of the TOP500 lists. In 2006 there was a single accelerated system on the list, 8 two years later, 17 in 2010 – still only in about 3% of the systems. But by 2015 more than 20% (103 entries) had GPUs. In June 2022 a third of the entries (166) were accelerated. Their fraction of the total performance is much higher.

We now have the term *accelerated computing* to encapsulate what is becoming a common environment of heterogeneous hardware compute elements. This goes

---

[1]There were, and are, the non-numerical codes used by the intelligence community on HPC systems.

beyond accelerators for 64-bit floating-point numerical simulations. Reflecting the changes in the nature of workloads, there is a whole host of new attached processors that target the AI space. one of their key features is the use of significantly smaller word size for such application types as machine learning.

The hardware evolutionary transition aside,[2] the bigger impact to what we think of as HPC is the addition of new types of workloads. That is, the expansion of the *content* of HPC. Specifically, the inclusion of *big data* or *data analytics*, and that of AI-based methods in the form of *machine learning* and *deep learning*.

Data Analytics

An obvious outcome of the increase in processing power and the accompanied growth in memory and storage capacity was that HPC systems could ingest, and produce, ever increasing amounts of data. In the past the attention was given to streamlining the data as a support for the compute part of the process. Memory bandwidth was a challenge, but not the size of datasets. But now, exacerbated by the prevalence of distributed memory (and storage) systems, much more attention had to be spent on managing data.

In HPC circles people began to talk about *data-driven* processing. The shape, structure, and size of the data determine the flow of the compute process; not the numerics.

The category of multiphysics applications, where several physical properties are simulated together (a simple example is climate modeling) became central to HPC. Gaining understanding and insights into what the results mean, and given the amount of the data, requires the additional step of analysis of the output data. And the field of *data analytics* was incorporated into HPC.

The analysis of ever larger and more complex sets of data meant that new tools and methods were needed with which to perform the analysis. Developing these capabilities is the domain of *Data Sciences*. It deals with processes and algorithms for modeling data, devising the questions that can be asked and how to expose predictive powers from data analysis.

There are different tasks given to the analysis phase of data that was captured by instruments or computed via a simulation model. One way to describe this is through a simple taxonomy ([77]):

- **Descriptive**. Show what was found. A common final product may use visualization tools.

- **Diagnostic**. An analysis that explains the "why" – an insight to the process behind the numbers.

---

[2]The topic of quantum computing is touched on in the closing chapter of the book, as it is not yet a mainstream technology.

- **Predictive**. Using the existing data to predict how the process examined is likely to proceed.

- **prescriptive**. The most useful and demanding level: Suggesting action to be taken.

Of course, not all levels of analysis are always sought or even possible.

While data-driven processing of scientific workloads became a part of HPC, it has always been the focus in non-HPC areas that involve large populations or other objects. These include social studies in academia and digital social networks and online commerce, as well as high-frequency stock trading. Those applications have also seen an explosive growth in the amount of data they collect and commensurate increase in the capacity to process the data in a timely manner. They now require high-performance system components for memory, storage, I/O, and interconnect (when running on clusters). Those and high-speed integer and logical operations. Therefore, even when not requiring much in the way of floating-point performance, their more demanding jobs benefit from HPC-level of system configurations.

Which is why some in the industry now refer to data analytics, in the sense described above, as *High Performance Data Analytics* (HPDA). We can think of is as the Venn graph intersection of scientific computing and enterprise-commercial applications.

The addition of data analytics to numerical simulations can be seen as the convergence of HPC with Big Data. This is just what HPC has become. It is not all about number crunching anymore.

HPDA adds speed and capabilities to computational data analysis. It makes *data mining* over large datasets efficient. This is critical for large-scale customer relationship management and resource planning, for example. High performance is a must for situations where the information is needed in real-time. This applies to frequency-trading in finance, of course, but also for online commerce and other business intelligence queries, as well as for telecom and intelligence. An accompanied benefit of access to high performance is having computational resources to perform error-checking (on data items or results) while processing a task, thus enhancing the reliability of the outcome.

One of the more demanding data analysis applications is finding connections and associations in very large datasets of objects or people. This is known as *graph analytics* (GA). It is a network analysis problem depicted as nodes (the objects) and edges (connections of nodes that may indicate relationships or just association). There is a brach of mathematics, *graph theory*, that is dedicated to the subject. It deals with types of graph and algorithms for answering questions about objects and connections.

In the world of computing GA is used to find out such attributes as the weight given to a node (think of identifying an 'influencer' in a social network), information about density in a network (a denser space of connected nodes is a sign of a 'community'), measures of the strength (or weight) of connections between nodes (can be used to quantify shopping preferences or to aid ranking of references for a search engine), and more ([78]).

GA output is typically described with the aid of visualization tools. It is hard to conceive how else to make sense of reams of numerical data about nodes and edges. We can easily imagine how, given access to the appropriate datasets, GA can be applied to tracking routes of currency transfers for compliance purposes and to detect fraud. Or, to uncover connections, otherwise hidden, between individuals suspected of illegal activities through telecom records, and to streamline supply chains and better manage utility grids.

Data Analytics involves arithmetic, but unlike numerical simulation of physical systems the computational load is driven and dominated by the tasks of managing data. DA methods are now used in HPC applications such as climate modeling and genomics. The same methods are employed by non-HPC applications that now require, due to the size of their datasets, HPC-class systems.

Machine Learning

Artificial Intelligence (AI) has been an exciting research discipline for decades now. In fact, back in 1950 the famous mathematician Alan Turing proposed a test he called the *imitation game* (now widely simply called the *Turing Test*) that will establish if a computer program exhibits 'intelligence'. The test applies only to the use of language (as opposed to, for example, inventing a flying machine). There are several nuances involved, but in essence an observer moderates a two-way conversation conducted via written notes. The test is for the observer to figure out if one of the conversants is a computer program. The machine is deemed intelligent if the observer cannot tell which side of the conversation is machine-generated.

There is no consensus over some details and the interpretation of the test devised by Turing. It certainly looks like examining only one aspect of what we would call intelligence. And it seems passing the test may be more of an act of imitation than one that involves thinking. Until recently no program passed the Turing test – there was no success above the level of randomly selecting one side of the conversation, but some say a Google chatbot passed the test. Be that as it may, progress is made, and here we consider AI's relationship to HPC.

I bring up the Turing test here only so we can do away with the term 'artificial intelligence' when it comes to our use of some of its aspects in computing. Thomas Sterling makes the case against the use of 'artificial', since 'artificial' is understood as 'not real'. Whatever degree of computer intelligence we achieve, it *is* real. He suggests calling it *machine intelligence.*

What we consider *intelligence* exhibited by computer programs takes many forms – chatbots, language translation, speech and face recognition, chess playing etc., but in HPC-world we associate "AI" mostly with *machine learning* (ML). It requires large amounts of data to be useful and effective.

Machine Learning is getting a lot of attention now and is applied with various models and for multiple purposes within and outside of HPC. But the idea behind is not new in the AI research community. The term itself was coined by the AI pioneer Arthur Samuel in 1959 - over 60 years ago, well before computers were able to make real use of the concept. Samuel should be remembered for several other achievements: He worked with Donald Knuth on the development and documentation of TeX, the precursor to LaTeX that many of us use today. Samuel is also credited with starting the ILLIAC IV project (see Chapter 6), though he left it to join IBM where he created one of the earliest *hash tables* programs. Samuel defined ML as "The field of study that gives computers the ability to learn without explicitly being programmed."

There is, of course, software involved. But unlike 'classical' HPC, where the program operates on the data and produces results, then repeat it on subsequent datasets, ML software uses initial dataset so the app can *learn* how to respond to other datasets of the same type. This first procedure is the *training* phase of the process. It fixes how the app will process from here on. If it is found that the results when acting on a previously unseen data are not good enough, then the 'training' can be repeated using another dataset or with some 'hints'. The material difference here compared to past programming is that for numerical simulations we would fix the code, where as in ML-land we just expose the app to new data to improve its performance.

ML is useful when we want to digitize operations that we do naturally but that are hard to program. With ML these tasks can be done on a large scale and fast. Think of face recognition or natural language translations or speech-to-text capture.

We can look at ML as a set of methods for performing data analysis where we don't only gain insights but create an object that can then function on its own when given new information (data not seen before). As with data analytics, ML outcome can be descriptive, predictive, or prescriptive.

The practice of ML programming is very different than that of numerical simulations.

There are several approaches for creating an ML system, chosen depending on the type of data and the desired learning. The most common method is *supervised* training. This is when data items are labeled by humans and the machine is taught to identify items by these labels. Think of learning to recognize specific objects – buildings, cars, cats, faces – in a set of images.

An *unsupervised* machine learning is when the data is not labeled and the program seeks to find patterns or connections among the data items. It often results in unexpected findings of correlations or groupings to be interpreted by humans.

There is a more interactive methods usually termed *reinforcement* which is a trial-end-error process where the machine is 'told' when it makes the right decision. It is useful, for example, in training autonomous vehicles.

Computationally, much of ML models are programmed in languages such as Python, R, Java, Julia, even LISP. Not in Fortran or C. Most of the algorithms used are not familiar to practitioners of numerical simulations. Though, linear regression, for example, that is used to create *best fit* lines that show relationships between variables, is not unfamiliar. Then there are *classification* algorithms such as *logistic regression* for discrete values, *Decision Tree* or an ensemble of decision trees known as *Random Forest*, *Support Vector Machine* (the vectors are coordinates in n-dimensional space and the algorithm separates 'clusters' of vectors), *Naive Bayes*, *k-Nearest Neighbors (kNN)* (associates an item with a group based on its distance from other items, by some metric). An algorithm called *K-Means* is applied as unsupervised solver for figuring out the number of different clusters. More algorithms exist, and surely more variants will be invented.

Often mentioned in conjunction with ML is the term *Deep Learning* (DL). DL itself is often used interchangeably with *artificial neural networks* (ANN), which is another important computational model for AI applications. The 'deep' in DL indicates a neural network with multiple layers, and it is an *unsupervised* method of machine learning. Important challenges in our digital world, such as autonomous vehicles, natural language processing, and image recognition, require DL-level solutions (see, for example, [79]).

The software tools used for ML, and AI in general, came from outside of the HPC community. It originated in companies whose products involve 'big data' that needs yo be studied and acted upon. One of the best-known examples of such a software tool is Google's TensorFlow. It is an open source library of tasks that are common in machine learning, inference and neural networks. *Tensor* is the mathematical term for multi-dimensional array and is used here in recognition of the multi-variable datasets the ML models are applied onto.

The AI-driven innovation did not stop with software. There are a number of new specialized accelerator-type of processors for ML and DL. Google, for example, developed an accelerator called Tensor Processing Unit (TPU) to support ML training. It excels in linear algebra operations. In particular, small word-size dense matrix multiply (it contains 65K 8-bit multiply-add units). These operations dominate the ML training phase in neural networks learning models, and the TPU speeds them up relative to CPUs and GPUs.

The above is certainly an incomplete introduction to machine learning but highlights how different a space of computations it is compared to our past 'classical' HPC. For more on ML and its models see these short introductory articles: [80], [81] and [82].

HPC and AI

It was quickly understood that ML techniques can be useful not only for social networks and tracking of online shopping habits, but also for gaining insights and increasing the predictive power of HPC applications. These include weather forecasting and climate modeling, spread of diseases and drug discovery, genomics, astrophysics, even fraudulent transactions, and more.

Just as data analytics is at the intersection of scientific and commercial applications, there is a subset of AI-type applications that resides at the intersection of big data and HPC.

To capture how AI is adding to HPC scientific applications I turned to a report from three DOE science labs. The report "AI for Science" ([83]) begins with:

"From July to October 2019, the Argonne, Oak Ridge, and Berkeley National Laboratories hosted a series of four town hall meetings attended by more than 1,000 U.S. scientists and engineers. The goal of the town hall series was to examine scientific opportunities in the areas of artificial intelligence (AI), Big Data, and high-performance computing (HPC) in the next decade, and to capture the big ideas, grand challenges, and next steps to realizing these opportunities."

By 'AI' the authors meant more than strictly AI (machine and deep learning), but also data analytics and automation. Essentially, methods that are added to the numerical simulations of physical systems. Though focused on DOE needs the scope of disciplines examined covers most areas of HPC. Turns out there are opportunities for 'AI' assistance in Chemistry, Materials, and Nanoscience; Earth and Environmental Sciences; Biology and Life Sciences; High Energy Physics; Nuclear Physics; Fusion; Engineering and Manufacturing; Smart Energy Infrastructure.

There is also a role for AI in Computer Science, Software, and Imaging. Even in driving Hardware design and with instruments that produce large amounts of data. The report recognizes that investment and progress are needed to gain a better understanding of the foundations of AI in order to utilize its full potential.

Some details on the use of AI methods in the application areas covered in this book are included in chapters dedicated to them. Weather and Climate (Chapter 27), Engineering (Chapter 28), and Life Sciences (Chapter 21).

An example of where AI can, and will, assist in basic science, taken from the "AI in Science" report, is in High Energy Physics. Particle colliders, in particular the Large Hadron Collider in Geneva, produce experimental data in amounts that can reach exabytes when digitized; and do so very fast. Simulations for testing theoretical hypotheses by comparing to experimental data also produce similar amounts of data – that's the idea of the *digital twin*, the complete replica of reality digitized. Cosmology studies are already being helped by the use of AI tools. Deep learning and classification methods are used to estimate redshift data, feature extraction, and more. In particle physics machine learning techniques are used to identify events and particles and estimate energy levels.

Looking forward, physicists identify challenges that can be helped by AI: Reconstructing the history of the universe, cosmic structure formation, and uncovering "new physics", among other topics. With AI methods they hope to answer some foundational questions such as: Is the Higgs boson a composite or a fundamental particle? Are there particles not predicted by the Standard Model? Are there particles that may be the constituents of dark matter?

AI models are seen as indispensable in the quest to understand the make-up of the universe, including its dark matter and dark energy and its early moments. No less than understanding the universe we live in and peering into its future.

The take-away from this chapter is that what was the traditional HPC – by and large, numerical simulations – is being augmented by computational tools that are very different. And it impacts system designs and software components.

CHAPTER 25

HPC in the Cloud

Delivering HPC Cycles from Remote Datacenters

N OT only the content of HPC has been changing over the last decade or so. A new option for delivering HPC cycles has been evolving. It was brought about by the emergence of Cloud Copmuting, which was mentioned before in the context of weather forecasting and life sciences. Cloud computing has become a factor in the HPC ecosystem that is worth looking into. First, how we got here with respect to HPC and 'Cloud'. Second, what it means to HPC.

Cloud Computing is a shorthand name for the practice of using, via the internet, a remote datacenter of networked servers for processing and computing. This mode of delivering computing services was not conceived with HPC in mind. The ingredients that enabled it were developed over time, and somewhat independently. The Internet providers, in an effort to satisfy online gamers and video streaming, are offering bandwidth rates that made it practical to move data back and forth between the user's location and some remote site. Concurrently, also unrelated to HPC, communication giants, social network companies, and online content providers showed the world that present technologies – processors, network, storage, power consumption management, make it possible to construct very large datacenters. Quite a few of these datacenters had aggregate flops count well above that of the top HPC systems (but, of course, not having the high-end components and interconnect necessary for parallel HPC jobs).

Then there is the business case for cloud computing. Access to online presence became essential for all aspects of commerce. But not all businesses wanted, or could, invest in acquiring and maintaining systems in-house. Small and medium size businesses also did not need a system of the size needed to process their workload full-time. Bigger businesses who had their own systems are having periods or circumstances when their compute requirements peaks and exceeds their in-house system's capability. Being able to outsource computing services makes sense to plenty of users.

DOI: 10.1201/9781003038054-25

So, we have the hardware technology and the business case. What is still needed is the mechanism to make on-demand remote compute a reality. Essentially, the software envelope that drives the process. Looking back, there were several projects and ideas that preceded cloud computing, and though they don't resemble it, included aspects that were adopted into the cloud stacks.

A Bit of History

A few years after the birth of the Internet and the world wide web, the notion of Peer-to-Peer (P2P) computing gained popularity. That happened in the late '90s and came to the forefront with the music-sharing app Napster. The idea was that computers, specifically individual computers, can communicate and interact with each other directly. Music sharing made P2P popular, but there were many other productive applications developed by quite a few startups in that period. They emphasized collaboration and trust on the web.[1] P2P, as a fad, faded away, but it sewed the seeds for inter-computer collaboration ([85]).

Going beyond file-sharing, the tools that allowed computers to interact directly in a decentralized, distributed, network environment, led to envisioning collaboration between computers where many computers contribute to a single task. The idea was that there are applications that can be divided into chunks of work units that can be done independently of other chunks. When the individual computer completes such a subtask it forwards its output to a computer assigned to manage the distribution of work and collect the results. The 'worker' computer can then disconnect from the joint activity or get another work-unit to process.

Thus was born the concept of *crowdsourcing computing*,[2] and several interesting scientific computing projects. No supercomputer involved, but the combined performance of the participating desktop computers was certainly in the supers' class. This mode is not strictly P2P, where there is a one-to-one interaction, because here we have one-to-many and many-to-one exchanges. Crowdsourcing benefits and draws from the reality that today's home and office computers have considerable amount of spare cycles that can be harnessed for the good of us all.

Perhaps the best known example is SETI@home from the Berkeley SETI Research Center that operated from 1999 to 2020. SETI stands for *Search for ExtraTerrestrial Intelligence* ([86]). The participants downloaded radio telescope data and ran a program over it that looked for signals or signatures indicating that an intelligent entity is behind it.

Similarly, there is the Folding@home Consortium that partners with a dozen of high-tech companies (yes, employees also contribute idle cycles on their company's

---

[1]I got involved in P2P for a short time and wrote a book about it with the subtitle of "Technologies for Sharing and Collaborating on the Net" ([84]).

[2]These days there are mobile apps that resort to crowdsourcing in the sense of collecting information and passing it on via the app. Typical of this kind is the navigation app Waze.

equipment). Its name refers to the 'folding' aspect of protein simulations, which is at the heart of the endeavor. As their mission statement says ([87]): "We exploit the biological insight these simulations provide to inform drug discovery and other efforts to combat global health threats." The consortium was also an active participant in addressing COVID-19 (see Chapter 30). Folding@home is by no means the only project using crowd sourcing, or "citizen science" (See, for example, a 2020 survey from the NIH National Library of Medicine on such projects for "open innovation in drug discovery" in [88]).

There are more examples of crowdsourcing. For example, PrimeGrid is an organization that use volunteers' computers, mostly home and desktop PCs, for finding prime numbers of different forms. Apart from the interest to pure mathematicians, prime numbers are often used in strong encryption of data ([89]). It turns out there is a live list of dozens of active distributed computing projects ([90]) under various categories – Science, Life Sciences, Cryptography, Financial, mathematics, etc. There is no official 'standard' for these crowdsourcing projects but many of them build their application on a software platform called BOINC. BOINC stands for *Berkeley Open Infrastructure for Network Computing* and is run from the University of California, Berkeley ([91], [92]).

We see that the internet allowed the development of protocols and tools for direct sharing of content between computers, and then for the creation of ad hoc collection of computers who all collaborate in processing a large-scale scientific computations, albeit such that no communication is needed between participant 'worker' computers. A next logical step, starting at the same period of mid-to-late '90s, was the idea of applying the sharing (of content, data) and collaboration (of work, computing) to a set of supercomputers. In particular, the U.S. DOE national labs have several high-end HPC systems and so do major universities. Computer resources are distributed when considering the set of available systems. They are not similar in terms of architecture and their components – such as memory, storage, visualization, software stacks, etc. The idea was to enable HPC systems connected via a network to jointly perform computational tasks. Conceived by Ian Foster from the Argonne National Lab and the University of Chicago, and Carl Kesselman from the University of Southern California, the concept was named *Grid Computing*, a reference to the power utility grid with which it shared some conceptual similarities of regulated resources 'on the grid' that can be arranged to be accessible by consumers.

Grid Computing is the means to process a computational task as a coordinated use of shared resources from systems that are networked but operate under different administrative regimes. The targeted workloads are large-scale scientific computations that require tight coordination and synchronization in applying different types of resources (compute, storage, visualization, etc.) that is greatly more demanding and stringent than that of crowdsourcing projects.

Organizationally, interested institutes and individuals created the Grid Forum in 1999. It became the Global Grid Forum in 2001 to include participants from Europe and Japan, and in 2006 became the Open Grid Forum (OGF) after merging with a private sector group – the Enterprise Grid Alliance. The forum, in its various incarnations, dealt with proposals and 'standards' for tools that enable grid computing.

To help users of grid computing focus on the science and not the IT aspects, Foster, kesselman, and Steve Tuecke (from Argonne Lab at that time) created a non-profit service called Globus ([93]). When I met Foster and Tuecke in 2001, while researching P2P, they articulated the Globus (and Grid Computing's) mission as addressing interactive coupling of computers and smart instruments, distribution of computations among networked supercomputers (and even desktops), and collaborative design in virtual spaces. To achieve these goals they defined protocols, and developed tools and libraries that offered resource management, data management and access, application development environment, and data security measures. Clearly, ingredients found today in Cloud Computing services. Indeed, one of Globus' present partners is AWS (Amazon Web Services).

Now, some 20 years later, Globus emphasizes managing the movement and sharing of data files and kit tools for development of applications and gateways. Researchers in hundreds of institutes use Globus to deal with data, IT teams apply its tools for storage access, and app developers use it to automate data workflows.

Foster and Kesselman narrate the history of grid computing in a recent paper – see [94]. In some ways Cloud is simpler than Grid computing, especially since it is contained within a single administrative domain.

HPC and Cloud Computing

Cloud computing was built on concepts and tools that had a strong imprint of and from the scientific computing environments: Remote processing, virtual machines, job schedulers, and more. But at its start its target users were not from the HPC universe. The datacenters that offered cloud services included many racks of servers with groups of them interconnected. They mostly ran Linux. But they did not have the tight high-performance networks and components, nor the software stacks that are necessary for any serious HPC workloads.

The HPC ecosystem was never a static one. However, until the last 10-15 years HPC has always been associated with specialized facilities that were not available to the masses. Still, remote processing was a fixture of HPC for many years. National centers, government labs and universities, provide HPC cycles to thousands of researchers; some of them globally. We saw that Grid Computing enables both remote and multi-site access to HPC cycles.

Cloud computing borrowed and built on ideas and tools originated in HPC. Can the HPC ecosystem benefit from Cloud computing? Can the 'Cloud' model be an option for delivering HPC cycles?

Let's first consider the similarities and the differences between HPC's Grid computing and Cloud computing. Both deliver compute cycles to remote users. The big difference is that Cloud delivers from a facility under a single management, where as Grid established collaborative processing among multiple domains, allowing variety of architectures. That means that in Cloud the job runs as it would run on an in-house system (with only inputting the data and retrieving the results crossing administrative regions). In principle, this is similar to a researcher submitting an HPC job to a remote national lab's system, which is the most common mechanism of high-end HPC. Grid computing use is less common and involves breaking up the job into major tasks – not necessarily equal tasks, but likely defined by function: compute, data manipulation, storage, visualization.

Grid computing is distributed and decentralized. Cloud computing is based on centralized client-server relationships. The Grid concept demands interoperability support. The Cloud does not.

Differences aside, the Grid developers and HPC, in general, demonstrated, before there was Cloud, how to manage users' request for resources and secure data movements for remote access.

Cloud computing came about from the largely correct assumption that large datacenters can deliver compute needs more economically compared with each user or company acquiring their own systems. This model also saves the users having to have in-house staff and expertise to manage such systems. It allows on-demand processing while eliminating idle time of on-premise resources. And it offers a resource for periods of high demand, whether seasonal or unexpected, so the consumers don't have to maintain systems that are not fully utilized most of time.

Given the potential and real advantages of Cloud computing, it is reasonable to ask if this model of delivering compute cycles and data handling can be applied for HPC needs.

In fact, in some sense the HPC community was ahead of the Cloud providers. Supercomputers, by definition, are far and few, and the scientists and public sector researchers have to access regional or national computer centers remotely for years. Still, a user has to belong to an agency or apply in advance, submitting a proposal to get a budgeted access to such a facility. A number of agencies with operational missions are having trouble at times providing timely cycles at moments of crisis and high demand.

As mentioned above, initially Cloud providers built datacenters that were not tailored for the tight high-performance requirements of HPC. With the increased maturity of the Cloud delivery and configuration mechanisms, and that of the privacy and security during data transfers and execution in the datacenter, it became clear that if parts of the resources were to be configured with clusters such as those used for HPC then the Cloud model can serve this sector too.

Indeed, in the last few years public Cloud providers have been hiring HPC experts and constructing HPC-grade capabilities. We see that in the offerings of AWS (Amazon Web Services), Google Cloud, Microsoft Azure, and Oracle cloud services, to name a few. Now these Cloud providers offer 'instances' that are specifically for HPC jobs. Within the Cloud ecosystem a new flavor of companies that help endusers use the Cloud was created (Rescale is an example, but there quite a few others).

That is not to say that Cloud is replacing the traditional HPC in-house ("on-prem") systems. As with most options we encounter, there are pros and cons to use of HPC in the Cloud. Consider some of the benefits Cloud can offer to HPC users and sites:

- Where there is an on-prem system, it can be kept just sufficiently large for the regular load of jobs while occasional or periodical needs for timely processing that is beyond the in-house capacity being directed to the Cloud on a on-demand basis.

- For small and medium size companies, a possible operational model is having a modest system in-house for development a and testing and running the full-scale production codes in the Cloud with its vast amount of resources.

- The scale of a Cloud provider's datacenter allows it to track technology more closely, more frequently, than the typical refreshment cycle of on-site system.

- The 'instance' in the Cloud can be tailored to the application's need, where as the on-prem system is optimized for a broader range of application types.

- There is likely to be cost-saving of expenses such as ISV software licenses, data storage, and data access during execution when using the Cloud due to its pricing negotiating power and features such as parallel file systems and the per-use pricing of software.

- Cloud computing often makes possible access to IT-level of HPC expertise, thus reducing level of in-house staff. This is especially relevant for the smaller user entities.

There are potential cons, though:

- Cost: Creating on-demand HPC instances for just a short execution time will make the per-compute-unit cost higher than it would be on-prem.

- Performance: Beware of the instance's interconnect performance for parallel jobs. Use of Virtual Machine to package a job with its special software environment (from the home system) may also affect performance.

■ Managing data: Moving data back and forth is time consuming and costly, but so is keeping it at the rented space remotely. Optimizing the process is no trivial matter.

■ Security: Commercial and governmental data requires more stringent security measures when not on-prem. This adds complexity, cost, and risk.

The conclusion most organizations draw from all the above is that the Cloud model has benefits and drawbacks. In other words, it should be use when appropriate. When an on-prem system is possible, or required, it is advisable to keep the Cloud option available for some apps and/or some times. This is known as *Hybrid Cloud* computing.

CHAPTER 26

The NCAR Models
Suite of Models for Research and Applications

HAVING described the momentous changes that occurred to HPC in the 2010s, we now turn back to how HPC is applied. This chapter and the next are dedicated to weather and climate modeling.

Funded by the National Science Foundation (NSF) the National Center for Atmospheric Research (NCAR is located on a mesa above the Colorado town of Boulder. The center was established in 1960 at the initiative of scientists from departments of meteorology at some top universities who were concerned at the reduction in funding and interest in atmospheric sciences after robust activity during WWII. They also realized that the complexity of the subject is better addressed at the national level. NCAR's stated mission was to address the fundamental atmospheric processes by assembling a large scale research facility for this purpose. It was to do so through an interdisciplinary approach not possible at a university department.

The center has a long history with supercomputers. The list of flagship systems at NCAR reflects the ups and downs of HPC system architectures and the companies that manufactured them. From Control Data's scientific mainframes in the 60s, to Cray Research vector processors through the mid 80s, and massively parallel Connection Machine in late 80s early 90s. An early cluster as a main system appears in '92 (by IBM). Another MPP-class machine, the T3D, from Cray Research in '94. A late representation from the mini-super era, a Convex system by HP in '97. Followed by SGI Origin and a distributed memory server-based from IBM. This started close to two decades of several IBM systems – Power-based, including BlueGene/L and a cluster with x86-architecure processors, with one break of a Cray system (2010-2013) – also an HPC x86 cluster. Continuing with the same genre of architecture, an SGI system was installed in 2017, shortly before SGI

DOI: 10.1201/9781003038054-26

was acquired by HPE.[1] And HPE is delivering the next NCAR supercomputer in 2023 – this one a Cray design (see [95] and [96]).

NCAR offers a view of how weather and climate models have evolved, even though, or perhaps because, it is not an operational forecast center. NCAR is a research institute, and it does now much more than atmospheric research. It is a choice institute for exploring the set of models and computational tools that is needed to study the Earth System. We looked at WRF in some detail, but it represents only a piece of the puzzle, and much of the development work has shifted to more complex modeling systems.

There are four major models at NCAR ([97]):

- Community Earth System Model (CESM) – a global climate model.

- Weather Research and Forecasting Model (WRF) – the robust model described in Chapter 15.

- Model for Predicting Across Scales (MPAS) – featuring a flexible hexagonal mesh.

- Whole Atmosphere Community Climate Model (WACCM) – modeling the atmosphere up to the thermosphere (up to 375 miles above earth's surface).

But that's not all. There are several, more specialized, models at NCAR: Solar Models for exploring the Sun's impact. A variety of Water Models – land surface, hydrological, water resource management, and some linked to WRF. Regional Atmospheric Chemistry Models for exploring issues of air quality and its chemistry.

The scientific disciplines served by the NCAR models are categorized as Climate and Global Dynamics, Mesoscale and Microscale Meteorology, and Atmospheric Chemistry.

Here we bring personal observations and insights related mostly to MPAS and CESM.

The core model's development group of WRF has moved on and now focuses on MPAS (the Model for Prediction Across Scales). The model was a joint development project with the DOE's Los Alamos National Lab (LANL).

Bill Skamarock (who we encountered before – see page 124), a senior scientist and section head in the Weather Modeling and Research division of NCAR, is the lead scientist for MPAS. Skamarock got his start in weather models, in the early '80s, at Stanford University working on an adaptive mesh for a weather model and running on the CDC Cyber 205 that was located at the navy's research facility

---

[1]As fate would have it, I joined SGI in time and for a role that had me deeply involved with this acquisition by NCAR

in Monterrey, CA. He arrived at NCAR in '87, and is still there. He talked to me about what makes MPAS a modern weather code:

"In some ways MPAS is a continuation of WRF. WRF was designed as a regional model. Its predictions are good for few hours to a couple of days. That was its niche, and it's a big niche. MPAS represents a better path to global model than WRF. This is why we're making the transition. If you look at the weather centers, their models are getting now to grid spacing of about 10 km. The centers also run ensembles to get an estimate of the uncertainty, and those run at about 20 km grids. The global models are coming down to the scales that we have been doing research on for the last 30 years, and we are moving now to the cloud scale.

"My sense of what's going on in NCAR: We're trying to consolidate atmospheric models. In particular, bring together weather and climate. Meaning, we want to run a weather application inside a climate system. Right now, MPAS doesn't have an ocean model that it can be coupled to. And that's becoming more and more important, even for weather. ECMWF and the U.K. Met Office will soon be running weather models that are coupled to an ocean model. They can show that it improves the forecast. When you increase forecast days, this becomes important. Therefore, at NCAR, we're bringing MPAS into the Earth System model that has atmosphere, ocean, sea ice, land ice, and lakes parts. With that we can do weather prediction research within a single earth system model.

"MPAS is brought into the climate model CESM (Community Earth System Model). And the reason is that MPAS solves the non-hydrostatic equations. It does not make the hydrostatic approximations, so we can simulate clouds. The other dynamic core in the climate model is hydrostatic – which does not show any vertical momentum. It assumes the atmosphere is hydrostatic, and, therefore, cannot simulate convection. That is why we bring MPAS to CESM. This way MPAS get used in the context of a coupled model.

"Also coming is a new Chemistry model. It helps with air quality calculations. It also helps with weather climate predictions because it is used for aerosol simulation, cloud forming, radiation, etc. We can see the difference between simulating with clear air and with aerosols. The model brings all this together – chemistry and air quality, to the research community. What's happening at NCAR is that the overlap between the disciplines (weather, climate) is growing larger and larger."

MPAS has "Across Scales" in its name. It refers to the fact that the horizontal view of the grid (enveloping the surface of the earth) can have variable resolution. The mesh is finer over regions where this will contribute to more accurate outcome because of greater variability in that region. For example, transitioning from flat plains area to a high mountains range will benefit from smooth refinement of the grid. In the past weather models used grids that followed the longitude and latitude directions, creating approximate squares or rectangles, ignoring topographical consideration. This also caused unequal spacing as we get closer to the poles; in fact, it generates a numerical singularity at the poles. Modern grids are more complex, but

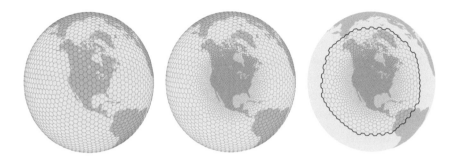

Figure 26.1: MPAS Mesh Options. Left: Global uniform meas. Center: Global variable resolution mesh. Right: Regional mesh. Source: Bill Skamarock et co., ©UCAR.

address the numerical instabilities. MPAS uses a hexagon-shape cells which remove the unwanted asymmetry of the 'longitudinal-lattitudal' grids. Physical quantities can be defined on edges between cells or at the cells' center (See Figure 26.1).

As Skamarock explains it: "The scheme allows for a different mesh density in different parts of the globe. It is a static mesh. The other way to include varying mesh is to do grid nesting. Starting with a coarse mesh, a finer mesh is created by subdividing cells. In this case the transition from coarse to fine is abrupt. It brings up all kinds of issues. In MPAS it is an unstructured grid in the horizontal and the transition between cells is very smooth. So the problem associated with wave refraction and reflection is minimized.

"With the MPAS scheme when the calculation moves to a coarse part of the grid it gets absorbed smoothly. It does that better than nested grids would. Generally, the fine part absorbs the computational errors from the coarse part and smoothes them."

As Skamarock point out, MPAS is a cleaner code compared to WRF. Importantly for today's systems, it has a GPU version. He predicts a bright future for MPAS: "Not for us to say if it will stay a research tool or become operational. Maybe both. The Weather Company is doing so right now."

More on the role of MPAS and CESM in the modeling space:

"We are bringing the MPAS dynamical core into CESM. It is one of several other dynamical cores in there: For the atmosphere, a spectral core, a spherical harmonics one, and others. For each run you choose one of the cores available, depending on the application. For climate we use the spectral elements core; for weather – the MPAS core. Maybe in the future they will use MPAS core for climate too. MPAS has a core for atmosphere, and another core for Ocean. And others for land and sea ice. They all use the same horizontal mesh, share operators and structures. The ocean model is used inside DOE's climate model, called E3SM

(Energy Exascale Earth System Model). They are using the MPAS' ocean core and also the land and sea ice core. DOE, for a number of reasons related to its mission and priorities, branched out with their own climate model. They needed a separate system. They still use the atmospheric spectral elements core that is in CESM. They share components but not at the same repository, so the models will diverge. However, the MPAS components sit in a shared repository used by NCAR and DOE. It is an open repository shared by the developers."

The story of WRF is an example of the broader questions of when and how a research tool is transforms to an industrial-commercial operational tool, and if and when it ends its life when more modern new tools are developed. In this case and at this time, WRF is a research model that finds usages by commercial interests, and was even used operationally in weather centers for regional forecasts. After 20-some years its user community growth and research work continue unabated.

The development of MPAS has been going on for more than several years by now. CESM has been around even longer. Yet, WRF's user community is much larger. The newer models have advantages – they are global models, use grids that are better for the numerics and local refinements, etc. But WRF has not been replaced and is still very popular with researchers and even operationally (see Roy Rasmussen's declaration page 125).

It appears that transition to newer models will occur in time, but it will be a long process.

That said, the users are looking for more. The main requirements are for the model to be global, and for it to describe the whole earth system, not just the atmosphere. But the product has to be well tested and stable. As Rasmussen expresses it:

"If I'm going to run a global model, I'm going to have to interact with the ocean directly. CESM has that. MPAS doesn't have an ocean coupling. That's why they want MPAS integrated with CESM. I'm waiting for it to happen and be well tested. I'm on the user side so I'm waiting for it to be production quality. To be able to see how it captures the snowfall in the Andes, for example. Maybe it's close, we'll see.."

We now turn from the NCAR example to a more general look at how HPC is enabling the modeling of the earth system.

CHAPTER 27

Modeling the Earth System

The many facets of simulations of atmosphere, oceans, land, and ice

T HE role of HPC in weather forecasting and climate modeling is a prominent theme in this book. Not only has this been the application area I was close to for many years, but it is relatively easy, compared to other fields, to evaluate its progress against advances of the HPC systems. The content of the models evolves, more physics and chemistry is added, but the object they simulate – the earth – remains constant.

Much progress was made from the early days of supercomputers when weather forecasts for the next 2-3 days were run on coarse grids to the high fidelity 10-14 days forecasts on very fine meshes nowadays. From atmospheric-only models to atmosphere-ocean-land ice-sea ice coupled models, and from near-term forecasts to long-term climate predictions.

More concisely, the state of the art evolved from crude simulation of layers of air to a very complex fine-mesh modeling of the whole earth system from the depths of the oceans to the upper layers of the atmosphere.

Trends in Weather/Climate Modeling

Much is known about how to improve the two main tasks of meteorologists and climatologists:

- Achieve more reliable forecasts for more days.

- Generate high-confidence climate predictions with means to investigate mitigation and adaptations ideas for climate change.

DOI: 10.1201/9781003038054-27

It is obvious that finer grid will result in higher accuracy (up to a point; see later). And because the amount of computation goes up by an order of magnitude when the horizontal interval is halved (the time-step is also shorten for numerical stability and more vertical layers are added), while the forecast results are time critical, this requires commensurate increase in computing power. Yet, the trend continues, as we show below.

In fact, some global models now have fine enough resolution so that in areas where the land's features are smooth there is no more a need for a regional model there (though small scale features such as convection over oceans are still beyond their capability).

Data Assimilation

Just as important is improving the initialization data for the forecast runs. There is an increased focus on *data assimilation* (as described in Chapter 4). The amount of timely observed data has increased over time and the instruments improved. In addition, the computation time given to the assimilation phase has gone up. Getting better starting values involves computing the equivalent of several hours of forecast. Since the data used has to be the latest possible, this has to be done as late as possible. And, therefore, be computed fast, requiring high-end systems. Together, more timely data and more data data manipulations are a major contributor to the improved fidelity of forecasts over a longer period.

It turns out that the reason the two big European weather centers, ECMWF and the U.K. Met Office, have a better forecast accuracy record than other centers, is that they do a better job at the assimilation phase (in addition to the strength of their models). They predicted recent hurricane paths better than the U.S. models.[1]

The data assimilation phase is getting more attention and prominence in the workflow of weather forecasting. Here is how Bill Skamarock, from NCAR and a developer of MPAS, describes its importance for severe weather situations, where predicting small feature phenomena is critical:

"We're not looking for a particular thunder storm. We're looking to see if the environment produces conditions for thunder storms that have tornadoes or heavy hail associated with them. The model tells us that a thunder storm conditions will exist in an area. It can tell us that for the next 2-3 days. The models are not looking for a deterministic prediction of a phenomenon but at the statistic probability of it occurring.

"A lot of people, NOAA and others, are trying to warn of severe storms 1-2 hours in advance. Success has a lot to do with the data assimilation phase. Looking

---

[1]Recently (as of early 2020) the U.S. National Weather Center, NWS, moved to 4th place in its fidelity of data assimilation. The model of the Canadian Meteorological Service of Environment Canada is also doing better.

at the operational centers, the number of people working on assimilation is 2, 3, or 5 times bigger than the number of people working on the atmospheric model. And the amount of compute time put into assimilation is also very large. Data assimilation is the most important part of the system."

Ensemble

Ensemble mode of running a forecast has moved from research to production and is ingrained by now in the major weather centers. At ECMWF ensemble has been operational for over 30 years. Other centers followed later. Reducing the interval between grid points and doing a better job on initial data help the average accuracy of the model, but do not eliminate the risk of occasional large divergence from correct values, due to the non-linearity of the governing equations, and it can happen even when the input error is very small. The idea is to run multiple copies of the model concurrently, with each member of the ensemble with small changes to the initial data and the model, commensurate with the assumed model error. Investigations showed that extracting an outcome of the multiple sets of results produces a better quality forecast. This is true even though the ensemble runs a at lower resolution than a single copy would run (consider that doubling the grid spacing allows about 10 ensemble members for the same amount of computations). For example, the U.K. Met Office, in 2022, made 17 perturbations to the analysis and, with the unperturbed version, runs an ensemble of 18 members ([98]). Ensemble mode provides a 'spread' of outcomes, corresponding to the spread of inputs. It needs to be interpreted and the techniques for doing so are probabilistic and still maturing.

Skamarock observes that, often, the ensemble's members don't exhibit enough spread. To remedy, modelers resort to stochastic methods designed so the spread is indicative of the actual error and is correctly propagated over time. "It is an active area of research. The spread is telling you something about the uncertainty. For example, when tracking hurricanes and cyclones by looking at predictions of different models, sometimes the paths are bunched together, sometime they're all over the place. This tells us about the uncertainty of initial conditions, and how it develops over time. Within a model, members exhibit a spread, but comparing different models also produces a spread. That is because each model has different error characteristics, and combining them tends to give a better prediction."

The spread and its underlying uncertainty result in probabilistic prediction, which adds to the difficulty of communicating a clear forecast to the public.

The ensemble concept is useful for climate modeling too, where, because of the much longer periods simulated, the uncertainties and possible divergence from a single run can grow much larger and result in inaccurate predictions if not checked.

As an example, the climate lab at NCAR produced the 'large climate ensemble' with some 50 members and simulated about 100 years. As expected, over the period simulated the members all dispersed somewhat. But climate ensemble allows investigation of interesting scientific questions of possible future scenarios. Whereas a weather forecast with much divergence may be useless, for climate, looking at broad averages, it would still be useful. Mathematically this is so because climate is a *boundary conditions* problem (beyond decadal simulations), not an *initial value* problem that weather forecasting is.

Performance and Precision

In another example of history repeating itself the possibility of using single-precision (32-bit) arithmetics is back. Recall (Chapter 5) that the U.K. Met Office opted for single-precision, with its double compute rate, on its Cyber 205 in 1980. Though single-precision is still applied in some models, most have been using 64-bit operations only. This is changing. In recent years, with the emergence of ensemble mode of forecast runs, we see leading centers turn to single-precision. Doing so creates memory space for more members in the ensemble, and even increases performance. The European premier center, ECMWF, established that such practice comes at no cost to the quality of the forecasts ([99]).

John Michalakes,[2] the WRF developer, talks about the need for faster computing elements:

"For operational models to have value the simulation needs to run about 150 times faster than real-time.[3] The problem cannot be solved by more parallelism only, since the time dimension will always be sequential. In addition, the time step needs to be shorter when the spatial resolution is finer to maintain numerical stability. But we can be smart about structuring the grid. For example, for a U.S. forecast it makes no sense to put a very high resolution grid over China."

The trend away from deterministic to probabilistic forecasting allows adding ensemble members to provide a more accurate picture of the uncertainties involved. Ensembles allow for more efficient parallelism. The computational advantage is in spending computational resources on more members rather than on higher resolution. Of course, adding members can reach the point of diminishing returns. In addition, we do need very high resolution for phenomena such as convection. That is the tradeoff between high resolution and more members in the ensemble. A subject of an active debate within the weather forecasting community.

Grids

As is the case with any other simulation of a physical space, much attention is given to the construction of the grid, or mesh, that defines the locations where

---

[2]Michalakes wrote a recent review chapter on the relationship between HPC and weather forecasting[100].

[3]Some say 120 times faster is sufficient.

values are computed. In the early days the grid was formed along latitude and longitude coordinates. This works fairly well for regional models that are not close to the poles. For global models developers and numerical analysts had to go to great lengths in dealing with the condensing of grid points as we get closer to the poles and with the actual singularity there. For the last few decades the modelers constructed more sophisticated meshes that required changes to the discretization schemes.

Skamarock explains: "What is driving the developments of these grids are two things: One, to get certain properties in the solver. Conservation of certain physical quantities (such as angular momentum), and the other consideration is how they map to emerging architectures – GPUs, in particular."

Several models adopted a *hexagonal* grid. When drawn on a sphere representing the globe it looks like a soccer ball (see Figure 26.1). The RIKEN global model that did the groundbreaking run with a 0.9 km resolution ([101]) also uses a hexagonal mesh together with its dual mesh. A Dual grid is the one generated by connecting the centers of the original grid cells (longitude-latitude grid is its own dual). Connecting the centers of the hexagonal grid results in a *triangular* mesh (such a grid is used by the German weather service, DWD). The two dual meshes can be used within the same model, with some quantities placed at the hexagon's edges and some in its center (a node of the triangular grid).

Yet another 'modern' grid is the *cubed-sphere*. It is derived by taking a grid on the surface of a cube and projecting it to a sphere, resulting in six curved faces. This grid is used by NASA and NOAA, for example. One of the dynamical cores of NCAR's climate model, CESM uses spectral discretization, which can be thought of as a finite elements grid. The ECMWF model mentioned below has been using spectral representations for decades now.

So, there are two competing issues that developers wrestle with: The accuracy of the solution, and how well it maps to HPC architectures; that is, its efficiency. In the literature you often see how fast the model computes on the system. Accuracy is often disregarded. The efficiency really needs to be the accuracy relative to the cost. When making a choice it should be based on the model's accuracy versus the cost. Given the initial data it becomes a numerical analysis, or applied math, matter. This is why one center runs one way, and another takes another way, depending on how they weigh these aspects.

Related to grids and solution efficiency is the choice of numerical method for the solution of the differential equations. In the introduction chapter I mentioned the *multigrid* (MG) method where one cycles up and down several levels of coarser and finer grids. It got my attention when Skamarock told me that "multigrid is making a comeback" since I worked on adapting a multigrid solver to vector processors in the early '80s ([102]) collaborating with the 'father' of MG, Achi Brandt from the Weizmann Institute of Science in Israel. The reason, Skamarock points out, is that it scales nicely, whereas alternatives such as *conjugate gradient* methods do not. Indeed, the U.K. Met Office uses MG in its *Unified Model* ([103]).

On Programming Languages

While Fortran is still the language of choice for the numerical simulation, the model complexity makes it advantageous to make use of higher-level languages and tools not written in Fortran.

An example, typical of other centers too, was given to me by Tom Clune, a computational scientist at NASA Goddard, when we talked about Fortran in general (Chapter 32):

"The next-generation data assimilation system, to be common to NOAA, NASA and other agencies, has been written from scratch in C++ using templating. There will be a large C++ layer at the top of our call stack. We are also seeing, because of the increased importance of accelerators and the long lead time between Fortran releases, that we are likely to start seeing other solutions to the 'under the hood' codes. For our dynamical core for instance, we are counting on a technology called GridTools, a set of software libraries that comes out of the Swiss weather center – MeteoSwiss."

GridTools ([104]) uses C++ code and templating to recognize objects such as stencils and is able to perform high-level optimizations such as fusion of loops. Its next generation will be written in Python. With that it will be possible to write computational kernels in Python and have those, basically, spit out templated C++ in the end for a variety of operations.

More Applications

There is an ever-growing versatility in how models and their derivatives are applied. We should not think of weather models as only providing forecasts in our daily lives. It is important to repeat here that there are numerous ways they are applied that are mostly hidden from the public at large. Some examples:

The *renewable energy* sector: Local, mesoscale models are used to design wind turbines, and run during their operations – giving alerts and forecasts on local wind conditions. Similarly, specialized high-resolution local models assist utility providers of solar energy by providing forecasts of cloud cover.

Another emerging use of derivative models are those tailored to model waves. They provide the means to predict surges from storms still distant from shore. And they are used in the design of off-shore structures and alerts afterwards.

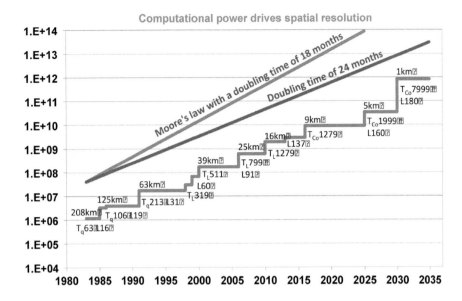

Figure 27.1: The progress in the degrees of freedom (vertical levels, grid columns, prognostic variables) of the ECMWF operational global atmosphere model in comparison to Moore's law. Source: Christoph Schär, ETH Zurich, and Nils Wedi, ECMWF in ([105]).

ECMWF and Earth System's Digital Twin

ECMWF, the European Center for Medium-Range Weather Forecasts, has been a premier weather center from its inception (as introduced in chapter 5). I reached out to Peter Bauer to learn about the current state-of-the-art in earth system modeling. Bauer, previously the Deputy Director of the Research Department at ECMWF, is now Director Destination Earth in ECMWF's Programmes Department.

Bauer showed an interesting chart that ECMWF have been updating over the years – Figure 27.1 below. It maps the forecast's model degrees of freedom against the measured sustained performance of their HPC system starting at 1984 and until 2018. The rate of *sustained* performance increase corresponds to the '24-months' version of Moore's Law. The chart extrapolates into the future to indicate when the goal of 1 km horizontal resolution might be reached. The starting value of 208 km interval on the surface and 16 vertical layers evolved to 9 km and 137 layers in 2018, and the expectation is to reach 1 km and 180 layers by circa 2030.

The slope of increase in the delivered performance reflects more than just computing with a larger number of grid points. Weather models evolve and more physical and chemical processes are introduced over time. In addition, finer resolution allows simulation of fine-feature phenomena. *Cloud resolving* is an example of a compute-intensive process that is important for more precise local and global forecasts. There is a delicate trade-off calculus going on between higher resolution and the computational content of the simulation.

ECMWF's mission, and it is in its name, is to provide medium-range (10-14 days) weather forecasts to its European member states. Monthly and seasonal forecasts are now also included in its mission. How about climate modeling? – Bauer explains:

"Weather forecasting is our mission indeed. All the member states were very keen to assign a mission to us that is weather forecasting, and actually *medium range* weather forecasting. So it is not short range weather forecasting or climate prediction. But over time these boundaries have blurred. Of course, if you do medium range forecast, then you automatically do short range forecasting also. But increasingly, we have extended even our medium range into the sub-seasonal and seasonal timescales. Seasonal predictions are for up to 9 or 13 months ahead. And parts of our model are actually used by others for climate modeling. For example, in the European Consortium model called EC Earth the atmospheric component is nearly identical with the one we use for weather forecasting. In short, even though ECMWF's mission is not climate prediction, parts of our system are actually used for it – by others. What we do, though, is climate monitoring, and that refers to as re-analysis. Re-analysis is a way to use the data assimilation system to perform analysis of the existing state since we have observations that are needed for data estimation. A forecast model is needed as well. The analysis is used to produce forecasts. But re-analysis is about using the same system to run a consistent analysis with the same system over decades, even a century, to achieve a good climate monitoring record of the past evolution of the climate. So that is not climate prediction, but it is at least climate monitoring. These are the best records of past evolutions of climate that exist today."

The *Integrated Forecasting System* (IFS) of ECMWF is a model targeted and tailored for different purposes, referred to as *components*. The atmospheric component, that which produces the daily forecasts, includes a 10-day high-resolution model, a coarser 15-days ensemble, a coarser-still 46-days ensemble, data assimilation model (single run and an ensemble), and seasonal forecasts – monthly runs for the next 7 months, and quarterly runs for the next 13 months. There is an ocean and sea-ice model, an ocean-wave component, and a set of re-analysis runs ([106]).

The European member states have access to all of ECMWF's raw data and products. Most run ensemble systems for their local forecasts using ECMWF's output to determine the boundary values for their country's forecast run in what Bauer calls "a cascade of models." There is a subtle point here: Even though the statistical average produced by the ensemble is shown to provide a more accurate forecast than any single run, that data averaging cannot be used for boundary values of regional runs because the ensemble mean is not a physically consistent set of values. It is not a deterministic evolution produced by a model. The member states can use the center's output of the high-res model, or, more typically, the output from each ensemble member to make their own local ensemble-based forecast.

Asked about how weather modeling changed over time, Bauer highlights several areas: "We have invested in numerical methods and in physical parameterizations. As we increase resolution you also have to adjust the parameterizations. The parameterizations have to be adjusted to the scale of the distance between grid points. We also learn more about the physical processes. We understand better how they actually work. There is a constant cycling of research into operations where we try to reflect the latest knowledge of how the physics works and implement that. Also, in the last 10 years there has been a lot of progress regarding model coupling. We used to run only atmospheric models. Increasingly, everybody is running global-scale systems coupled to an ocean model because of the importance of momentum, heat, and gases exchanged at the interface."

The coupling, or integration, of models to better simulate the earth's physical reality does not stop there. Added are components that model ice over land and over seas, and snow and ice sheets. And also characterization of surfaces in terms of vegetation and how it changes with the seasons, types of soil and the effect of precipitation on it.

ECMWF has been using spectral methods for solving the differential (Navier-Stokes) equations. Where other centers moved to, or stayed with, grid-point stencil discretization, spectral methods involved transformations to the frequency domain, and then back to the spatial coordinates' values. It's been thought to be too computationally expensive and that it'll get worse with higher resolution because of the global communication aspect in spectral domain solution process. But, Bauer states: "So far, with the combination of the spectral method and the semi-implicit time stepping, we see that we can run quite high resolutions in the spectral space with very large time steps, which make our model actually faster than any other grid point model in the world today, even down to a kilometer spatial resolution. The spectral method was declared dead many times in the past because of its computational cost, but it seems that it's actually quite efficient, at least in the way we implemented it."

Weather centers measure their model's *skill*; that is, how well it forecasts. It can get complicated. Which quantity to measure? at what location? how to average over time in the day? There is a methodology agreed on that averages absolute value of the error. The message from the findings is that accuracy improved over time, and it can be measured. The combined contributions from observations, modeling, and compute power leads to the *"one day per decade"* progress. If in the '80s we had a certain ('good') level of skill at day 2 of the forecast, then 10 years later we were getting this level at day 3. According to ECMWF we now get a good level of skill with high-res model for 7 days of forecast, and achieve that level for 10 days with the ensemble.

I asked Bauer to what he attributes the status ECMWF gained as the premier weather center in the world: "There's no silver bullet. ECMWF has been a leader in medium range weather forecasting since we issued our first forecast, for over 45 years. It is the combined effect of investment in different areas and making sure that the investments are applied very quickly." He went on to list the top factors that contributed to the center's leadership status, emphasizing the institutional culture of a significant research function and a quick turnaround of scientific findings into operational skill:

- Data assimilation: Adapting quickly to data from new instruments and new data types.

- Numerics: Being aggressive with increase of grid resolution while maintaining stable and efficient dynamical core.

- Physics: Constant improvement of physical parameterization and capturing the effects of convection, turbulence, clouds, etc.

- Coupling: Adding ocean, ice, and land surface effects on weather.

- Computing: Always a top system relative to other centers, and with sufficient allocations for research.

The culmination of modeling the earth system is the creation of its *digital twin*, a concept we saw in engineering and in life sciences. Indeed, such an ambitious project is underway (of which Bauer is its director at ECMWF). Destination Earth (DestinE), commissioned by the EU, is a partnership of three organizations: ECMWF, the European Space Agency (ESA), and the European Organization for the Exploitation of Meteorological Satellites ([107]).

By pushing the limits of climate science and computing the program's goal is to have, by 2024, the tools and building blocks necessary for constructing the earth's digital twin. It will support Europe's Green Deal actions on climate change and the environment. ECMWF will provide two digital twins: One on weather and environmental extremes – for risk management, and the second on climate change adaptation – for analysis and testing of various scenarios.

These 'models' will deserve the term *digital twins* because they include all of the physical environment's contributing factors – atmosphere, sea, hydrology, ice, and land, and when they do so at a level of resolution currently unattainable. And, importantly, because they will include societal sectors related to energy, water, food, health and risk management in the modeling framework, and an interactive system where users can play with workflows, data and models. An earth digital twin can be applied in three temporal modes – learn and validate from the past, assess present situations, and, of course, predict future changes (Figure 27.2).

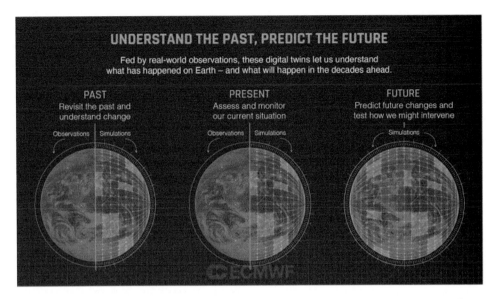

Figure 27.2: Earth System Digital Twin: Revisit the past to understand change. Monitor the present. Predict future changes. Source: ECMWF, Destination Earth Initiative.

AI Methods to Support Modeling

The field of weather and climate modeling is one of the HPC disciplines to adopt machine learning techniques.

At the ISC, the annual European HPC conference, a 2021 keynote presentation by Xiaoxiang Zhu from the Technical University of Munich was dedicated to the use of machine and deep learning methods to manage earth's observation data. ML became a necessary tool when the fleet of satellites of the European Space Agency's Copernicus program began producing hundreds of petabytes of data. To get an idea of the scale of the problem Zhu explain that an area of 10 x 5km is represented by about 50M pixels, each undergoing an intensive computational process which can only be done with the aid of a high-end HPC system. The amount of data garnered from remote sensors requires, and the available compute power allows, going beyond classical neural networks and applying *deep neural networks* that are more complex and include more layers ([108]).

At NASA Goddard, the center for NASA's climate modeling, they began to introduce neural networks in the data analysis phase. Further out is the idea of using AI methods for optimizing the physics columns in the vertical layers of the model. This is still far out in time, but the center started to save models and data for training an AI-type model.

This is but one example. Many other weather and modeling centers are introducing AI methods to their climate research and forecasting workflow.

Interviewed about their reactions to a report by the IPCC, several HPC experts mentioned the growing role of AI ([109]). Rick Stevens from Argonne National Lab touts its potential in the downscaling process, as well as in interpreting both observational data and the model's output. Thomas Sterling opines that AI will assist in correlating atmospheric data to chemical processes affecting climate dynamics.

Some researchers are going further than just augmenting the numerical weather prediction (NWP) models with data mining and neural network techniques for better forecasts, and ask if *deep learning* (DL) methods can completely replace NWP ([110]). The authors conclude that borrowing machine or deep learning methods from other application fields is not sufficient in the case of predictions based on weather data. However, they propose that a research direction could be one that starts with the weather as a big-data problem to which some DL methods are applied, discarding with the traditional flow of data assimilation, numerical modeling and the output processing.

We are clearly not there yet, and maybe never will be. However, Stevens and Kate Evans, the division director for Oak Ridge National Laboratory's Computational Science and Engineering Division, see AI methods included in the model itself. They see the future models as a combination of numerical simulation with machine (or deep) learning which will result in lesser uncertainty and better understanding of the processes involved.

Auroop Ganguly, from Northeastern University in Boston, investigates climate resilience (related to climate adaptation mentioned below). In the annual supercomputing conference in 2022 (SC22) he stated that the subject's complexity demands going beyond modeling and simulation and applying other methods: Physics-guided ML and computer vision; probabilistic graph ML guided by engineering principles; agent-based models guided by social science theories.

At ECMWF there is an active research into development of machine learning methods with neural networks to emulate components of the numerical model, such as the radiation scheme. The argument for this is that ML methods would be computationally less expensive, both in number of operations and the ability to use lower precision (see [111], for example).

Earth system modeling is a stage for one of the most impactful example of the convergence of numerical physics-based simulations and AI methods.

Climate Change and Society

One of this book's main themes is the impact HPC has on society, through the applications it is used for. Climate modeling is one of these applications, so it is worth noting that there a segment of society that denies what science tells us via HPC.

With all the talk about climate change and how to address it, we cannot avoid mentioning that not only there is no full acceptance within our society that it is an urgent matter. The issue has been politicized. Mostly in the U.S. In particular, there are those who are skeptics or deniers of either that there is climate change, and/or that human activities contribute to it. Roy Rasmussen from NCAR addresses me when he tells this story:

"I hope you can explain to people the climate change issue. I stick to the science and models. I had the experience with a Wyoming rancher, I said: 'you listen to the weather forecast every night; right?' and he said: 'No. I don't believe to anything the government says'. I was flabbergasted. The forecasts issued are excellent for the next 2-3 days. You're a rancher. Why wouldn't you listen to it?"

Of course, it is even harder to make people believe in climate models' predictions about what would occur many years from now. Instead of starting with the future, a convincing case can be made looking to the past. Validate the model by testing it against known outcome, by simulating the last few decades, for instance. Test it with and without processes and impacts of human activity and show that the 'without' comes out way off. Again, as Rasmussen explains it:

"I say to people 'you see the weather forecast from weather models. I use the same model to do climate predictions. Run it for 10-30 years, average the weather, and that's climate'. To validate, we run the model from some point of the past, say 40 years, and show it predicts what we know has happened. We use WRF over the Continental U.S. (CONUS) for that. I think WRF is just perfect for the question you posed. That's exactly what I'm doing. I'm hiring post-docs to analyze the last 40 years. People also test models by removing the man-made impact and see the predictions don't come out right. They do the runs with and without industrial pollution. We only get agreement when emissions are included. That's probably one of the strongest arguments for climate change we have. By now people agree there is climate change. But some say it's a natural cycle. We know there are cycles. That's why we have to run multiple models to see there is a real signal (of other effects)."

The Sixth Assessment Report (AR6) ([112]) by the Intergovernmental Panel on Climate Change (IPCC) emphasizes that now climate change is much more clearly upon us, and not a future prediction. This has impacted the direction research is taking. It is a shift from *mitigation* only to an equal effort spent on *adaptation* – there was a separate working group for each (the third was about the modeling itself). The difference between the mitigation and adaptation is beautifully expressed by a saying attributed to Lord Nicholas Stern, Professor of Economics and Government at the London School of Economics and Political Science:

"Adaptation is managing the unavoidable, and mitigation is avoiding the unmanageable."

Climate modeling continues to rely on access *at-scale* to the most powerful HPC systems available. Not only for running existing models, but for basic-research level of work to better understand processes in nature that could not be simulated before. And it turns out improvements in climate predictions require running climate models in high resolution. This is not for a few-days' forecast, but for simulating decades or more. One of the main drivers for this requirement is *clouds*. Simulating clouds is fine-feature phenomenon, and it is critical to understanding the impact of rising temperatures. Cloud cover patterns are evolving with temperature changes, and clouds possess two opposing properties: They absorb heat radiation rising from the earth's surface, while they also reflect solar radiation. Working out the fine-features physics and the balance between both processes is essential for accurate prediction of future average temperatures.

As far as the society at large: The sentiment within much of the research community is that their job is to continue to focus on the factual and numerical findings and explain their impact. Some feel it is not their job to 'sell' their concern to the public; others contend they are the most qualified people to do so.

CHAPTER 28

HPC, Cloud, and AI for Engineering

Engineering Adapting to Modern HPC

HERE we return to the more current state of affairs in the use of HPC for engineering – in the automotive and aviation industries. This includes the use of current modes of computational components and the inclusion of data analytics and AI, as well as the development of driverless cars.

Computational Platforms

Sharan Kalwani explains why there is little to no use of GPUs by third-party engineering software vendors:

"In the last five to eight years GPUs have become a permanent feature of HPC. Not just because of AI apps, but because they very good at math on a large number of cores. However, there is a business wrinkle, which is why there aren't many CAE, CFD software apps taking advantage of GPUs. Most of these codes in the automotive industry are now third party codes. The same is largely true of the aerospace industry (but they do have their own transonic and supersonic codes because they are very proprietary to their aerospace designs). GPUs are only implemented in some of these companies, because of how the license fees are structured. They start to become enormous when the GPU component is added to it. This is because the use of GPUs would force a major re-write of the codes, an effort that the relatively small software companies cannot afford without a substantial increase in the license fees."

The re-write of the codes for GPUs could have potentially be justified if the application can then be executed at-scale – on many thousands of cores. But this

DOI: 10.1201/9781003038054-28

is not the case. The CAE codes don't scale well beyond, roughly, a thousand cores. In fact, most of the CAE licenses are for a single node.

The GPUs do have a role in the engineering domain. There are used, within relatively small clusters, for graphics and visualization, as was the original role of these devices.

As for Cloud Computing, legal and technical issues limit their use by the automotive and aero industries. These have to do with intellectual property concerns. Therefore, they rely on in-house computing capabilities. But, they do take advantage of cloud computing for short periods of peak demand of computing or for short-duration projects. Doing so cuts time-to-completion of the projects and is beneficial to the business even when the cost of the remote processing is a little higher than doing so on the premises.

The economic argument for using cloud computing plays even for military contracts. In 2021 Rolls-Royce were awarded the contract to provide the engine for air-force bombers ([113]). They won by digitally modeling the (old) wing and pylon with the newly design engine. In the process, James Ong from Rolls-Royce says, it was the government that "pushed us go to Microsoft Azure to establish this government-approved cloud with its security features, to ensure all the requirements will be met." This anecdote is a testament to the advances made by public cloud computing providers both in terms of high performance computing capabilities and of data security measures.

Ong captures the trade offs caused by the limited compute capability in the commercial sector and their dependence on HPC: "whatever simulation we do, the goal is to get a solution back in under 12 hours. So, many times we sacrifice model size or accuracy by simplifying the model to accommodate this design cycle. We need HPC to help us eliminate that barrier. High-performance computing is certainly the key enabler to help us continue to reduce the time to getting the design out of the door. This really is our vehicle."

Aviation

Ong talked about the design of turbine jet engines in Chapter 14, and the challenges of high-quality simulations of the CFD part of the design. The simulation capabilities for the structures part of engineering are more robust relative to the state of fluid dynamics. According to Ong, "We're able to model the whole engine structure. It is feasible to model every bit of the structure. The model we put together has about 67 million nodes. Using the finite elements method this amounts to about 200 million degrees of freedom." Five years ago it would have taken the team at Rolls-Royce seven to eight weeks to get the results of the run back. Working with the software vendor of LS-DYNA, they now use a much improved implicit-direct solver implemented on a cluster, and able to get a simulation run done over night.

Indeed, in 2014 NASA commissioned a study of the state of CFD and a vision as to how to improve matters. The report, *"CFD Vision 2030 Study: A Path to Revolutionary Computational Aerosciences"* ([114]) was authored by a team that represented the relevant disciplines: Aircraft design and manufacturing, turbine engine manufacturer, researchers and developers of CFD methods, software development architect, and mechanical engineer. The findings, in the words of the authors, "were obtained with broad community input, including a formal survey, a community workshop, as well as through numerous informal interactions with subject matter experts." The report should be taken to represent the consensus among the members of the CFD community, not just those of the named authors.

Here is an excerpt from the report's conclusions:

"Despite considerable past success, today there is a general feeling that CFD development for single and multidisciplinary aerospace engineering problems has been stagnant for some time, caught between rapidly changing HPC hardware, the inability to predict adequately complex separated turbulent flows, and the difficulties incurred with increasingly complex software driven by complex geometry and increasing demands for multidisciplinary simulations."

We see that some 20 years after the first fully digital design of a commercial airliner (verified via a physical prototype before going to production, as described in Chapter 14), the CFD community is not satisfied with the state of the art concerning turbulent flows.

One of the major findings in the report is the observation that because the hardware landscape is undergoing a paradigm shift in the form of heterogeneous massive parallelism and new technologies, the application developers need to come up with new algorithms that can take advantage of these new capabilities.

To address the CFD challenge for aerodynamics simulation, namely *turbulence*, the authors determined that computational algorithms require major improvements in numerics (solvers), error estimation, and discretization techniques. Related to these items, and a necessary tool for addressing them, is the stage of mesh generation, the grid upon which the simulation is calculated. The mesh does not only have to be locally refined (where turbulence occurs), but also adaptive, shifting as the simulation progresses. The NASA report identified these topics as ones that have not been sufficiently invested in.

Then there is the added concern of the ability to manage the much larger amounts of data generated by higher resolution simulations, and analyzing data coming both from simulations and experimentally. This is an example of the changing face of HPC we introduced in Chapter 24.

NASA and its partners are acting on the vision's recommended actions. Turns out there is a team, with close association with AIAA (American Institute of Aeronautics and Astronautics) that tracks the roadmap that was drawn in order to realize the CFD 2030 vision, and report annually on its progress and status. They are the "Roadmap subcommittee." A 2021 update report ([115]) looks at the state of the main technical domains identified for needed progress: high performance computing (HPC), physical modeling, algorithms, geometry and grid generation, knowledge extraction, and multidisciplinary analysis and optimization (MDAO).

For the HPC item there are goals (or, rather, needs) specified for CFD on massively parallel systems and for CFD on revolutionary systems (such as quantum computers). For the former there is a gap of three orders of magnitude to bridge by 2030 which the authors consider very challenging, but not out of the question. As far as quantum computing goes, in the words of the authors, "While progress has been made in the field of quantum computing, this technology has yet to penetrate into aerospace CFD." The areas related to applying HPC – algorithms, grid generation, data management, and MDAO – appear to be on track to meet the desired milestones.

The 2014 *Vision 2030* report sought to lay a path for revolutionizing aerosciences. That was followed by another NASA-commissioned study that resulted in a 2018 report – *Vision 2040: A Roadmap for Integrated, Multiscale Modeling and Simulation of Materials and Systems* ([116]), that is more specific and targeted in its objectives. Pushing forward the concept of modeling all the components of a system in a single, holistic, model, as we have seen with the earth system, the vision is for a simulation that spans the scales from the exterior of the body, to the structural aspects, and into the insides of the materials it is built from. Looking ahead more than 20 years out, the authors – with input from many subject-matter experts – envision these foundational changes:

- Integrated design of systems and material (disconnected today)

- Development stages seamlessly joined (segmented today)

- Common tools, methodologies and standards across the community (domain-specific today)

- Material properties determined digitally (empirical today)

- Product certification mostly by simulation (physical testing today)

AI methods feature in the report as necessary participants in achieving the 2040 vision. For example, selecting an optimal solution from the mixture of considerations and parameters for materials, manufacturing, and design will be done by an autonomous agent. This is an example of where AI tools are integrated into a complex HPC-powered design flow.

The multi-disciplinary scope of aero modeling described by *Vision 2040* is analogous to the *digital twin* idea encountered before as applied to life sciences, physics, and weather/climate modeling. Indeed, at Rolls-Royce they invoke the term *Industry*

4.0, for the 4th Industrial Revolution, to mark the dominant stature 'smart' technology applications have taken in the engineering sector. They refer to big data, machine learning, virtual reality, and global connectivity, where *digital twin* technology combines them together to form a powerful tool. In their words, referring to these technologies ([117]): "At their core Digital Twins are virtual replicas of physical devices, products or entities created by combining data with machine learning and software analytics to create digital models that update and change alongside their real-life counterparts."

The digital twin continues to update through feeds of on-board sensors' data. Maintenance becomes less probabilistic-based, more deterministically timed.

Automotive and Manufacturing

We saw the role of HPC in the design of vehicles through applications in structural analysis, fluid dynamics, crash analysis, noise and vibrations simulations, and more. These days, while driverless cars are still experimental and under development, and as an interim to fully autonomous vehicles, the automotive industry is investing in what they call *Advanced Driver Assistance Systems* (ADAS).

The automotive industry is gradually implementing the tools that will eventually allow hands-off driving managed entirely by instruments. The main systems, still driver-assistant, are lidar (laser-based detection system), camera, and radar ([118]). The development of the devices, that obviously need the highest standards of safety and performance, requires HPC-level simulations. The virtual design process that needs to capture such varied facets as temperature, visibility, weather, traffic, surface conditions etc. can only be expressed as a multiphysics simulation model.

Autonomous vehicles are not going to be 'HPC systems on wheels', even when possessing a large amount of compute power when totaling the many processing units scattered around the vehicle. But HPC is a means to its development.

Automotive and aviation are two prominent examples of where HPC is a critical tool for the manufacturing sector. There are others. This broad area is commonly referred to as Computer Aided Engineering, or CAE.

The U.S. Department of Energy created the Advanced Manufacturing Office (AMO) that makes use of DOE's HPC capabilities to accelerate innovation in manufacturing processes, with emphasis on energy efficient solutions. The AMO can point to success stories in several sectors ([119]). Examples include metal additive manufacturing for light weight parts for aircraft, aluminum casting that reduce cracking, next-generation LEDs, energy-efficient process for drying paper pulp, improving yields of converting molten glass to solid fiber, predicting behavior of light-weight material for vehicles and aircraft, energy-efficient design of spray dryers, and more.

All these gains are made possible with the use of the DOE's National Labs supercomputers.

CHAPTER 29

Two Scientific Anecdotes: LIGO, Fusion

The Invisible Hand of HPC

THERE are numerous examples of how computing served as the third-leg of science. Below are two examples – one of purely scientific interest, the other with huge implications for society: Gravitational waves and nuclear fusion. HPC played a role..

LIGO and the discovery of gravitational waves

One of the most beautiful examples of how HPC helps scientific discovery has to be the story of detecting gravitational waves generated by events such as a most powerful collision of two black holes. Such waves were predicted soon after Einstein published his general relativity theory. This was followed by decades of devising and funding an experimental instrument to test the theory, then failing to detect the expected waves, until complex and massive computations on supercomputers narrowed down the search space and wave frequencies. It guided the design of sensors-detectors for these elusive waves. And now we have a validation of the theory and numerous detections of black hole and neutron star formations.

The project that brought it all together – an observational experiment, is known as LIGO. It stands for Laser Interferometer Gravitational-wave Observatory. Its main components were two antennas, one in Louisiana and one in eastern Washington state, tuned to listen to wave frequencies computed to be correspond to events that involved black holes. Two sites were needed so analysis can eliminate any local 'noise' and validate the extraction of the extraterrestrial waves. 40 years in the making and over $1B in cost (funded by the National Science Foundation), one day in mid-September 2015 a special wave pattern was recorded that produced a chirp sound. It corresponded to two black holes colliding some billion light-years away

DOI: 10.1201/9781003038054-29

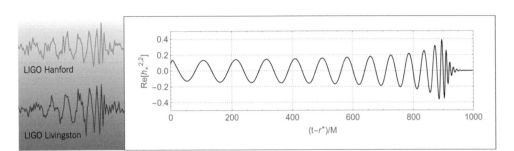

Figure 29.1: Gravitational waves signature. Left: As recorded at Hanford and Livingston, 2016. Source: LIGO, Caltech, M.I.T., Simulating eXtreme Spacetimes project. Right: As computed by the Cactus framework prior to 2011.

(and that many years in the past). It was a triumph of bold theoretical courage (yes, there were scientists who were 'black holes skeptics') and experimental ingenuity, aided by complex computations that made the discovery possible.

John Shalf, the Computer Science Department Head within the Computational Research Division (CRD) of the DOE's Lawrence Berkeley Lab (LBL), worked on the Fortran programming and the running of the code that computed the range and signature of the frequencies of the gravitational waves that would indicate presence of black holes. He recalls the exciting day when it all came together (for him):

"I was waiting for my flight when, over someone's shoulder, my eye caught a chart of wavy frequencies in a NYT article. It looked familiar. I suddenly realized this is the signature pattern that our LIGO calculations came up with 25 years ago. Sure enough, the article was about the LIGO experiment and recent observations. I was excited and asked the guy – "can I get closer; I don't want to freak you out." The observed profile was almost identical to the computational prediction. To me it was pretty amazing. People think the computation is fast and we're done. Well, this was 10 countries, multiple institutes, over 5 years. It's not like watching Star Trek."

Indeed, the report published in Physical Review Letters included over 1,000 authors – all scientists who contributed to the findings. More on some of the key physicists behind LIGO can be found in the article Shalf mentioned [120].

Figure 29.1 shows the computed and the observed wave forms. Close to the end of the computed wave form we see the 'chirp' signature.

HPC and the Demonstration of Fusion Ignition

At the closing days of 2022, the U.S. Department of Energy (DOE) announced ([121]) a major experimental achievement at one of its labs. Scientists at Lawrence Livermore National Lab (LLNL) managed, for the first time, to perform a controlled nuclear fusion ignition, meaning that more energy was produced than what was put in to create the heavy hydrogen (one that has a neutron added to its nucleus).[1] A successful ignition, the theory says, at the right setup of materials and environment, will generate a self-sustaining fusion accompanied with a continuous release of energy.

The 2022 achievement is a big deal, even if only confirming the ignition stage and far from a commercial volume application. It is the first time, after more than 60 years of trying, that the theoretical assertion was confirmed by experiment. The research and development that led to the recent success involved lasers, optics, materials, and designs of experiments. It also involved computer simulations on supercomputers. The importance of HPC in the fusion-ignition achievement cannot be better stated than by how Brian Spears, a principal investigator at the National Ignition Facility (NIF), put it ([122]): "Our laboratories have two pillars of excellence, in simulation and high performance computing, and in large-scale experimentation, and with this, the net result was just a fantastic demonstration of what HPC can do for us." Since 2009 the NIF team has run hundreds of thousands of simulations that informed the experiments, with success finally at hand.

There is a side story here of another example of how AI assist in the transition from simulations to experiment. The NIF team developed a cognitive simulation model – one that simulates human problem solving and mental processing, where the input was the data from prior experiments, to create experimental predictions that informed corrections to the simulation models.

While providing a new capability for the DOE's mission of Stockpile Stewardship, for the rest of us the fusion ignition milestone is a promising step toward deploying fusion as a source for clean energy and the corresponding reduction in the use of carbon-emitting sources of energy.

---

[1]Fusion is nuclear-level process where a heavier nucleus is created by *fusing* two lighter nuclei.

CHAPTER 30

The COVID-19 Campaign

HPC and the Fight Against the Coronavirus Pandemic

M OST of this book was written in the midst of the worldwide COVID-19 pandemic. Many of the institutes owning HPC capabilities, whether their mission included life sciences or not, made it their high priority to make their system and personnel available to the search for a vaccine, and perhaps a cure, for COVID-19.

Indeed, as early as the first half of March 2020 all around the HPC community people asked "How can we leverage the nation's and the world's most powerful HPC capabilities and resources to accelerate understanding and the search for cure? And, how can we provide those resources to help COVID-19 researchers worldwide to advance their critical efforts?" The answer in the U.S. was the creation of *The COVID-19 High Performance Copmuting Consortium*. It was set up within days as a partnership between the public sector (U.S. government Office of Science and Technology, DOE, NSF) and the private sector (IBM) offering a single point of no-cost access to computing resources and technical support. A process for submitting proposals was quickly established, researchers applied – often dropping what they were doing before, and HPC facilities prioritized resource allocations to the fight against the pandemic.

More organizations and institutes joined the effort. Among the early ones were Rensselaer Polytechnic Institute, Microsoft, Google, Harvard Medical School, MIT, Amazon's AWS, and NASA. And, of course, the DOE national labs and NSF-sponsored HPC facilities. It is estimated that already at the start of the consortium more than 300 petaflops were made available to its projects.

It did not take long for the membership to grow to over 40. Members came from academia, industry, government agencies, and software companies. And it went international with member organizations from Korea and Japan, as well as collaborations with initiatives in Europe and Australia.

DOI: 10.1201/9781003038054-30

By the end of 2022 there were 114 active projects and the consortium offered machine time from a reservoir of over 600 petaflops. The projects cover several aspects of the response to the spread of the virus. Here are some notable achievements:

Treatment of the disease (COVID-19): Molecular-level simulations of how the virus' (SARS-COV-2) interacts with membranes. Chemical synthesis of over 300 assays delivered to researchers resulting in 20 showing potency against the virus. Predicted compounds passed on to experimental teams led to discovery of inhibitors to viral protein with a couple entering clinical trials. Predictions from some 35 billion simulations of the effects of combination of drugs led to 10 drug pairings predicted to target the disease. Simulation model of airflow in ventilators enables splitting a single ventilator across several patients allowing for their varying needs. Analysis of differences of gene expression among patients helps plan their treatment.

Spread of the virus: Modeling and analyzing potential mutations of the virus and how they will impact tests, vaccines, and treatment. COVID-19 transmission model, for California, that forecasts hospitalizations, ICU beds, and mortality. Simulations of aerosol transport that include turbulence resolving at classrooms helped planning re-opening of schools.

Improved diagnostics: Development of new version of the gene-editing CRISPR to better identify viral genetic material.

Among the lessons learnt from the consortium experience, that turned into a recommendation, was the need for a *National Strategic Computing Reserve*.

The more complete story of the consortium's history and accomplishments and details of its activities can be found in [123] and [124].

Crowdsourcing was described earlier as sharing some of the technologies that led to grid and cloud computing. Some of its applications are HPC-class in reach even when not tied to HPC-class systems. The methodology came in useful in the war against the coronavirus. One such project, named COVID Moonshot, was initiated in March 2020 (contemporary to the formation of the consortium described above). Its open-science premise and sharing of genetic information of the virus aimed at accelerating the discovery of anti-viral therapeutics that cannot be patented. In less then a year there were some 14,000 compound designs, and the project transitioned to a collaborative drug discovery mode. Candidates for clinical trials were identified by 2022. When the need arose for a large system for simulations it was the *The COVID-19 High Performance Copmuting Consortium* that provided the machine time on supercomputers.

Another crowdsourcing project, the Folding@home Consortium (mentioned in Chapter 25), joined the 'COVID war'. They applied the protein-folding simulation software to the virus, to gain understanding of how the viral proteins work and, from there, how to design therapeutics to stop their invasion.

There are many more examples across the world of how HPC facilities and people came forward to help society in a time of crisis – the coronavirus pandemic, in this case.

VI

Wrap-Up and Outlook

Some HPC Themes and What the Future Holds

CHAPTER 31

P is for Performance
The metrics, the debates, and its importance for HPC

PERFORMANCE, or rather, *high performance*, and how to achieve it, is a central theme in HPC. Some would say it is the whole point of the HPC enterprise. For as long as there were computers, and especially supercomputers, there were claims and counter claims about which system is the most powerful for its time. True, in our competitive commercial world there was the 'bragging rights' motivation. Setting this aside, it is nonetheless useful to have some metrics that would enable measurements to ascertain which system is best suitable for a certain workload. It is useful, and not just for procurement purposes, to quantify performance differences between systems. Whole books could be, and were, written on the subject. The subject is close to my heart since all of the first half of my career was about how to make applications run faster on supercomputers.

Here I will touch on some aspects of how the *performance* concept plays out in HPC, or, indeed, in the world of computing systems.

Performance Has No Universal Units

Performance in the context of HPC is a nebulous concept. Intuitively, and quite naturally, if a code runs in a shorter time on one system than on another, we would say that the first system performs better (on this specific code). Similarly, we would say that a higher performance was achieved when a run finishes sooner after some code modifications on a fixed system. So, we can say that *time-to-solution* is a possible measure of performance, with values expressed as units of *time*.

Any single measurement of performance cannot stand alone for it to be useful. It has to be qualified. It only applies to the specific code, or some mix of codes, used for measuring it. When comparing two systems based on such a timing test, we need to look at the resources applied in each case. This can be generally seen as a comparison of the systems' sizes. Or, more commonly, comparing the cost of the systems, and then *normalizing* the measured timings against a cost unit. This is the *price-performance* (PP) metric, expressed as a ratio and, with some care, allowing the consideration of scaling the system in question up or down. It would make it a more valid comparison than just the timing data. But it does have its limitations. Consider two environments:

The workload consists of a mix of many jobs, none of them very large. One solution is a single large system. The other is made of multiple small systems. The latter would have a better PP value because of the integration and interconnect that make the large system what it is. What the organization gets with the single large system is more flexibility and ease of managing the job stream. A scaled-down sample of jobs determining the PP value is not, by itself, helpful for a selection decision.

Another environment has a workload that includes one or more mission-critical jobs that can run in a reasonable time, or at all, only when taking all of the large system's resources. Here the PP comparison makes sense only between systems of roughly the same size. Thankfully, this condition is satisfied in most HPC procurements. There are cases when the performance of a critical job is of such importance that the system with the shorter time-to-solution will be chosen even when its price-performance ratio is worse.

The bigger point here is that *timing* alone is of limited use. It applies only to workloads with characteristics profile that closely resembles the test measured. It is not a universal identifier of the system. A timing test of a different workload profile is likely to result in a different conclusion about the performance of the same two systems compared before.

A question I will address below (in the last section) is whether it is possible to create tests that will represent something like an 'average' HPC workload, or at least its numerical simulation characteristic.

All of the above, while well recognized within the HPC community, does not make up for the desire for a simple, even if oversimplified, single figure to rank systems by. First, we can move to a metric of *speed* if the number of instructions or operations, executed in the time measured were known. The number of operations divided by the measured time gives us a measure of how fast the system performs. It is a 'normalized' number that hides the size of the job that was measured and allows comparisons of the same workload profile at different scales. It is a more 'universal' measure.

But the above *speed* metric is still unique to the test run. So we resort to what is known as *theoretical peak performance* (TPP) that stands for the calculated value of the most floating-point operations per second, or FLOPS, (also written as flops and, generally, assumes 64-bit arithmetic) the system can perform if there is no delay of waiting for data and while employing all the *add* and *multiply* arithmetic functional units available.

The TPP is a relatively easy figure to calculate for any given system. It is also quite misleading as a measure of performance. We are counting only floating-point operations with the idea that this count is an indicator for how fast the numeric calculation can proceed. That is of academic interest but is of little practical use. There are two main problems with it: One, the flops count ignores integer and logic operations that also move the algorithm forward. This is in most numerical simulations a small correction to the flops count. But the other instructions missing from the count – those that deal with moving data, communications, and system overhead – make up for most of the execution time for most applications. The TPP measure does not help us understand why the actual performance is what it is. It does not provide any guidance about what to do to extract a higher performance from the system. It is of little help in comparing systems' performance on a given workload.

We could, for instance, figure out the memory bandwidth of the system and use it instead. In fact, as it turned out, for many HPC applications the limiting factor, and bottleneck, for performance is memory bandwidth. Therefore, it would make sense to rate a system by the memory metric. But that would be a non-representative metric for other applications, though perhaps less so than the TPP figure. It could be said that the memory bandwidth is not adequate because the focus was on arithmetic, which produces results we look at. But a memory bandwidth score is a better indicator of systems' relative performance of HPC codes.

To wit, it is very common for a production code to run at the range of 5-10% of TPP. This is a fuzzy figure, and arguably even optimistic. The statement requires a fairly accurate count of the floating-point operations executed during the run. Or, at least, an estimate that relies on an operations count of a heavily used kernel – such as the solver part of an application. But there is at least one solid data point; or rather, a set of points. The TOP500 site added an optional test to its better known linear algebra (HPL) test. It is called the High-Performance Conjugate Gradient (HPCG) benchmark (see also the Benchmarks section below), that represents a more realistic form of a solver – one that requires sparse algebra calculations and is memory intensive. None of the top performing HPC systems in the world can achieve a flops count on this test that is above 3.6% of TPP as of 2021 (check [2]).

Another metric of performance that is receiving much attention in the last 20 years or so is related to *power consumption*. Facilities of large systems are now often optimized for power. There are limits on how much power can be delivered to a site. There is a cost aspect, of course, as well as environmental considerations. Calibrating the CPU clock speed is one way to optimize for power. For a given fraction of reduction in power consumption, the performance loss is a smaller fraction (the relationship between power and performance is one of exponential vs. linear). For memory intensive codes somewhat slower compute speed may hardly change the run's time to completion. In fact, the TOP500 site mentioned before also keeps a list they call the "Green500" which ranks the top systems by their ratio of gigaflops/watts when running a standard test.

TPP numbers do provide a simple indicator of the size of the system in terms of CPUs and GPUs. We see that limiting the performance discussion to the count of theoretically possible arithmetic operations says very little about the computational benefits the system delivers on applications of interest. That said, the systems ranked high by the flops count, in general, still do run jobs faster than systems with lesser flops count, even if not in the most cost-effective manner, because they have more processors and thus more channels to memory – or more layers of memory when the flops come from GPUs. Importantly, those high-flops systems can execute larger jobs. The above, true in many cases, can be shown to have counter examples. The supercomputing history is strewn with cases of high-end and medium-size systems where the lesser-flops system outperformed benchmarks and mission-critical applications when data movement was the performance bottleneck. Examples include the Fugaku running the TOP500 conjugate gradient benchmark, and past performance of fluid dynamic codes on Japan's 'Numerical Wind Tunnel', 'Earth Simulator', and the K computer. There are more such examples from the '80s and '90s – the era of diverse architectures. For more on applications' performance vs. peak flops see a study by Gustafson ([125]).

It should be noted that while computer architects recognize the singular importance of a system's memory bandwidth, increasing it is much harder than adding more compute power. Technology allows upping the compute speed at a rate that far exceeds that of increasing memory bandwidth and reducing latency, and it is tempting to boost the flops rate at the cost of lesser investment in the memory system. The hope has always been that software will lessen the efficiency gap through innovative algorithms and code optimizations. Much of the HPC community is engaged in this enterprise.

Scaling and Speedup Through Concurrency

The performance topic can also be viewed via the concept of *speedup*. Hardware-based speedup in the early days of HPC was mostly 'generational' – applying faster and denser circuitry to the next-generation single processor system. Speeding up a single CPU has run its course when it hit the power and heat barrier. Prior to that the rate of speedup from denser circuitry has slowed down, and the architects started to resort to increasing the number of functional units and processors. Parallelism at the system level (multi server nodes) and concurrency within the CPU chip (multi-core) became the main avenue to raw hardware speedup.

We saw how Amdahl's Law is used to estimate the potential system's speedup when some components are sped up and some are not (see page 37 where it was applied to vectorization):

$$Speedup = \frac{1}{(1 - f) + \frac{f}{s}}$$

Where f is the fraction of the time that was sped up, and s is the speedup multiplier.

The formula was also applied to when a portion of the code can be parallelized. One limitation of the expression is that it applies to a fixed-size problem. In practice, quite often the purpose of a newer and bigger system is to run a bigger size problem of the same application. For instance, when the larger dataset problem run on a new system at the same time the smaller size problem executed on a previous system, we can look at the ratio of the datasets sizes as a *speedup* factor.

John Gustafson, at the DOE Sandia Laboratories at the time, addressed the theoretical estimate of speedup in parallel systems where the problem's size is allowed to grow with the number of processors. The article "Reevaluating Amdahl's Law" was published in 1988 ([126]). The new speedup formula is now known as *Gustafson's Law*:

$$S = s + p * N$$

$$\text{where } S = \text{Speedup of the whole program,}$$
$$s = \text{Fraction of the execution time done serially,}$$
$$p = \text{Fraction of the execution time done in parallel,}$$
$$N = \text{Number of processors,}$$
$$and \quad s + p = 1$$

The expression above has the number of processors as an explicit parameter. Implicitly, it assumes that when N changes the performance of each processor does not. The assumption is correct when the amount of work given each processor is constant, which is the case when the problem size scales with N. The serial part, s, is a constant for the application, and is assumed to not be dependent of N, and therefore independent of the problem size. This is often a correct assumption for numerical simulations, where the solver can be parallelized and s represents initialization setup that is not affected by the size of the datasets.

Gustafson's Law applies nicely to cases where the users wish for higher fidelity solution that can be achieved with higher resolution of the discretized grids. This is certainly the case for weather forecasting models, auto and plane design, and more.

The case of fixed-size problem speedup, covered by Amdahl's Law, is often called *strong scaling*. When the workload at each processor remains constant, the speedup as described by Gustafson's Law is referred to as *weak scaling*.

The reader may share with me (and others) the thought that calling scaling due to increase in problem size *weak* is belittling, if not disparaging. There is nothing weak about running a much larger problem size. The use of the terms *staring* and *weak* is attributed to the IBM marketing apparatus who, with the introduction of the many-processor BlueGene, had to backtrack from the Amdahl's Law interpretation of scaling.[1]

The ideal case of parallelism relative to a single processor is a speedup factor that is equal to the number of processors in a parallel system. Or so it would seem. In fact, there are cases where we can do even better than the supposedly ideal best-case. This phenomenon is called *Superlinear Speedup*. How can this be possible?

To understand superlinear speedup we start with the reminder that when nodes are added resources other than CPUs are also added. The memory isn't just proportionally larger. It is also distributed. The interconnect often has more paths per processor. The total amount of cache memory also multiplied. These resources can assist in gaining extra performance.

The cache item is significant for memory intensive codes and fixed-size problem (or where its size grows less than the increase in the number of nodes). When each processor gets less data to process, more of the data (all of it in the extreme case) resides in a cache for relatively longer period of the execution, eliminating much wait time for RAM memory access.

Distributed memory, with its independent memory banks, reduces wait times for problems that would hit on data elements sharing a memory bank. A more robust interprocessor network can reduce congestion of data exchanges by providing relatively higher number of alternate paths. And then there is the case where a new parallel version of a code is done optimally, but the single-processor version used for base-timing has not been improved.

Superlinear speedups do happen; but rarely. Typically, for a fixed-size problem and a modest level of parallelism. When it does happen it also has to make up for overhead introduced by parallelization, such as communication and exchanges of data between processes.

The issue of applications' scaling is pervasive in HPC. As computer technology advances and as user sites acquire new systems the new compute power may be used to run the same application faster. It may be used to run a more accurate version – with finer grid and/or more content, at the same execution time as before (as is the case for weather forecasting models). The "ExaScale Computing Study" report ([72] discussed in Chapter 23) identifies four categories of applications by their scalability properties. Assume a new system is X times faster, a perfect scaling can be:

- Application that remains unchanged but applied to X times more data executes at the same time on a system X times faster. Example: Weather model on a finer grid.

---

[1]Similarly, calling a perfectly parallelizable code *embarrassingly parallel* is making such an application somehow less worthy, rather than a fine opportunity.

- The unchanged application completes X times faster. Example: Hurricane weather model.

- Unchanged application and data size, but X times more time-steps executes at the same time as before. Example: Climate model simulates a longer period.

- Application with X times higher resolution that requires more calculations at each point, with unchanged execution time. A most demanding category. Example: Ocean model where fine features such as turbulence are included, or weather models with cloud resolution added.

Of course, the scaling by X can be achieved with a combination of the above. A mix of more data points, complexity, and time steps. And, needless to say, in real life we rarely observe such perfect scaling.

Performance as an Optimization Process

Performance as discussed so far applies to reducing execution time of a single code. This is certainly the right methodology when considering a critical production code – a weather forecast application in a weather center, for example. In particular, the single-code optimization approach is necessary for the class of applications known as *Grand Challenges*. These are problems that are deemed both very important for progress in their respective fields and that addressing them computationally stretches the capability of current systems. Such problems are often called out when making the case for the next generation of supercomputers.

There is another way to think of performance. Optimize the productivity of the system, not that of an individual code. Most institutes, even those with a mission-critical code or two, have periods where a the system has to process a mix of user jobs. The metric for performance now becomes the ability of the system to process more of these jobs in a given time period. This is referred to as *Throughput Performance*.

Of course it is still desirable that individual jobs run fast, but now the considerations are different. Consider a case that when doubling the resources for one job it will run 50% faster. Very nice for the single job, but it is very likely that the system's throughput will be higher if those additional resources were used to run other jobs, even when the single job completes later that it would have had if given more of the system.

The central mission of a computational facility defines its performance metric, and the staff would employ optimization strategies to address it. It could be algorithmic, coding, software libraries, parallelization, and compiler optimizations for the organization's mission-critical application. Or, it could be a focus on job scheduling and managing workload techniques when the goal is to optimize throughput. In most real-life situations the compute environment is a mix of the two. The facility may schedule periods when the single-app performance is paramount, and other times when the system is optimized for throughput. It may associate relative weights to the conflicting performance targets when selecting a new system and when assigning resources for improving performance.

As a practical matter, there is an argument against pushing the optimization of utilizing the system resources to its limit. Smooth operations benefit from some excess capacity. Unexpected critical demand and failure of some components can be handled with less disturbance to other jobs.

The considerations above apply to the system after it is handed to the users and the support staff. However, the performance campaign starts before that. It is a given that HPC systems are designed for performance in the sense that they employ high-performance components – CPUs, memory channels, memory DIMMs, large memory systems, accelerators, storage and file systems, and high-bandwidth low-latency interconnect. The optimization process here is putting together these components such that the facility's workload runs on it efficiently in the sense of minimizing bottlenecks that hold up execution speed. Quantifying the performance is done by benchmarking and is done within the constraints of cost and power consumption.

There are situations when facilities have to optimize their power consumptions. One strategy, as noted above, might be to set the CPUs clock speed at a lower level than the maximum possible and to not allow burst rate when this option exists. If such an environment is desired permanently then the organization would also save on the cost of the system by purchasing a slower variant of the CPU.

One interesting form of optimizing for performance is to find the 'sweet spot' between the job's execution time and the resources it used. For example, when the memory bandwidth cannot keep up with the compute speed it is possible to deploy less CPUs without affecting (much) the time to solution. In this vein, here is an anecdote as told by John Gustafson:

"I heard the following from AMD: They use supercomputers for fluid dynamics simulations to model turbulence in the Formula 1 race cars, how well they corner, and so on. Now, everything in Formula 1 has specifications and limits. The car can't weigh more than this and other mechanical constraints. And one of the limits is on the peak flops in the supercomputer they use. So AMD took one of its multicore processors that had eight cores, they turned off seven of the cores. And it still had all the memory management units and all the bandwidth going into the chip. But now the peak speed was down by a factor of eight. It barely hurt the performance at all, of course, because all microprocessors are starved for operands, especially when they're doing computational fluid dynamics. And so, that allowed them to put eight times as many processors in their system (each with one operational core). And they got a very high fraction of peak, because now they achieve system balance."

We can think of the HPC performance issue in terms of an analogy to cars and driving. The car manufacturer produces hardware optimized to a mix of performance and safety, with some parameters that can be tweaked by the driver. The car's performance and handling are set until they are changed. These are analogous to CPU and OS and libraries settings of the HPC system. But, then, the driving experience is also affected by the driver's skill. Similarly, the programming skills of the enduser – how well the code is structured to allow the compiler to better utilize the resources available, will determine the application's performance. Both activities involve tweaking what one is given and exercising the skills one possesses.

The *Efficiency* Metric

When concluding that the theoretical peak performance (TPP) is usually of little use by itself, we noted that most HPC codes perform in practice at a very small fraction of that "not-to-exceed" figure. But perhaps this TPP measure is still useful in conjunction with additional information. The most common application of this idea is the metric of *efficiency*. That is, the fraction or percentage of TPP that has been actually achieved by an application or some workload. The efficiency metric is a useful starting point for drawing conclusions about the fit between the code and the system.

Whereas TPP is an easily calculated value, getting the actual flops performance of a run requires some careful attention. We need the operations count of the run and the execution time in order to derive the efficiency. The 'floats' count can be sometimes estimated to a close approximation if most of the execution time is spent in a solver or a small number of loops. A better approach would be to invoke a *performance monitor* tool that counts executed instructions and groups them by type. And then run the application without the tool to get timing without the tool's overhead. For many codes this would provide a reliable measure of the efficiency as understood here. But there are cases when different datasets result in very different execution paths, and, therefore, in possible different efficiency values.

Whether the efficiency measure meets or exceeds, or fails to meet, some expected value depends on several factors: Does the balance between system attributes such as computational speed, number and type of functional units, memory bandwidth, interconnect latency and bandwidth, I/O performance, etc. match the characteristics of the code? Is the code optimized (vectorized, parallelized, instruction-scheduled) for the system?

In other words, is an observed low efficiency due to the fact that the system was designed for applications of different profile than the one examined, or because the programmer failed to extract all that can be had?

Another source for uncertainty is the expected efficiency itself. It is hard to profile an application accurately. It is not just the operation counts, or the memory bandwidth, or the communications requirements, but how they all interact and overlap.

With all that said, an efficiency number, if done half decently, is still a more constructive measure than a performance value that states the average flops rate of the job. Given the calculated peak computational performance the efficiency measure provides a quick answer as to whether one should look for optimizations opportunities, be them code changes, compiler options, choice of libraries, or even changing the numerical algorithm or the datasets structure.

There is another measure of efficiency that is of particular interest in today's computing: the *efficiency of parallelism*. It is a measure of how much is gained in speeding up the time to solution when adding processors (nodes) to a parallel application. This efficiency metric is applied almost always to a fixed size problem (though Gustafson's Law that stipulates increasing the problem with the number of processors can also be tested by varying N in the formula). Users have the problem size (dataset) they wish to run on a large system. For jobs that may run many times it is useful to determine an optimal level of concurrency. If, say, doubling it results in a speed up of 5%, then perhaps this would be a wasteful use of resources. For 50% gain it may be tempting to go with the larger configuration. This process corresponds to Amdahl's Law, and if the time spent in the sequential portion of the code is known, then one could figure out the useful concurrency cut-off without experimenting.

When we talk about efficiency and 'waste' of resources we need to consider the human factor too. Take the cost of person's time and the costs of computer hardware and its operations 50 years ago, and compare them to present time. The computing costs has gone down by a factor in the millions. The professional's time is now worth considerably more. So having resources underutilized is worth it if their availability on-demand increases people's productivity. Back to the car analogy, Gustafson reminds us, that we often drive alone in cars that can hold several passengers. He also notes that at any moment most of the grand piano keys are sitting there idle when music is played. We should look at computer resources the same way.

Benchmarks

Benchmarks in computing are codes (and sometime system software routines) that are meant to represent a workload, an environment, or profile the demands of a computer facility. Benchmarking (the verb) is to measure a system's performance with one or more of these codes. The verb is also often used as shorthand for the process of modifying the codes or the setup for a better performance, what we called the *optimization* process.

Benchmarking is a field fraught with misconceptions, arguments, subjectivities, misleading information, and more. Let's start with a nod to a 1991 tongue-in-cheek article penned by my past colleague, the now-retired numerics scientist David Bailey (we were together at NASA Ames). The title says it all: "Twelve Ways to Fool the Masses When Giving Performance Results on Parallel Computers" ([127]).

Bailey lists sins of omission and of commission in how benchmark results on parallel systems are sometimes reported with the purpose of making them look better than they deserve. Here a few of them: Quote single-precision results (when they are faster) without mentioning and compare to double-precision results. Quote the timing of a kernel or solver, possibly with some assembly code, as if it were the whole application. Run the test on one configuration (size) and write down an estimated projection the performance of a larger system. Compare a parallel code to an unoptimized scalar code, or compare to an older version of a code run on an older system. Use algorithm that results in a higher flops rate (when this is the measure asked for) even when the execution time is longer.

No one is claiming these transgressions are common; only that such cases were observed. The 'offenders' come from the ranks of researchers parallelizing codes and from vendors trying to make a sale. Being misled by the latter can be avoided by careful instructions in the Request for Proposal (RFP) and diligently verifying of the reported results. Spotting the pitfalls in published papers or conference talks requires detailed attention and some skepticism.

Bailey's *Twelve Ways to Fool the Masses* aside, benchmarks are a useful tool and serve several purposes. The most common of which is the selection of a future system. Other applications of benchmarks include measuring historical progress of processors and systems, and evaluations of parallelizations and other code optimizations. In its most general terms, benchmarks are used to compare systems' performance under some constraints and for specific workload environment.

Benchmarks can be classified into several categories:

- Standard Community Benchmarks

- Procurement Benchmarks

- ISV Benchmarks

Standard Benchmarks

By *standard benchmarks* we mean those tests that are available to all (not necessarily managed by Standards bodies, though there are some of those too). They tend to be fixed over time, and modified almost always only to accommodate new architectural features.

There are many open source benchmarks in use within the HPC community. Probably the best known is the one used to create the TOP500 list ([2])- the High Performance LINPACK (HPL). LINPACK is a library of mathematical routines for linear algebra developed in the '70s and its core are the BLAS – Basic Linear

Algebra Subroutines. In the passage of time LINPACK has been largely replaced by LAPACK, which is better suited for current system architectures. This is an appropriate place to introduce Jack Dongarra, a central figure in the development of LINPACK and LAPACK, and one of the founders and authors of TOP500.org. Dongarra is now a Distinguished Professor at the University of Tennessee. His research team maintains LAPACK and other numerical software collections in the *Netlib Repository* (netlib.org).

Dongarra put forward the *LINPACK Benchmark* where a system of dense linear equations is solved using the Gaussian elimination method with partial pivoting to get LU decomposition form of its matrix representation. It was defined for several matrix sizes. HPL (introduced in 1991), the high-performance variant of the test and the test to rank the top HPC systems, was designed to let the target system demonstrate the best performance it can achieve on this problem. The problem size can be chosen to fit the system. Dongarra explains the motivation was "to have a benchmark that could scale over time and remain relevant in tracking performance." This aspect is consistent with the notion of the weak scaling principle behind Gustafson's Law (1988). But there is more: Code changes are allowed. Even replacing the LU factorization scheme, as long as the operation count is not reduced. However, there is a strict bound on the accuracy of the results. The HPL benchmark is one where the performance can be close to TPP. High parallel efficiency is achieved by spreading blocks of the large dense matrices to each node, allowing for plenty of local computations and infrequent inter-node data exchanges. Within the node it is possible to achieve high efficiency by further dividing the arrays to blocks that fit in the processor's cache. Most of the time is taken by a highly optimized matrix multiplication library routine, so that there is no stress on the memory bandwidth either (though the memory system is typically filled to capacity). Of the top 500 systems not many are below 50% efficient. The median for that group is around 60%. And systems with particularly high performing interconnect, exemplified by some of the Cray XC40's in the November 2021 list, for example, achieve over 90% efficiency!

Of course, it is not a complete application, and, in fact, most real applications involve sparse matrices for which the possible compute rate is much reduced. But that's not to say that this LINPACK benchmark has no value.

The TOP500 list, published twice year since 1993, is coming on to its 30th year anniversary. It provides a historical record of the advances and architectural transitions of the top HPC systems. The data the lists provide show the ups and downs of computer vendors, the distribution of top systems across continents and countries, and much more.

The TOP500.org group added in recent years HPCG and Green500 that were mentioned above, so now there is an historical record of the top systems' performance on a test that more closely represents the numerical solution of a real physical system, and of the energy efficiency of systems, respectively.

There are quite a few less known standard benchmarks that are not applications-based. Their purpose is to measure specific features or components. For a comprehensive list of standard, open-source, and specialized benchmarks see [128].

Procurement-Targeted Benchmarks

The message from the previous section is that if an organization or a user community want to select the best system for their budget, they better do the measurements themselves using a representation of their workload.

Therefore, when it comes to procurements, the request for proposal is usually accompanied by a set of benchmarks that are used for the evaluation of a future system. The tests may include some of the 'standard' benchmarks discussed above, but the critical test, more often than not, is an in-house production code or several such codes.

This is a good place to bring up an issue that has plagued benchmarking in general, and very specifically when procuring a new system. It is the verification of the results; the output of the tests. For many HPC applications the bit-wise result may have a dependence on the order of the operations. Different vendors, or a new generation of the same vendor, may implement details such as rounding differently. Parallelizing applications or even changing the scale of parallelization may affect interactions of variables and operation ordering.

When is the output deemed correct? Do all values have to replicate old ones exactly, up to the least significant bit? If the last bit or two are different, is the old one 'better'? If one vendor's results are closer to the reference results, should that deserve more credit?

It is hard to verify computed results against the actual physical phenomenon, because much is not amenable to direct measurement and the of accuracy of the measuring devices is also limited.

Perhaps for simplicity of process, or to avoid appeals by competing vendors, a number of user sites insisted on bit-for-bit correctness relative to existing results. This includes some major weather centers' past procurements; but others too. This requirement came at a cost. Vendors had, at times, resorted to relaxing compiler or library optimization options in order to comply, with accompanied loss of performance. And there was never any evidence that the slightly-different results were any less valid.

The right approach, often taken now, is to define, and declare, an allowed tolerance level of how close the new benchmarked results are to the reference numbers.

ISV Application-Specific Benchmarks

Independent Software Vendors (ISVs), mostly in the areas of engineering (CAE, CFD) and life sciences, keep and maintain their own benchmarks. They are somewhat different than the benchmarks covered above. The ISVs' interest is in how their app improved from one software version to the next. Here, the problem to be solved remains constant and the impact of the code changes is measured. The test compares versions of code on the same compute platform.

User facilities in a procurement cycle may want to use the ISV result when their app is an important part of their workload. Often the procuring party can only quote what the ISV provides. There is never access to the source code, which is proprietary, and no opportunity for the hardware vendor to optimize it. Competitive information between vendors' systems is often hard to come by.

This class of benchmarks is a useful and necessary tool for the ISV, and while its data is important to some users, they have only a limited influence on it and limited access to it.

Multi-Component Metrics and Predictive Models

So far we saw benchmarks and performance measurements in general as just telling us *what is possible* in terms of maximum capabilities. As such they are not helpful in estimating what an application or a workload will perform like on a given system in advance of actually measuring it.

People have long sought to relate the system's performance to their applications, and better yet, to be able to predict, or just estimate, what their performance would be on a candidate future system. The starting point is the notion that a big portion of the HPC application space can be captured as falling into one or more of a relatively small number of categories. such a category represents a computational method with its arithmetic, logic, and data access patterns. Next, find or create a simple code that exhibits the category's characteristics and profile and that can be studied on multiple platforms and also analyzed for its use of system's main subsystems (compute, memory, communications, storage).

Continuing this line of thought, the workload for a target system can be, in principle at least, characterized by the fraction the computation types take of the total, and the information then used to optimize the design of a future system by attempting to reflect the balance called-for among the system's components. Or, sometimes, the analysis shows where software and application optimizations will pay off.

Much thought has been put in over the years by individuals and teams in creating tests, frameworks, and metrics that will be applied to benchmarking ways to manage systems (job scheduling), select future systems, and design more 'balanced' systems.

The DARPA 2008 HPCS report ([51]) recognized contributions in this vein. They discuss noteworthy efforts done up to that point. Some examples: A cleverly devised code that provides a performance graph as it executes, while measuring the error-reciprocal (improved 'quality') of the solution. It is scalable and adaptable to different precision levels of the arithmetic. Interestingly, a number of existing benchmarks' results match different points on this benchmark's 'quality' performance plot (Gustafson and Todi, [129]). John McCalpin created the STREAM benchmark that measures the memory performance when executing simple operations. It measures *machine balance* – defined as the ratio of peak floating point operations to memory operations, per cycle. The more flops per memory op, the higher value for the 'balance' metric. When conceived, in 1995, it correlated well to vector performance. Systems with a higher 'balance' performed vector codes proportionally better ([130]). Based on Gustafson's work on application signatures and machine signatures, a team from the San Diego Supercomputer center (SDSC), under the leadership of the late Allan Snavely, developed a methodology for ranking HPC systems according to machine metrics – several ways of data access, interprocessor communication speed and latency, compute rate – and applied them to HPC applications ([131]).

The discussion above about the *efficiency* as a metric correlates well with the notion of *balanced system*. A system well-balanced for a particular workload achieves high efficiency mark. It is important to realize that when we talk about balance among the system's components, it is always qualified by the codes being run. Therefore, there is no absolute, or universally accepted, 'balanced system'. Though, it is generally true that a system that has a higher memory bandwidth relative to compute rate than another system is better balanced for HPC workloads.

These approaches provide insights, are relatively complicated, and of limited predictive value.

The Dongarra team, mentioned above, is also credited with the development of what is known as the *HPC Challenge* benchmark. The benchmark is made up of a set of seven separate tasks, some of which involve multiple components, that are meant to measure different aspects of the system, and was developed for the DARPA's High Productivity Computing Systems (HPCS) ([51]). The computational aspect is represented by three tasks: HPL, library routine for matrix-matrix multiplication, and FFT (Fast Fourier Transform). The memory system is tested for sustained bandwidth and its performance of random access updates to memory. And the network system is measured for total capacity, and its bandwidth and latency (more details in [132]).

The HPC Challenge Benchmark provides more information related to a system's maximum capabilities, having added tests for the memory and the communications systems. Is still does not include I/O and storage measurements. This is a more complete picture of the system's characteristics, but is of little help in how the interplay of these subsystems is reflected in the performance of real applications.

The benchmark is a collection of 'peak performance' measures. These may be useful only if the execution profile of an application is fully understood.

There were other benchmarks that provide multiple numbers as output, with the hope that it will be possible to pick one component or more to fit a given workload profile.

An oft quoted example is a set of fairly small codes put together by a group of computational scientists at NASA Ames Research Center. The NAS[2] Parallel Benchmarks (NPB) started with a novel approach to benchmarking the authors suitably called "paper and pencil" benchmarks. The problems to measure were presented as an algorithm – with words and equations, letting the testers write the code as they see fit. The first release, created in the early '90s, years before the HPC Challenge benchmark, specified 8 problems that together cover a broad spectrum of methods used for simulations of physical systems. Three of them are 3D PDEs solvers using multigrid, conjugate gradient, and FFT. Three ways of solving non-linear PDEs. A sort problem, and an 'embarrassingly parallel' (one that requires no inter-processor communication). The non-linear PDEs problems are what we may call 'pseudo-applications', the other five are more like 'kernels'. Later version added a couple more problems. Reference implementations, in multiple languages and programming models, were added in later years. For more on NPB see [133] and [134].

A discussion of multi-component approach to performance evaluation of HPC workloads cannot be complete without recognizing and highlighting the work done in the EE and CS department of the University of Berkeley in the early 2000s. It culminated with a seminal technical report titled "The Landscape of Parallel Computing Research: A View from Berkeley" ([135]). The 11 authors include the computer architect David Patterson, as well as Katherine Yelick and John Shalf who we met before. Having addressed questions about applications, hardware, and how programming models bridge them, they tackled the issue of how to measure success when examining the impact of transitions and changes to the combined hardware-programming models-applications complex.

Just as at NASA, where the NPB was composed of several short codes that represented the *types* of computations that amount to their mission-critical aeronautics applications, the Berkeley group sought to identify kernel-size mathematical and data structure problems such that each "capture a pattern of computation and communication common to a class of important applications." They came up with what they called the "seven dwarfs" – after the seven numerical methods put forward by Phillip Colella, the applied mathematician from LBL:

■ Dense Linear Algebra

---

[2]NAS stands for NASA Advanced Supercomputing, a division at NASA Ames and the flagship HPC facility of NASA

- Spase Linear Algebra

- Spectral Methods

- N-Body Methods

- Structured Grids

- Unstructured Grids

- Monte Carlo Algorithm

Perhaps not surprising, all the NPB problems correspond to a unique 'dwarf'. The above HPC characterization was the group's first step. They were exploring parallelism in general, and it was applied in areas other then HPC, the main ones being machine learning, databases, and graphics and games. This expansion of the inquiry space resulted in six additional 'dwarfs' ([135], page 14). The purpose of the 'dwarfs' was not that of benchmarking in the sense of selecting or comparing systems, but for decisions about system architecture matters such as balance between components, communication patterns, processor features, and more. But one can conceive of defining specific problems to be computed for any or all the 'dwarfs' and using them in actual measurements of systems to estimate performance for a class, or classes, of computational tasks. Indeed, I submitted such a proposal as a white paper for the first IESP meeting in 2009 ([136]).

Knowing the profile of a site's HPC workload in terms of its 'dwarf' types and relative weights in the workload, and having them represented as (relatively) small codes, can provide a useful means for comparing systems for that particular site.

Mathematicians' Contributions to Performance Gains

The question of how much of the performance advances in HPC came about not just from hardware technologies and computer architecture, but from innovation in numerical algorithms is not an easy one to answer. Attempts to quantify the contribution of new and improved algorithms to performance are open to some level of subjectivity and selectivity in what is presented. With that in mind, one such study was done in 1991-92 by the *High Performance Computing and Communications Working Group* reporting to the *Committee on Physical, Mathematical, and Engineering Sciences* of the *Federal Coordinating Council for Science, Engineering, and Technology* ([137]). The working group was made up of over 20 highly respected individuals from some 10 U.S. Government agencies with stake in HPC. Its main thrust was the *Grand Challenges* of HPC – those applications that are critical for the well-being, progress, and security of our society, and that are hard or impossible to execute well on the then-current supercomputers. Part of the analysis involved the algorithms and their impact on performance – the aspect explored here.

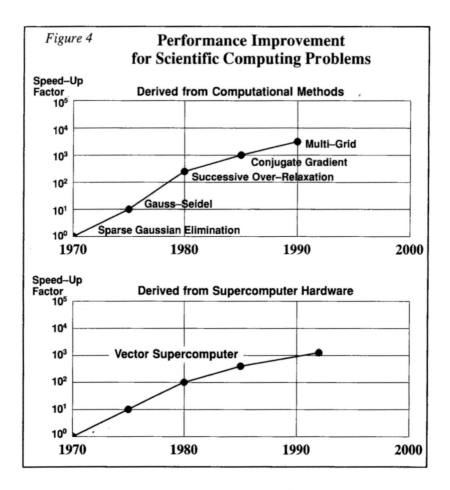

Figure 31.1: Performance Improvement for Scientific Problems Due to Hardware
and to Computational Methods. Source: Grand Challenges: High Performance
Computing and Communications. Office of Science and Technology Policy. 1992.

Fig. 31.1 is a visual presentation of the working group's findings. The curve from
the hardware-derived speedup and that of the speedup derived from computational
methods over 20 years look similar. But they are drawn on a logarithmic scale,
and, in fact, whereas the hardware curve shows a factor of 1,000-fold speedup, the
algorithms curve is shown to have improvements that add up to a speedup of 3,000
times.

The message conveyed was that numerical algorithms and computational methods contribute to speedup over time at least as much as hardware technologies and
innovations. The authors did not provide information about how the chart was
constructed. There are reasons not to take the data points as definitive.

Consider the hardware chart: The speedup according to TPP from 1970 to
1980 was just over 20x (from the CDC 7600's 36 megaflops to Cray X-MP's 800

megaflops). By 1990, we had a vector processor (Cray C90) at 16 gigaflops TPP. Memory improvements were slower than those for the computing speed. It looks like the hardware speedup for that period was less than half of what is claimed. (The rate of hardware speedup has increased from the mid '80s on.)

The computational methods chart is more vague. What problem and what size dataset was used? How were the hardware changes over time been taken into account? Was the change of workload profile taken into account? Or was it a rough estimate based on paper and pencil analysis? – The latter is likely: Take a problem, perhaps Poisson's equation, for example, and do an operation count according to the method used. It is hard to account for data movements, though.

The point made via the chart is valid and important: algorithmic innovations of numerical methods are a main source of performance improvements beyond what is provided by hardware technologies and system architecture.

On the Relationship Between Architecture and Algorithm

In the early days of the vector processors people discovered that after a code has been vectorized, when run again serially (on system with no vector instructions), it often ran faster than before. Today we find, similarly, that parallelized code often runs on a single processor faster than it did before. This is mostly due to changes about how data is organized and accessed, and a better use of data already loaded.

But optimizing for a specific architecture is not always helpful for other architectures. An example: Iterative methods are in common use for solving partial differential equations. They involve repeated updates to each grid point through values of its immediate neighbors. To correctly get a new value from values of the previous iteration, in vector form, the grid is often split into two, in what is known as the *red-black ordering*. Each point is surrounded by points of the other color. The vector code updates the points of one color, then updates the other color. Each new value is computed by values of the other color that are all from the same iteration step. The procedure does not follow the order of operations in a sequential execution, but it is more sound for numerical consistency. In a typical sequential code a new value is placed into the same location the previous one was in. As a result, most new values are computed from some 'old' values and some 'new' ones. The red-black method works well for vectorization. However, running the code now serially on a microprocessor degrades performance since the scheme creates redundant loads of the data because microprocessors load a cacheline at a time. John Shalf, now at the DOE Lawrence Berkeley Lab, laments: "I spend years at NCSA to undo red-black codes for microprocessor clusters."

The Case for Fat Nodes

Contrary to early predictions, Frontier, the first exaflops system, was built with fewer than 10,000 nodes. That is, much more performance was packed into a single node than was expected 10 years earlier. We know that this was due to use of multiple GPUs in each node. This is not the classical fat node that was made of symmetrical configuration of 4-way or 8-way general-purpose processors. Frontier's nodes have one host processor managing four GPUs. But from the application perspective it is a fat node (though each GPU has hundreds of computing elements and not all the memory is accessed directly by the GPUs).

We mentioned the notion of fat nodes in Chap 23 as a design point of the most recent exascale systems, an outcome that was not expected 10 years earlier. Here is why fat nodes are beneficial to applications' performance:

The larger local memory configured with the computationally more powerful node means that a bigger chunk of data resides and is computed within the node. Less data needs to be transferred between nodes. This is sometimes referred to as the *surface-to-volume effect*. Many physical simulations are three-dimensional, and each node is given a 3D 'block' to process. The size of the data and amount of computations goes up as n^3, while the data exchanged between nodes are the surface points which is proportional to n^2. So, the bigger the chunk of data in a node is, the more it computes locally, and the less the relative size of data is that needs to be moved between nodes.[3]

Programmatically, even in the days of 4-way fat nodes, the larger memory allows for larger MPI tasks on the node. Reducing the network traffic was a performance boost since the networks were not keeping up with the compute power. The same argument applies to the GPU-based fat nodes. Only now the application has to be written so that fine-level data parallelism is exposed for effective use of the GPUs. What is left is task-level MPI exchanges between the nodes, which lightens the inter-node communications. The key is exposing data-parallelism where possible.

The smaller number of nodes also enhances the reliability of the system. There are less connected building blocks in the system. Users of such systems can expect to be able to run applications uninterrupted by system failures for days and even weeks.

---

[3]I am rephrasing Mark Seager. He used this rationale in designing the 4-way Itanium cluster – Thunder, described in Chapter 18. Fat nodes designs before the wide adoption of GPUs did not catch up in the past due to their premium cost.

CHAPTER 32

Fortran: The Coarrays Story

Expressing Distributed Memory Parallelism: In-language or
Library; MPI wins

FROM the early days of supercomputing there was always Fortran. To this day it is the language most suitable to numeric calculations. The language has evolved over time to include more capabilities, to abstract more complex data structures, to accommodate parallelism. And also to stay relevant when other languages, mostly C and C++, gained popularity. Having arrived at the end of our HPC journey, it is appropriate to visit the *state of the Fortran language*.

The next three chapters, including this one, are dedicated to this end.

My start at HPC was by learning FORTRAN IV on the CDC 7600, and Fortran has always been the software workhorse enabling much of scientific computing. A testament to the durability of Fortran and its evolving nature is the 1982 quote by the British computer scientist, winner of the Turing Award, Tony Hoare:

> *"I don't know what the language of the year 2000 will look like, but I know it will be called Fortran."*

What follows is not an exposé on the evolution of Fortran. But, rather, some perspectives about its present place in HPC and parallelism, its relationship to other emerging languages, as well as the dynamics between the language guardians and the user community.

To this end I sought out a few of the people who play, or played, a role in defining the Fortran language on its standards committees (by which I mean both the U.S. body – known by its former shorthand J3, and WG5 – the abbreviated name of the international committee).[1] A mix of old-timers and some who arrived

---

[1]To be more precise: The U.S. Fortran Programming Language Technical Committee is formally called INCITS (InterNational Committee for Information Technology Standards) PL22.3.

on the scene more recently. They have all dedicated, and are dedicating, much time and energy for the advancement of Fortran:

Ondřej Čertĵk, a scientist at the Computational Physics and Methods group of the Los Alamos National Lab.

Milan Curcic, a scientist at the university of Miami and WRF[2] user and contributor.

John Reid, past convenor of WG5 ('99-2017); now retired, but maintains a desk at the Rutherford Appleton Laboratory in the U.K.

Tom Clune, senior computational scientist at NASA GSFC; leads software infrastructure team for weather and climate modeling.

Jon Steidel, compiler engineer at Intel; previously at Cray Research, SGI, Cray, SRC.

Steve Lionel, current convenor of WG5; a.k.a "Doctor Fortran." Retired from Intel; previously at DEC (Digital).

Damian Rouson, Group leader at Lawrence Berkeley National Laboratory and President of the Sourcery Institute, and lead developer of OpenCoarrays.

Two of them, Ondřej and Milan, belong to the new generation of Fortran advocates. John, Jon, and Steve represent the 'old guard'. And Tom and Damian are in-between. Steve and Jon are, or were, compiler writers. The others come from the user community; the people who program in Fortran.[3]

The next chapters in this series contain perspectives from members of the Fortran standards committee, and observations about current Fortran-related activities and the status of the language. But first, to set the scene, let us go over an episode that exposes the issues of Fortran's role and evolution – The *coarrays* idea and its journey to be included in the Fortran standard. Of course, we should be careful not to over-generalize based on one episode in the language's history. There are some generalizations to be made, and examining other aspects – such as data structures, data typing, and memory allocation – would shed a different light on the committee's work. It is just that expressing parallelism is so central to HPC.

It was formerly called ANSI (American National Standards Institute) X3J3, and still referred to informally as J3. It is the U.S. representative for the international standards committee – ISO/IEC JTC1/SC22/WG5. One can see why people use the shorthand of J3 and WG5. I refer to the bodies interchangeably as the Fortran Standards Committee, or just the standards committee.

[2]The Weather Research and Forecasting model described at some detail in previous chapters.

[3]The presence of two convenors among the listed above reminded me that shortly after arriving to live in the U.S., working for Control Data at the Colorado State University (CSU) campus in Ft. Collins, Co., I was fortunate to have met the late Jeanne Adams from NCAR. Jeanne was the first to lead a Fortran standards body, first known as the Fortran Experts group, then as the WG5. She was later the chair of J3.

Some Context

A 2020 paper[138] by three long-time Fortran practitioners – John Reid, Bill Long, and Jon Steidel – captures the aspects of the Fortran evolution that relate mostly to the expression of parallelism. Fortran's birth was at IBM in 1956.[4] Designed for numerical calculations it became the main and dominant programming language for HPC. Fortran could not remain static. It had to adapt to the evolving computer architectures. It had to be done in an orderly fashion. Standardization began through the American Fortran committee, J3. People from other countries joined the committee and its working groups.

First there was the identification of independent vector operations that can be streamed. Then came the multi-vector-processors. Elements of arrays and vectors can be split between processors, where each pipelines the operations. The significant feature for the programmer was that the processors shared a single memory system. Each processor 'saw' all the data. Care must be taken, but sharing the address space made it possible within the existent language syntax, with additional means of communication between execution threads. Once we arrived at distributed memory systems, the days of the MPP (Massively Parallel Processing) architecture and later of the clusters and accelerators, things got more difficult. Expressing parallelism across images of operating systems, over interconnect network between compute nodes, and with data distributed among multiple memory systems became the programming challenge for HPC.

It is understandable that while the system architecture was evolving, and *changing*, a solution to the reality of distributed data that did not impact the foundations of the language was more convenient and expedient. This approach resulted in an external library that contained the tools for manipulating distributed data in a parallel program. The Message Passing Interface (MPI) is the library most in use today. The *library* approach was (and is) attractive in that it offers a solution that can be easily adopted by multiple programming languages. It can be more easily experimented with as people learn what works best for large scale parallelism. And it was developed and implemented faster than a language modification, through the Standards, would have allowed.

The Coarrays Story

There is, however, a non-library approach that is now part of the Fortran language. It followed a number of previous solutions that never quite caught on; at least, not universally.[5] The paper cited above [138], titled "History of Coarrays and SPMD Parallelism in Fortran", set out to tell the story of a powerful feature of

---

[4]There are different accounts of Fortran's birth date. An IBM history page puts Fortran's start at 1954, with a commercial availability at 1957.

[5]I am referring to such techniques as microtasking, macrostasking, compiler directives, block directives in HPF, Fortran extension in the Vienna Fortran, auto tasking, and more.

Fortran – Coarrays, that was first proposed (invented) by my friend from our days at Control Data, Robert Numrich. Numrich, at Cray Research at the time (early '90s) and involved with the Cray MPP systems – the T3D and T3E, set out to remove complexities of expressing parallelism for the application user and the compiler.

The story of how the coarrays idea was introduced into Fortran is interesting to me for a couple of reasons. One, it provides a vivid example of the *codesign* idea described in Chapter 23. Second, the process of realizing the idea allows a critical look at the pros and cons of *"move fast and break things"* vs. orderly committee work.

Thinking about the challenges of parallelizing codes for massively distributed memory systems, Numrich asked a question along the lines of *"how can one processor access data from another processor if the address on the remote processor is known?"* Cray Research was a 'full service' vendor, but a small enough company, and Numrich, from the applications and software side of the house, was able to interact with the hardware group. There he found from the interconnect team that, indeed, there was a sequence of instructions that allows a processor to point to, or 'touch', an address in the memory of another.

What remains is to actually *know* the remote address of a given data item or variable. This would typically be the starting address of an array or a subset of an array. Fortran had a statement called COMMON block. Variables placed there are available to all the routines where the block is declared. Now, in the SPMD (Single Program Multiple Data) programming model, each processor gets an identical image of the program to be executed with its own portion of the data. All that is needed, then, it to declare the arrays that are to be *common* to all the processors – that is, a *coarray*, and place them in COMMON blocks. The memory mapping of the data is identical in all processors – this was known as Symmetric Memory Processing (SMP, not to be confused with Shared Memory Processing), so any processor can figure out the address of any element in a coarray on another processor.[6]

Armed with these ingredients, Numrich defined one-sided communication functionality of PUT and GET – for placing and retrieving data between processors. There is more to the model and the story, but the important lesson here is that innovation moves faster when people from different disciplines and skillsets collaborate.

That was in 1991. Numrich formalized his ideas with a proposal of a variant of Fortran he called F⁻⁻. The name designation is a tongue-in-cheek play to contrast the C++ language. Whereas the "++" stand for adding complexity to C, the minus signs symbolize the simplicity proposed for expressing parallelism in

---

[6]As implemented later, the declaration of a coarray is sufficient for the compiler to place it into symmetric memory across processes (or allocate it on a shared heap). There is no need to put the coarray into a COMMON block

Fortran.[7] Numrich continued to develop his ideas and conduct early experiments of implementing them. Early work was done with Fortran 77 on the Cray-T3D. The foundations and the proof-of-concept work done by Numrich are captured in his '97 paper titled "F⁻⁻: A Parallel Extension to Cray Fortran"[139]. At this stage, he was able to get to a partial implementation by taking advantage of special hardware of the T3D/T3E. Though lacking the elegance and clarity of the later coarrays standard, it provided an undeniable proof of concept. It took a summer intern from the University of Minnesota, assigned to Numrich almost as a favor, to accomplish this feat. The intern incorporated into the Cray compiler some existing assembler code, that had been designed for other purposes outside the compiler group, which took advantage of specialized hardware in the T3E.

It was during that period, from the early to mid '90s, when MPP and distributed memory solutions took center stage, that Cray Research became less sure-footed about programming models for their compiler. Various approaches were tried out, none successful. This is another factor that explains the hesitancy in promoting and pushing for the coarrays SPMD model.

So Numrich proceeded without support from the compiler team: "It took me a while to convince myself that the GET/PUT model worked on the T3D. It worked because the T3D had something like a global address mechanism, and the systems people put common blocks at the same address in local memory on each processor. So I wrote a little bit of assembler code, called GET, to set a remote address for moving data around. I still remember the day that I was able to run the NAS Parallel Benchmark codes on the T3D without any compiler support. Many people inside Cray told me over and over again that my idea couldn't possibly work until I showed them the results for the NAS codes. Even then, management didn't recognize it. In fact, for over a year, the only way people could benchmark the T3D was by using my little bits of assembler code. I only needed a GET function, but people insisted on an additional PUT function. Then the whole thing evolved into the SHMEM Library."[8]

He justifiably concludes: "With my little GET function, I saved the Cray T3D from failure."

Resorting to a library solution was out of necessity. However, Numrich found a way around it even back then. He recalls: "I always wanted the GET/PUT model to be implemented as part of the language rather than implemented as a library. Since the compiler people wouldn't help me, I wrote another little bit of assembler code that converted the local address of a variable into a remote address for the same variable on another processor. Using a Cray pointer, an extension to the

---

[7]Turns out Numrich credits a colleague, Geert Wenes, with helping him choose the name for his variant of Fortran. I came to respect Geert when we were colleagues at Cray many years later.

[8]SHMEM stands for "shared memory" and was a library for one-sided access of distributed memory in parallel programming. The acronym was later interpreted to stand for "Symmetric Hierarchical MEM."

language before pointers became part of the Fortran language, that was returned from this function, I could write communication code in Cray Fortran with no compiler support. It worked really well."

It was later, in the mid '90s, close to the time of SGI's acquisition of Cray, that the compiler group got more involved. Fortran 90 was stable enough, and its array features allowed a less specialized implementation of the model. It showed promise in terms of ease of programming and performance.

Jon Steidel recalls how his involvement with coarrays started: "Bob Numrich had his shared memory library, which became SHMEM. One day he walked into my office, and started writing on the whiteboard. And he asked, '*What do you think about this syntax for doing the SHMEM stuff?*' And I looked at it and I thought, '*This is pretty cool. This is pretty clean.*' He was strictly focused on Fortran 77 at the time, and I was working on the Fortran 90 compiler. So I taught Numrich about Fortran 90 features, and he got more excited."

Numrich recalls that the reason he got excited was because F90's derived types removed the restriction of symmetric memory allocation for a co-array object. The ability to reference a non-local allocated component within an F90 derived type was something that no other proposed language extension was doing. In other words, every processor could see all of the remote memory whether allocated symmetrically or not. In addition, it did not require data buffers for data communication that included computations, as was required in MPI-1's two-sided message passing, for example.

Irene Qualters, now at LANL after a stint at the NSF, managed Cray's software at the time. She met with Steidel: "And she said, '*We've got our proprietary model*', which was the CRAFT compiler at the time. It was very much like HPF where you had explicit data distribution, and we've got Numrich's F^{--}. And she said, '*I want you to figure out what direction we should go.*' And so I met with a few other people, and my recommendation was that we go ahead with F^{--}. I gave the nod to F^{--}, and had somebody in my group implement the syntax for it."

F^{--} was later implemented as an extension to Fortran 95. It was then that Numrich named it Co-Array Fortran (CAF). That was done on the Cray compiler for the SGI ORIGIN 2000 and the CRAY-T3E. It is worth noting that it took 3-4 Cray developers just about a week to create this compiler version. An example of an early use of CAF is given in [140], showing results of a multigrid solver. In general, the authors claim, CAF's performance compared with MPI was almost always as good for small numbers of processors but invariably scaled better to large number of processors. To be sure, this was a comparison of one application against one vendor's implementation, and not a general finding about the superiority of one tool over the other.

This F^{--} project was done somewhat skunkworks-style. Indeed, Steidel concludes: "We almost got fired for it."

Still, at that period John Reid collaborated with Numrich to define the extension to the language more precisely. Their work, captured in [141], was used as the definition of coarrays for at least five years.

Numrich relayed another anecdote that shows the benefits of co-design: After SGI parted ways with Cray, which was passed to Tera, the Seattle-based company that later named itself Cray Inc., the Cray people built the Cray X1/X1E. This machine actually incorporated the co-array model into the hardware address space. Addresses on the Cray X1/X1E included a node number in its higher bits. Reference to a remote node just required changing these higher bits to the remote node number. In his words: "It was the best approach to distributed memory ever designed, and it was all done to support the co-array model."[9]

The broader background to the story is the question of whether parallelism is expressed within the language, be it Fortran or C or any other, or by calling on an external library. Cray had coarrays in its compiler from the late 90s, but no other compiler vendor did. Though the Cray implementation served as a model for the standards, it was not until 2008 that the details for the standard were agreed on. Implementation by other vendors would take a couple more years. The result is that the MPI library model is now, essentially, 'the law of the land'.

That's not to say that the coarrays model is dead. People other than Numrich, including John Reid who sat on the Fortran Standards Committee, found the concept and the proposed syntax attractive.

As John Reid says: "I got involved in coarrays because it struck me as a much nicer way to express parallelism. It is safer, too, because the writer's intention is so much clearer."

And the feature took a different path, that of a formal addition to the language. Getting the coarrays model into the official Standard Fortran turned out to be a long and arduous journey, driven mostly by the enthusiasm of John Reid and Bill long. Reid did not propose coarrays for Fortran 2003 because the changes already agreed, including object orientation and polymorphism, were huge (see Chapter 34). Proposals were written and debated. Not all the committee members agreed on the need or the urgency of this model. There were delays due to disagreements over syntax – which did evolve somewhat over time. The coarrays model made it in for Fortran 2008, which was approved in 2010. More was added to coarray Fortran, described as Fortran's *parallel execution model*, in Fortran 2018[138]. The coarrays approach to parallelism in Fortran, with emphasis on how it is applied, is captured in Numrich's book[142].

Steve Lionel recalls: "It was in 2008, my very first international Fortran meeting in Tokyo. I liked coarrays, but there was a lot of objection to it from vendors who didn't want to do the work to implement it. And it almost didn't make it into the

---

[9]Unfortunately, this product was not a commercially successful, and it was the last proprietary processor from Cray.

language. I think I was a deciding vote for keeping it in the language. They voted by country, and it won over by one vote. Nowadays, the vendors have kind of accepted coarrays. A lot of it comes down to the fact that we are a small committee, and we had been working inefficiently."

To recap: Numrich conceived of the coarrays model in '91. He first called the compiler F^{--} but by 1998 changed the name to Co-Array Fortran (CAF). It has taken until '97 for Numrich to be able to talk and publish papers about the model outside of Cray. One factor holding him back was the unfortunate coincidence of the negotiations with SGI about acquiring Cray, because SGI's focus at the time was on systems with globally coherent virtual memory. And Numrich's model targeted distributed memories, although his model worked just as well, if not better, for a globally coherent virtual memory. Cray's management was also not supportive of the idea. Numrich admits to having failed to make a strong enough case. But, more significant to the topic of the Fortran committee, was the fact that, as Numrich puts it: "I had no idea how to communicate with the committee." Thanks to the work of Reid and Long as committee members it finally made it into Fortran 2008, which was published in 2010. It would take compilers other than Cray's more time, measured in years, to implement coarrays.

It has taken the coarrays model between *15 to 20 years* from concept to implementation and general access!

The MPI Perspective

For comparison, let's take a look at the history of the competing approach – that of a library:

MPI (the Message Passing Interface)[143] is the most successful and the most used library for parallel Fortran codes on distributed memory systems. It came after several other, mostly vendor proprietary, software systems, including PVM (Parallel Virtual Machine)[144]. PVM, like MPI, was developed by a user community team. It was first written in 1989, with at least two more versions 2 years apart. A number of its developers became developers and advocates of MPI. MPI was designed and built by learnings from the early libraries that laid the foundations to the message passing approach to parallelism in distributed memory systems. Its developers' goal was to create a high performance and *portable* library that would eliminate the need for a high performant vendor-specific solution.

MPI grew out of a workshop organized by the late Ken Kennedy in April 1992, with the goal of creating a target for parallel compilers (the "Williamsburg workshop"). The workshop provided the motivation for the community to get together and define a library from scratch. The real work started at the November 1992 ACM/IEEE conference on supercomputing (also known as SC '92 and Supercomputing 92). The MPI Forum that was formed then conducted some eight working meetings in 1993, culminating with an MPI specification draft that was presented at the November '93 Supercomputing conference. It was finalized and released in May 1994 as MPI 1.0. By then the MPI project had grown to about 80 people from 40 organizations. All the more impressive that it has taken less than *two years* from conception to a released product.

In fairness to the Fortran committee, it should be noted that MPI, while a standardized interface, is not a formal standard. It should be more properly referred to as *specification*. MPI is not an ISO[10] or ANSI standard, but Fortran is. This allows the MPI developers more frequent releases.

To better understand the process that led to MPI and the thinking behind its design, I contacted one of its principal developers – William Gropp. Gropp is currently (2021) the director of NCSA (National Center for Supercomputing Application) at the University of Illinois at Urbana-Champaign. He is also a professor at the Computer Science department there. Gropp has been involved with MPI from its inception, and for the last more than 25 years.

Why was MPI designed as a library? – Gropp recalls: "We did think about whether we should be designing something that was compiled or was a library, but we decided it had to be a library because we didn't believe we could get the language developers and the compiler vendors, in particular, to cater to our niche community. So that was a compromise. But I think that that was an important, pragmatic one because it gave us portability."

Gropp clarifies that the reason MPI started with support for two-sided communications is that, at that time, all the production system supported two-sided, but not one-sided communication. He adds: "One-sided was starting to emerge, and MPI added one-sided in MPI-2, only a few years after the original MPI specification. However, this version of one-sided also had limitations, some a result of the limitations of the technology at the time, and others because of a desire to have a well-defined standard that avoided ambiguities. That actually turned out to be a problem, which was addressed in the MPI-3 revision of one-sided operations."

There has been a recurrent aphorism among programmers: *MPI is the assembly language of parallel programming*. That is, it forces the programmer pretty close to the hardware. Damian Rouson tells of the time he ran into Gropp at a conference a few years ago, and repeated quip. Gropp responded (allowing for possible paraphrasing here): "That's what we always intended it to be! The problem is that it took so long for parallel programming languages to come along, and domain scientists couldn't wait, so they embedded MPI directly in the application source code."

So, did Gropp and his colleagues intend for the MPI calls to be generated by the compiler of the parallel language of choice, and not to be called directly by the application programmer? – Well, yes and no. Gropp explains:

---

[10]ISO stands for International Organization of Standardization

"I did not see MPI as primarily a compiler target – it really is too high level for that. However, I did not and do not believe that most programmers should use MPI extensively within their applications. Rather, they should take advantage of its features for "programming in the large", such as its support for libraries, and limit MPI to the implementation of core communication and parallel computation operations, much as is done in numerical libraries that provide distributed computing abstractions."

He expands on the idea: "Most programmers should either use a parallel framework (e.g., PETSc or Trilinos[11]) and leave the MPI programming to someone else, or they should design good abstractions and put those into a few routines that implement those abstractions. What programmers should *not* do is sprinkle MPI throughout their program as if they could use it as assignment (store) or reference (load) of individual words. MPI has features that make it easier to write those libraries (including message contexts, support for arbitrary groups of processes, and data caching on key MPI objects)."

Gropp says this about high performance production codes use of MPI for parallelization: "In many cases, the applications are not using a big common library, but what they have done is they have defined the operations on their data structures, and then written their own application library (using MPI for the defined operations). So, instead of having thousands of lines of MPI calls scattered through the application, the actual MPI calls are restricted in a special communication module. The developers optimized very special data structures and very special sets of operations. And MPI gives them the tools to build the parallel communication part of the application for an arbitrary data structure."

This, in Gropp's view, gets to a core reason for the success of MPI:

> *"MPI makes nothing easy, but makes everything possible."*

Gropp emphasizes that there is no parallel data structure that cannot be built with MPI code. The level of difficulty is the same whether the structure is a Cartesian mesh or a dynamic, unstructured, weird graph. Just as the mantra above states. He believes that models that attempted to be simple could not cater to most real applications, because their data structures were some perturbations from the simple structures.

And Gropp anchors the view above with a *performance* argument: "A lot of this comes back to what is really the fundamental problem in my view: We want to program with operations on individual elements, but for performance, because of the latency to all layers of local to remote memory, we need to work with aggregates, and use algorithms that give us sufficiently large aggregates. MPI forces the programmer to work in terms of larger groups of data (you can do individual elements, but the

---

[11]Two examples of open-source collections of portable libraries, a toolkit, to be used as building blocks for scientific applications.

performance will be so terrible you won't) – this makes it often harder to write the first version of a program but *easier* to get to the performant one – because you start by acknowledging that, whether it is cache lines, RAM rows, network flits, or disk blocks, data in real systems moves in blocks."

As for coarrays, Gropp likes the fact it is a tool that can used to teach people to work with blocks of data, not with remote scalar variables or single elements. Though, he adds, the coarrays model "does make it very easy to write point-wise code that accesses data on other processes."

MPI, in light of the above, is really a library whose direct use can be ineffective. Instead of its aspirational use by parallel languages, it may have found its optimal place embedded in a framework the application is built around. Marc Snir, who is an emeritus professor in the department of computer science at the university of Illinois Urbana-Champaign and also a principal early developer of MPI, gave a provocative talk on the occasion of 25 years to MPI. Its title was "MPI is too High-Level/MPI is too Low-Level" [145]. Depending on the role played by MPI, people claim that MPI is too low-level as an application programming interface, but too high-level for communication run-time not exposed to the application. Snir goes on to explain that MPI lacks functionality needed to be an assembly language. And, yes, its communicator has features not needed for low-level protocol. The examples Snir provides convince me that Gropp is right in encouraging users not to 'get in the weeds' by sprinkling MPI calls throughout their application, but rely on a 'mediator' library as an umbrella for MPI use.

Beyond the issue of how to incorporate MPI into the application, the early years saw some "truly dreadful implementations" in Gropp's words. It has taken some years and efforts by the MPI developers and the vendors to get the good performance MPI users enjoy today.

There is talk about MPI being too big, or complex or difficult, and whether it should be replaced by something else. Gropp shrugs and says "I wish I knew what it should look like." He goes to highlight the foresight, at the '90s, of designing MPI to be thread-safe. But acknowledges that MPI did not, or could not, take into account what modern processors will look like; meaning the parallelism on the processor chip. Because MPI is thread-safe, modular, and composable, it was possible to marry it with OpenMP and with CUDA. These combinations became common programming models. This is not without its challenges. Each of the participants – OpenMP for multicore shared memory parallelism, MPI for distributed processing – does its job well. The challenge is with the interaction between the two standards. "Socially, it's difficult because it means that the two standards would have to agree on the interactions." Gropp points to some success in coexistence between standards: "Fortran is perhaps the best example where Fortran was made to support the memory model that MPI has."

Gropp points to the MPI-IO feature, and the lack of coarrays-IO, as a plus for MPI. He says a complete model should include I/O. "I should be able to open a file, put a coarray into it, and close it. And do so in such a way that I should be able to access the data correctly with a different number of processes, for example."

The criticism associated with MPI notwithstanding – complexity, performance issues, lack of compiler and runtime help, and more, the MPI project has been successful. Some years ago Gropp described the factors that, in his view, were fundamental to the popularity enjoyed by MPI (see [146]). He highlights six factors where MPI shines as the significant sources for its success:

- Portability: The most important property. MPI runs on most parallel platforms, and works with multiple programming languages.

- Performance: For small number of processors MPI allows effective placement of data in memory. For large systems it provides effective means to scale programs and algorithms.

- Simplicity and Symmetry: The number of routines matters less than the fact that MPI requires only a few concepts. That's the *simplicity* part. Symmetry is achieved by elimination of exceptions (by adding routines).

- Modularity: MPI supports component-oriented software, for a clean and maintainable service. A *communicator* ensures all communications are kept within a component, which make MPI a reliable library.

- Composability: MPI was designed to work with other tools, and exploit improvements in them.

- Completeness:Any parallel algorithm can be implemented with MPI.

Referring to the attributes above, Gropp summarizes: "Each of these is necessary in a general-purpose parallel programming system." He added that MPI is not perfect and suggests areas for improvement. Gropp wrote the paper in 2006. Much has been done since to improve MPI.

Lessons Learnt and Outcomes

The MPI vs. Coarrays is more than a Library vs. Compiler matter. Initially, it could have been seen as a debate about the programming model for expressing parallelism in Fortran (and possibly in other languages too). Message Passing vs. One-Sided Communication is one way of looking at the different approaches taken. As we saw, now MPI supports both. Then there is the *societal* aspect of who decides what the programming model and implementation would be. Should it be the compiler vendors and a few large users working with the Standards body, or should it be a broader and direct representation of the user community?

At the end it was the Message Passing approach, as library implementation, championed by one influential segment (the DOE labs) that became the norm. And perhaps the whole affair was more about the programming model than anything else.

We are left to wonder how a parallel Fortran might have fared had the coarrays model been promoted more vigorously from its inception and had been considered for the standard in the late 90s. Would it have accelerated adopting the model into the language?

What we do know, as Ondřej Čertjk tells me, is that the coarrays model is not being used at LANL, for example. The main reason for that is that the lab's users are not satisfied with the level of support it gets by the several compilers they use. That's probably true for the other DOE labs.

Similarly at NASA; at least at NASA Goddard. Tom Clune sees coarrays as a positive for some resurgence of Fortran use, in that it addresses parallelism very elegantly. Nevertheless, coarrays are not used for production codes there. A few years ago, Clune's group performed a prototyping exercise with coarrays that targeted a large legacy application based on MPI. They found they needed to work with arrays having different extents on different images, and this capability was not yet robustly supported in the compilers used in the study. Instead, they explored the use of coarrays in synthetic kernels, but did not find significant performance benefits in their computing environment. Although the NASA team understood that the need to use consistent array extents could have been avoided using F90 derived types, the necessary changes to demonstrate this within the targeted legacy application were well beyond the scope of the prototyping effort. While aware that other compilers were more mature in this respect, for example, the Cray compiler, the usefulness of co-arrays was being evaluated in the context of an application that needed to be supported on a system with Intel and GNU compilers.

Portability considerations may also be a reason for avoiding the use of coarrays in some organizations. IBM, for one, chose not to implement coarrays in its compiler. Nor is it implemented in the PGI compiler, now owned by NVIDIA. Even as it makes these compilers not fully compliant with Fortran 2008 and Fortran 2018.

That's not to say that there is no use of coarrays at all. Damian Rouson tells me of some meaningful projects he has been involved where coarrays was central. Meaningful because they were commissioned by government agencies such as the U.S. Nuclear Regulatory Commission, NASA, and NCAR. Rouson also sees interest in coarrays as a major reason people cite for attending his Fortran training courses.

One of the interesting fallouts of the Coarrays-MPI story is that some compilers found it useful (or convenient) to incorporate MPI in the implementation of coarrays. For example, gfortran of the open-source GNU Compiler Collection (GCC) supports coarrays in the language, but 'under the covers' the compiler generates MPI calls to execute the coarrays-related commands. In fact, it is the OpenCoarrays project[147], with its set of library routines, that makes coarrays support possible in gfortran. More on OpenCoarrays, led by Damian Rouson, in the next chapter. The Intel Fortran compiler also uses MPI calls to implement coarrays. This implementation choice comes with a possible performance penalty, due largely to MPI use of intermediate data buffers.

Users do not have to choose between the models. The use of MPI and coarrays can be mixed within a program. This allows developers to bring together routines that don't employ the same programming model. And it also allows a gradual migration from one model to the other.

One can only speculate on what HPC programming may have looked like if the library-oriented people and the compiler-language teams collaborated. One idea for a possible outcome of such collaboration was suggested by Numrich at a conference talk in 2009[148]. He suggested that the expression of parallelism could be abstracted into an object-oriented framework. From that abstract class each programming model might then implement its own functionality, since both Communicators and Teams, for example, can be descendants of the same abstract object. Other objects, such as collectives, could also be defined abstractly, and then be specified through inheritance. And so on.

Well, it turns out that the idea of object-oriented framework came up at the MPI Forum and it is now under discussion. As Gropp relays, the old C++ binding was removed so it would be possible to create and add a new one.

Everyone I talked to agrees that the coarrays model is an elegant, clear, and simple method for expressing parallelism within the Fortran language. From Numrich's experience: "Fortran programmers understood what a co-array meant without having to explain it to them because it fit precisely into the whole design philosophy of the Fortran language based on arrays."

We should also remember that the coarrays model was not without compiler-based competition. There were a number of other ideas for Fortran parallel programming models for distributed systems that have been tried in the first half of the '90s. They include HPF (High-Performance Fortran, from Ken Kennedy's team at Rice University), Fortran D (a data parallel Fortran also from Ken Kennedy's group), Vienna Fortran 90 (extension of Fortran 90), CRAFT-90 (from Cray Research), and a few others. Only the coarrays model survived among the models tried in Fortran.

That it has not caught on much seems to be the result of both the time-to-standardize and the preference of a library solution by an influential segment of the user community; especially the U.S. DOE labs.

With the passage of time, Numrich reflects that the investigation of the choice between message-passing and one-sided communication models, which is separate from the library vs. in-language debate, got cut short by the quick emergence of MPI.

An open source portable library offers more flexibility in the choice of the programming language, and provides a convenient vehicle for adding functionality and modifications as compared to doing so within the language. However, including a feature in the language allows for a better optimized code, better error messages, and faster compilation. Perhaps it is good that both options exist.

CHAPTER 33

Fortran Today

The State of the Language and Related Projects

BEFORE getting to the personal reflections of the Fortran's Standard committee members it is useful to go over some projects and activities that serve the Fortran user community, as well as summarize where the language is heading.

The expansion of HPC beyond numerical simulations of the physical world to encompass also data analytics and manipulations as well as applications in the AI space, brought about commensurate expansion of software tools in use. This includes languages and libraries. And, significantly, additional requirements for environments where multiple languages are applied to a single job, and for support of a richer set of data structures and data formats.

Recognizing this new reality the Fortran 2018 Standard defines a new layer that allows C and Fortran code to interoperate using more Fortran features. In the past it had been somewhat limited. For example, it was difficult to access Fortran allocatable arrays in C. Fortran 2018 added new elements to the language that enable doing all that in a standard, portable way.

Machine Learning (ML) and big data brought Python to the world of HPC. The Python ML programmer might need a computational kernel best done in Fortran. A typical implementation would have a light-weight C wrapper around the Fortran routine. C is sometimes used as a bridge between Python and Fortran to simplify access to Fortran.

Fortran is the language of choice for the numerically intensive kernels and solvers of what we may refer to as 'classic' HPC. Python is more suitable for analysis and visualization because it allows a more rapid prototyping and more convenient software for graphics.

This trend was reflected at the FortranCon 2020 conference, an international meeting that took place in July 2020 (virtually, of course). There were several presentations about interoperability between Fortran and other languages. Fortran

DOI: 10.1201/9781003038054-33

interfacing with C++, with python, and also with Lua (full content published at [149]). Beyond the Standards there is a lot of development work going on about how Fortran interfaces to these other languages. The Standards are only concerned with Fortran interoperating with C, but because C is such a versatile low-level language and so well designed, it allows Fortran to coexist with almost any other language.

The proliferation of languages other than Fortran in HPC happened despite adding those features that made the other languages attractive into Fortran. In particular, features for handling more data types and data structures. The Fortran Derived Type provides the functionality of the C Struct and the C++ Class (and their representation in newer languages such as Java and Python).

A major departure from the old-style Fortran came about with the addition of support for object-oriented programming style. Fortran 90 through Fortran 2003 saw incremental expansion of object-oriented programming facilities. Object-oriented programming became popular in the early '90s, especially for business applications and where more abstract and unstructured data was present. Today's Fortran supports most object-oriented operations imagined. This aspect of Fortran is crucial to the use of the language in processing that go beyond just number crunching with simple arrays.

Steve Lionel, as the WG5 convenor, gave a talk titled "Fortran 2018...and Beyond" at FortranCon 2020. He shows that whereas Fortran 66 was defined in a 39-page document (190 pages for Fortran 77), Fortran 2018 required 539 pages to capture all its features. This provides a measure of the increased richness (and complexity) of the language. Lionel summarizes the main additions in Fortran 2018:

- Parallelism: New coarray features for richer use cases

- Further Interoperability with C – numerous details for greater closeness of expressions and descriptions

- Other enhancements; but also declaring some features deleted or obsolescent

Lionel concluded with a list of about 20 small changes and additions planned for a future Fortran release, referred to informally as Fortran 202X. They have been kept small because the committee is determined that delay due to greater inclusion of features be avoided. Five larger features were proposed but not included for this reason or because much more work was needed on the details. One of them, *generics* – allowing *type* and *operation* parameters in functions and subroutine, was seen as so important that work should begin at once with a view to its inclusion in the following standard, informally known as Fortran 202Y.

The coarrays story led me to wonder about the dynamics and interactions between the Fortran user community and the body that determines how the language develops and evolves. My guess is that the pull and push among the bodies of developers, users, and commercial interests is not unique to Fortran.

I found out about FortranCon 2020 through an online article about a presentation that was given there. The talk described projects for creating a standard library and a package manager for managing a modern Fortran environment. These are the first two projects by a recently-founded (December 2019) organization called fortran-lang.org [150]. The organization's mission is to create a community and a forum for collaboration for Fortran programmers. Their current focus is on the development of modern infrastructure for Fortran. The presenter was Milan Curcic, who we met on the topic of WRF applications (Chapter 20). It was the Fortran connection that had me look for Curcic, only to find out about his WRF enterprise.

It started for Curcic with Fortran 77 that was taught to meteorology students at the University of Belgrade in Serbia[1] in 2006 (even though Fortran 90 and Fortran 2003 already existed). Fast forward to 2020 with Curcic now a researcher at the university of Miami. We talked about how Fortran is evolving now.[2]

The Fortran Standards Committee is made of a small group of people – a mix of representatives from compiler vendors and large user organizations. There was no process or tools for end-users whose company is not represented on the committee to provide input. Most Fortran users, if they even know about its existence, don't know how the committee operates. This has improved in the past a few years, but it is far from perfect. The vendors are more attentive to requirements of those large users, and are also subject to their own constraints of resources and costs when it comes to supporting proposed changes. But, Curcic and others argue, there is a broader community of users who do not have relationships with compiler vendors. There was no mechanism for these users in the U.S. to provide feedback and offer suggestions or demands (there is such a mechanism in the U.K., though).

This has changed in the fall of 2019, when Ondřej Čertĵk, the LANL scientist, opened a GitHub repository for proposals to be considered by the committee. In talking to Čertĵk, it becomes clear he wants to contribute in two main aspects of Fortran world: User community involvement and a vision for the language.

So, how did creating the GitHub repository work out? – Curcic describes it this way:

Within three months after the start of the repository there were already several proposals from the community presented to the committee for consideration. "This was a demonstration of how very easily and quickly a user from the community who knew nothing about the standards committee, can now submit feedback about the language." Čertĵk concurs that the user community was excited to see the establishment of the repository. They advised, though, that it is a long process from this first step of submitting a proposal. A proposal that would have to be revised, voted on, slated for future implementation (if accepted), and so on. A process that usually takes several years.

---

[1]The same department where the Eta model, later used as the North American Mesoscale Model (NMM) and a predecessor to WRF, was developed in the '70s.

[2]I spoke to Milan Curcic just before he joined the J3 committee as an alternate.

And that wasn't all. The participation initiative drove further developments: *Standard library*, and a *package management utility*[150].

The instigators arrived at the idea of working together on the Fortran standard library. That was in recognition of the fact that most programming languages have a fairly rich and mature general-purpose standard library with various utilities, collections, data structures, and so on. And Fortran doesn't. It only has a collection of intrinsics or built-in procedures and modules that are mostly oriented toward numerical work. There are several well-established and rich numerical libraries for almost any computational need, but the language contains no support for data structures such as dictionary or a linked list, for example.

Curcic: "The programmers would need to write their own. Of course, without coordination everyone has been rolling their own. There is no 'go-to' solution for some of commonly used utilities.

"So we thought, *'why don't we join forces, and work together on something that would provide general utilities, and we call it the Standard Library'.*" The idea was not to create a language Standard, but to offer a go-to library for people to use with the language. It is not done in isolation. The developers communicate and collaborate with several members of the standards committee who provide feedback on the development of this community Standard Library.

While working on the Standard Library, they identified a gap in the building of Fortran environment: The lack of Build and Package Management tools. Python, and other languages, have 'standard' tools for creating the software files, and package and distribute them. These are easily available to other users. Nothing like that exists for Fortran. Currently, sharing code means the recipient needs to literally take it and put the files into their application. There is no easy way to automatically build and link a code received from another user. That was the motivation for starting the project called '*Fortran Package Manager*'. At the time of this writing it is incomplete, but already useful to its users.

Initially, the projects target the open source Fortran, gfortran, from GCC and under the GNU umbrella. The goal is to extend the tools to other flavors of Fortran, as well as include support for non-Fortran dependencies such as C libraries and functions.

Čertjk sees this activity as a campaign that aims at *resurrecting* Fortran with a community of developers. See his summary of the current state of affairs in his blog [151]. I'm certain more projects will follow.

A better known, and longer in existence, is a project called *LLVM* (see [152, 153]). The project name is not an acronym. Started at the University of Illinois some in 2000, its webpage defines it as a "collection of modular and reusable compiler and toolchain technologies." In reality, its scope is broader than that implied by its own definition. The part that is of interest here, and central to LLVM, is a compiler backend and an intermediate representation language (IR) that, together, support compilation of any programming language. There are numerous subprojects under the LLVM umbrella, including at least two Fortran frontends:

An interactive Fortran, called LFortran, for interactive execution much like how people use Python, MATLAB, or Julia [154]. Both Curcic and Čertĵk are involved in this project.

And Flang, which is LLVM's Fortran frontend.

Quite a few companies rely on LLVM components. Especially on its compiler backend. They include AMD, Apple, ARM, Nvidia, and IBM. In addition, Intel has plans to use LLVM's backend for its compilers in the future. In 2005 Apple hired one of LLVM's co-founders, Chris Lattner, and formed a team for using LLVM as an integral part of its development tools for its flagship operating systems.

It is not likely that all will be migrating toward the use of a single community Fortran backend. As Jon Steidel puts it: "Most people building compilers off of LLVM create some proprietary optimizations, which go into their products but don't get upstreamed to the public LLVM repository."

It looks likely that LLVM will become a prominent component in the software layer in support of exascale computing, the era into which we are entering now. As touted on its site [155], its tools and breadth of collaboration across the labs and leading computer companies "made LLVM-based compiler technology the default gatekeeper to these [exascale-capable] systems."

Much was said earlier about the length of time it took to get coarrays implemented in compilers. That was, in large part, due to the fact that implementing the full coarrays repertoire in the compiler is hard. The *OpenCoarrays* project [147] was created to make that easier. It is, at this time and as a first step, an application binary interface for the gfortran front-end for the parallel programming features of Fortran 2018. Flang seems to be the project's next compiler target. I reached out to Damian Rouson to find out more.

Turns out the idea for OpenCoarrays originated in Italy. A graduate student in the University of Rome published a paper in 2014 with 5 collaborators from Italy, Germany, and the U.S., including Rouson, who now leads the OpenCoarrays development [156].

The project has a broader vision than just making coarrays implementation easier. Rouson: "One of the main goals of OpenCoarrays was to liberate Fortran source from direct reliance on one parallel programming model. MPI alternatives such as OpenSHMEM and GASNet can also be used and the switch doesn't require changing or even recompiling the source code. The decision about which one to use can be made at link time."

Rouson's goal for OpenCoarrays is perfectly aligned with Numrich's 2009 suggestion in [148] of an object-oriented framework under which different programming models can be implemented.

Rouson points to a couple of other practical reasons that keep OpenCoarrays separate and outside of any compiler, other than the freedom to choose the parallel programming model: The outdated development technology used, until recently, by the GNU Compiler Collection (i.e., gfortran). And the need for licensing flexibility in order to support multiple compilers.

All in all, despite the new types of workloads in HPC and the prevalence of other programming languages, the Fortran community is alive, well, and thriving.

CHAPTER 34

Thoughts from the Guardians of Fortran

Reflections from Those Who Do the Work

W HY has Fortran not managed to stay the absolute language of choice for HPC? – I can only theorize. One contributing factor has to be timing. Fortran has been a follower, not a leader, as far as adopting data structure and management capabilities. Add to it the time it takes the Fortran standards committee to deliberate and approve, and we end up with a time delay that drives users to seek solutions elsewhere. Other factors include the flexibility of programming in the more modern languages. Some are, perhaps, driven by what they see as a more elegant foundation of the syntax of newer languages. Or, the interactivity enabled by line interpreter compilers.

Perhaps this is purely of academic interest, and the community is, in general, quite happy with the current state of affairs.

It is best to hear from some of the people who are in the thick of all things Fortran. The conversations described here took place around December 2020 and early 2021. They show areas of agreement and a spectrum of perspectives.

John Reid lives in the U.K. and is the longest serving convenor of the committee so far (1999-2017). Though retired, he has an honorary position at the Rutherford Appleton Laboratory in Oxfordshire, U.K. His exposure to Fortran dates back to the '60s. He became convinced of the need for standards in the language in the '70s when his IBM Fortran codes could not run on other computers without changes. That got him to join the J3 standards committee in 1983, where he represented all the sites of the U.K. Atomic Energy Authority. They were concerned with protecting their investment in large Fortran codes.

I was interested in Reid's historical perspective.

DOI: 10.1201/9781003038054-34

As he puts it to me: "By 1983 Fortran was looking dated. I was keen to see improvements." But "There's always a conflict between people who want more features in the language and the cost of developing those features into a compiler." The vendor representatives on the committee were protective of their investment and preferred to keep old-style Fortran and make little to no changes. As Reid recalls:

"in 1988, the situation was really critical. We got into a total deadlock, that was finally resolved only at the ISO level at a meeting in Paris, because about half the J3 committee were in favor of a fairly major change to the language, and the other half were totally opposed to anything but a very minor revision. The rules that J3 operated upon at that time were that if a decision was made at a meeting to make some change to the language it couldn't be undone at a later meeting without a two-thirds majority. So, a two-thirds majority was needed to make any change and there wasn't a two-thirds majority either way. We were split about half and half. So we were in a deadlock.

"The long delay in the Fortran 90 standard was a disaster for Fortran. It made the committee determined to make the next revision (Fortran 95) small and make it happen soon."

"The idea was that it could be followed by a major revision. Here we had an overshoot. The committee was too nice: *'It is a major revision and this looks useful – we should include it.'* Fortran 2003 was too big. I was convener by then and I should have done more to counter this. But the role of an ISO convener is to achieve consensus, and all the additions had this."

Reid highlights the dilemma the committee has to contend with: Users complain the language is too big in terms of options and features. Yet, they lobby for this feature or another to be added. Of course, the existence of large and complex old Fortran codes makes it next to impossible to remove features. Even when outdated or rarely used. It is a standing requirement that these old codes continue to be supported in the language.

There is also a tension in what motivates changes in features. Reid uses the term *safety* to indicate the assurance that a code would do what the programmer intended, *and* that this intention will be clear to users years later. The other approach is to give the resulting performance a higher priority. He relates an anecdote from the 90s: "A group of Cray users were invited to a discussion with the compiler team on their plans for the compiler and associated tools. There were about ten of us. None of the others thought safety should play any part in the objectives."

As for the nature of features, Reid believes "Fortran should focus on what it's doing well, which is numerical computations, because this is wanted by many users."

Clearly, Reid represents those who prize, above all, code safety and stability. He summarizes his views on Fortran's past and future: The lesson learnt from past revisions is not to overload future revisions with too many changes. Allow compiler vendors to keep apace. In addition to performance, code safety has to be a priority factor to consider. Old superseded features are, regrettably, here to stay in support of old codes. Beyond that, the new generation of committee members and users have to decide what is important for the Fortran language.

Jon Steidel, from Intel, is another old-timer in Fortran-world. He played a role in the coarrays story as a colleague of Numrich and a participant in its proof-of-concept implementation. Having been a compiler developer for decades at several HPC companies, and a past and current member of the standards committee, I sought his perspective on Fortran's past and future.

Steidel is involved, on behalf of Intel, with the DOE's Argonne National Lab as they prepare codes for a future exascale system. He observed that whereas in the past, in the '80s, over 90% of the cycles were executing Fortran originated code, he estimates that now the fraction is down to around 60%. "Fortran was slow to get data structures in general. It wasn't until 2003 that we got Object-Oriented. And there was a lot of interest in C++ by that time because of its Object-Oriented features." He points out that though Fortran is superior, and the language of choice, for numerical simulations of all kinds, C interacts better with the operating system. Therefore, interoperability with C is important for data analysis tasks, for example. He recalls: "One of the things that I really pushed for in 2003 was to get some C interoperability stuff into the standard, because IBM, DEC, and Cray had it, but not in the same manner. I thought it was really important that we standardize that or Fortran probably would be in worse shape now than it is."

On the topic of innovation in programming languages, Steidel makes the interesting observation that for C and C++, addition and changes occur only after being implemented by at least two compilers. As for Fortran, "We go out and invent things that nobody has implemented. Maybe that's a problem.." For example, he points out, it took Intel 9 years to have a fully compliant Fortran 2008. To be fair, by 2020 they already have a fully compliant Fortran 2018.

The coarrays story is one facet of Fortran's struggles with expressing parallelism. We ended up with the majority using MPI for what can be called coarse-grain parallelism, and OpenMP for fine-grain – vectors and in shared memory regions. OpenMP is closer to the compiler, and now its interface to the compiler is in the standards. Steidel was around when this came about:

"OpenMP has its own committee. Its own rules. They aren't part of the ANSI or the ISO standards. It started back in around '97. There was an interest in standardizing shared-memory parallel processing. There was much effort, many meetings, and a lot of time put into thinking these things through. And there was actually a committee formed, but it disbanded. The DOE labs came to Cray and IBM and to Digital, and they said: *'Could you guys please standardize the way that we do parallelism?'*. And that was the start of OpenMP. Several vendor companies got involved with representatives from the labs. We would meet fairly frequently and hash things out. We came up with what was, basically, the Cray autotasking model with different spelling. OpenMP continued to evolve through the years. It is version 5.0 of the standard now. And it even contains the necessary syntax for GPU offloading."

The Fortran Standards committee is fine with letting handling of parallelism be done by the outside bodies of OpenMP and MPI.

Asked about direction Fortran should evolve Steidel joins others by highlighting generic programming. The language feature is also known as *templates*. It provides the ability to write a procedure once, then compile it with different data types as needed by the application. This is also a part of the increased focus on *high productivity programming* (see also the discussion about high-productivity in Chapter 19).

Tom Clune, from NASA, is another committee member from the user community. His role is that of a software engineer who advises and supports scientists-programmers. In his case, they are the climate and weather modelers at NASA Goddard.

Clune is not concerned with the level of Fortran-teaching in higher education institutes. The NASA recruits with a graduate degree in science arrive with sufficient knowledge of Fortran or can readily learn it on the job if they know other high-level languages, he states.

On the subject of object-oriented features, Clune points out that Fortran and C++ employ different models: "It's not like you can just write a simple thing in the standard to mandate how objects are interoperable between the two languages the way we could at the C-layer level." It is possible to achieve interoperability with C++ through a C wrapper, but it is tedious, and therefore not much used at NASA.

The topic above led to Clune expressing his perspective on the working of the committee:

"I have actually found that the committee itself is generally very open to communication from the broader community. But there's a difference between being open to that and successfully getting a paper through. You really have to have somebody on the committee that's championing it. It really does help to have somebody in the room for whom that's a key issue. But I'm much more concerned about the time frame. Not only that it wasn't until 2003 that Fortran had object-oriented capabilities, but it really wasn't until, maybe 2015 that compilers were robust enough to actually use them. It puts Fortran about 20 years behind other languages. This is partly because Fortran is a small community now. The vendors have very limited budgets to implement the new features. They can only introduce new features at a certain rate. If we want to keep our codes able to work with a certain compiler, we're oftentimes stuck with 10 years behind the standard in terms of what language subset we can use. As opposed to C++ where developers apparently spend a much smaller fraction of their effort diagnosing, reporting, and working around compiler bugs. This is frustrating."

Clune is of the opinion that whenever possible making smaller and more frequent updates to the standards will work better for the community. He observes that this is indeed the direction the committee is currently taking. At the same time, he realizes this is not always possible, and provides this example (that others have mentioned too):

"Fortran users have wanted something along the lines of C++ templating for a long time. We now have a mandate to work on that. It is considered to be such a big feature that we are aiming it for the standard after the next. It is referred to as F202Y (the next version is called F202X) – without committing to a specific year. It may be as late as 2030. Here I'm really looking at developing a feature that doesn't actually get robustly implemented until I retire. This is something I've wanted for a large swath of my career. I'm happy to help create it, but jealous of the fact that I won't get to use it."

When it comes to the matter of abstracting the language from the underlying system architecture, Clune agrees with the current philosophy, while conceding there should be some notion of distinct memory spaces as a way of addressing accelerators. Perhaps, a third level – beyond shared and distributed, to deal with hosted attached processors, such as GPUs. And, in this context, the question arises of how much control to provide the user in instructing the compiler. Not through non-standard extensions, but within the language. "There are language people that will say: *'the compiler is in a better position to decide the balance between vectorization and threading.'* I don't believe that is always the case. I think that the compiler is not even going to know what I'm doing with MPI. And it is not going to know that turning on threads is going to kill my performance because now it is oversubscribing resources. There should be a way for the user to provide that kind of guidance. And Fortran right now does not recognize that."

Ondřej Čertík from LANL observes that the Standards Committee tasks itself with defining a feature – how to express and how to use it. It does not comment on implementation. In general, the feature has not been tried and tested before it is published in the Standard (a concern expressed also by Steidel). He suggests that a better process would be to prototype a proposed feature. That will remove the estimation and guesswork of both implementability and usability. He says: "The committee standardizes things before they get used in practice. That's like putting the carriage in front of the horse."

The other point Čertík makes is that the Standards Committee should have some kind of an outreach program to promote a new version of the compiler. He mentions two elements: Advertisement directed at the users, and convincing compiler developers to implement early. He says: "To actually get the Fortran community to use new features you have to do more than just publish them. You have to advertise, and you have to ensure that it is implemented in several major compilers, with support for the users."

When seeking a vision for Fortran, Čertjk points to C++ as an example of what a vision may look like. It sounds more like an underlying design philosophy: C++ is to be rich and versatile enough so that user developers can create libraries for any processing task they wish. In contrast, for Fortran, he agrees with other Fortran guardians who are taking a counter view. For Fortran the language itself has to contain everything that is needed for numerical scientific computing (and not just for any possible processing task). "make Fortran the best language for scientific computing again." Čertjk fears Fortran will be delegated to the lower level of routines being called from programs written in other languages. Existing Fortran programs in production are being maintained, but there is not much new major applications being developed in Fortran. He worries that the newer languages that developers start with have rich enough set of libraries that reduces the need to call even the numerical libraries of Fortran. And, he suggests, Fortran will be helped by having a standard set of libraries for utilities and functions beyond the numerical.

A running theme in todays HPC programming is the use of multiple languages and the interoperability among them. For Čertjk, this matter was another reason he got involved with the Fortran standards: "It needs to become easy to mix and match C++ and Fortran, Python and Fortran, Julia and Fortran, and so on."

In my exchanges with Damian Rouson, he placed the OpenCoarrays project in the broader context of parallelism and parallel programming models. He points out that "One little-recognized fact is that the parallel feature set in Fortran 2018 extends well beyond coarrays. A surprisingly large fraction of applications can go parallel in Fortran 2018 without ever declaring a single coarray. For embarrassingly parallel applications,[1] collective subroutines, teams, failed images, image enumeration, synchronization, and error termination cover all of the application's algorithmic needs without requiring coarrays. Moreover, there are plenty of features in the language that are implicitly parallel, including array statements, elemental procedures, and the concurrent form of do loops. For all these reasons, I usually just talk about parallelism in Fortran 2018 broadly without necessarily tying it to coarrays."

Rouson, who teaches modern Fortran to researchers and scientists, further emphasizes the parallelism aspect of scientific programming, even if somewhat

---

[1]The term 'embarrassingly parallel applications' applies to a class of application where the distributed tasks require little to no communication during the parallel processing.

hyperbolically: "There's so much parallelism in the language that even the shortest standard-forming program is parallel. Or to put it another way, rather than asking the common question of how to parallelize an algorithm, we could start asking why we would ever serialize an algorithm. It's probably fair to say that nature is parallel in all ways that satisfy causality. Thus, when I teach modern Fortran, the courses usually start parallel from the first line of code and stay parallel throughout. It's an inherently parallel language now. Any Fortran code can be compiled in a way that it executes in parallel. The only question is whether the programmer is going to take advantage of that ability in any explicit way."

It is appropriate to round off the conversations with Steve Lionel, who is the current (circa 2021) convenor of the standards committee. His start with Fortran dates back to the late '70s and the compiler group at Digital Equipment Corporation (DEC).

Interestingly, it was then that the second Fortran standard – for Fortran 77, was published. Lionel notes that it would take 13 years before the next Standards version is published. The delay caused issues. Users demanded features and vendors had to implement them on their proprietary compilers. Of course, not all implemented the same features or in the same way. As a result, when the Standard came out there was re-work to be done, while preserving support of past implementations. That long-ago experience must be what is driving Lionel now, as the committee's convenor, to establish a process of more frequent and less burdened updates. It took 8 years between Fortran 95 and Fortran 2003. An improvement, but still too long. As Lionel explains:

"The Fortran Standards Committee has been a mix of compiler vendors and end-users. But in recent revision cycles the committee had not been soliciting input from the broad user base. Many of the members were familiar with what their customers were doing. The compiler vendor representatives tended to be from the support side. They were familiar with the types of programs that their customers were doing and the things they ran into, and they would make suggestions to the committee. And often these compiler vendors would create an extension to the language to make their hardware look better. Sometimes it would be as an actual language feature; sometimes it would be directives that would alter the way that the program was interpreted. Soon the compiler vendors started becoming more conservative about creating new language features because that was especially a problem if the new standard conflicted with the feature they created."[2]

Lionel repeats what Reid stated before. Fortran 2003 was a very big update, perhaps too much so: "Fortran 2003 was another big, big, big change. With

---

[2]John Reid adds that there was a huge outreach to users for the revision of Fortran 77 – there were 396 letters sent in the first public review and much committee effort was expended in considering them and replying. In recent years, it has become much harder to get opinions from users, but we have certainly been trying within the U.K.

polymorphism and a bunch of other features that had never been done in a Fortran before. But they were popular in other languages. So in addition to what customers or the users were saying they would like to do, the committee also looked at what other languages do and said: 'Gee. That would be a useful thing', or, 'That's something that people are asking for, so let's do that.' What made things worse was that the economy was contracting at that time. The Fortran compiler teams were getting smaller. There were companies that just stopped doing Fortran entirely. There was consolidation as when Intel bought the DEC team in 2001."

When comparing to other languages' updates, Lionel tells me, consider that the Fortran committee has between 15 and 20 members, with contributors from five countries, at most (U.S., Canada, Japan, Germany, U.K.), with only about 6-7 members doing much of the work. The C++ committee, for example, has some 250 members. Still, he is eager to reform, gently, how the committee operates: "When I became convener of the Fortran committee in 2017, the first thing I did was a public survey among Fortran users. We had 137 responses with hundreds of suggestions and we asked the users to rank them. We then got together, and went through all of those suggestions, plus those from committee members, and that's how we came up with the feature list for what would be the next revision, which we're calling 202X."

Fortran 2018 was actually published in 2018, unlike previous revisions which were named after the year in which the features were finalized (integration of the new features into a large and complicated standard document together with final publishing add delay measured in years). It adds clarity to the timeline of revisions, but only because that revision was previously called Fortran 2015. Following the 2018 publication and the user survey "We immediately started working on the next revision, planning a very short cycle, five years maximum. We have been very successful with that. We've got almost all the features designed already. Even with the pandemic slowing us down and holding virtual meetings."

Lionel's vision for the workings of the committee fits well with the GitHub repository described above, of which he is fully supportive. He highlights another advantage it brings about: "One of the things that's held us back in the past is that nothing happened between meetings. During the meeting there's this flurry of activity. People go up back to their hotel rooms and write papers and design features. We vote on them. We do that all week, and then we go back to our jobs and forget about it again until the next meeting. There are just three meetings a year. This is a terribly inefficient way of doing things. I think of the repository as an *'incubator for ideas'*. It serves to get ideas fleshed out so that we can get a running start on the next revision. It definitely makes things more visible to the end users, which is something that I was promoting."

"It has worked so far. I'm pleased that compilers are catching up with the standard. In fact, in the past compiler vendors were saying: 'The standards are just moving too fast. We can't keep up with it.' "

Lionel also has an ongoing blog series under the heading of "Doctor Fortran"[157]: "The blog is where I talk about Fortran standards activities and where I try to explain things. This is a way of keeping more in touch with the users."

We talked about Fortran's place among programming languages. Lionel is of the opinion that the standards committee cannot do much about schools not teaching Fortran anymore in the U.S. (not so in Europe and Asia). And that, anyway, there is no one language that is best at, or designed for, everything. Mixing languages where appropriate is just fine. That said, "Fortran continues to have its strengths in numeric processing. That's going to be important. I tell people to use the language they know and can maintain. And not to rewrite applications from one language to the other just because a language is out of favor. That would usually introduces bugs. There are new applications being written in Fortran, because Fortran is still very strong in scientific and mathematical engineering applications. Other languages have their place too. I don't think this is a zero-sum game. And my goal is to keep Fortran relevant for the people who want to use it."

Reinforcing the guiding principle of detaching the language from the hardware architecture the code runs on, Lionel says: "One of the things that we on the Fortran committee take a stand on is that we don't want to put into the language anything that looks like it's going to be temporary. Fortran is actually hardware agnostic. There is almost nothing in the language that's specific to hardware implementations."

For example, he believes GPUs use is temporary and will be over in a few years. This is based on the cyclical history of deploying attached accelerators. Anyway, "Our approach regarding GPUs and attached processors is that there is an existing model: OpenMP or OpenACC, to enable their use independently of the language."

This is consistent with a basic tenet of the Fortran standards committee: "Fortran's philosophy is unusual, in that it allows the processor – and by which we mean compiler and underlying OS and hardware – to do things in any way that it sees fit to get the desired result. Things like auto parallelism and vector processing and such, are invisible in the language. The implementation just makes it happen. We do try not to put restrictions in the language that make it difficult for accelerators to work. It's a different philosophy than that of some of the other languages."

And the justification, Lionel continues, is: "I would rather focus on language features that improve programmer productivity in terms of the development process and let committees, such as OpenMP and OpenACC, focus on getting the best out of the hardware if the compiler can't do it. There are tools that can analyze the running program and advise the user on what to do for a better performance. Compilers can tack on directives that say, 'vectorize this.', or, 'Don't vectorize that.'."

Again, the underlying assumption is that things like attached processors and even vector functionalities are all temporary.

Commentary

Here I offer my opinions about what I heard and observed. My major takeaways from the conversations described above, about the thinking of the Fortran guardians regarding its near-term future, are:

- New versions of Fortran standards to be published at the faster pace of one every 4 to 5 years.

- Desire and activity to generate a greater user community feedback and suggestions into the committee, as well as projects to enhance the language's usability and popularity.

- High productivity, of development and maintenance, gets more attention than performance enabling features.

- The language definition is independent of any hardware architectural details.

- Fortran continues to limit the scope of its target codes to numerical computations.

- Fortran for HPC is here to stay even if in a mixed-language environment. Interoperability with C is a central tenet.

The points above serve me as a roadmap for discussing my reactions to the conversations described in the *Fortran chapters*.

On the Workings of the Committee

Accomplishing the goal of a faster release cycle of the Fortran standards requires both that the amount of changes is kept in check and that committee work continues between meetings. The use of an online repository for proposals, open to the public, makes the goal feasible.

We saw that some people would like to see more promotion of Fortran. Lionel, the current committee convenor, would welcome it, but states that the committee has no funding or the tools for such an activity. Perhaps it is a role for the user community. Possibly driven by national research labs, government agencies with stake in the language, engineering schools; even organizations interested in higher productivity in computing. For example, it is likely that many recent graduates are not aware that much of what they do in C can be done in Fortran too.

Looking at the committee's past performance, with the coarrays story in mind, the verdict is not as complimentary.

By having taken until 2010 (when Fortran 2008 was published) to introduce parallelism intrinsically into the language via the coarrays model, the Fortran standards bodies, effectively, ceded the expression of parallelism to external libraries and other committees. Of course, that wasn't intentional, but what might be seen as a historical misfortune. Coarrays' use is far, far less than that of MPI and OpenMP (or OpenACC for GPUs). Not all compiler vendors feel compelled to implement coarrays. Most who do, employ MPI under the covers. Though not a part of the standard, there is a tacit acceptance by the Fortran committee (and users) that OpenMP is the tool to use for local memory parallelism and, with OpenACC, for GPU processing, and MPI for distributed (and shared) memory parallelism.

The outcome is that the details for expressing parallelism in Fortran are, in practice (because coarrays is not in common use), controlled by bodies external to the Fortran standards committee.

This is not a matter of status or recognition. But still feels awkward (to me, at least) that most programmers in Fortran, the language that accompanied supercomputing throughout its history, rely on tools external to the language for expressing parallelism – an aspect so fundamental to HPC.

The coarrays story is used here as an example in order to draw some more general conclusions regarding the evolution of Fortran. But it is not the only case from which lessons can be learnt, as was hinted about in the preceding chapters:

One such topic is the introduction of array syntax in Fortran 90. Influenced by the then-popular data-parallel paradigm, it turned out that the early implementations could affect performance negatively. It has taken a long time for the compiler vendors to fix the performance issue, after which many more of the users were willing to use the feature.[3]

Another important feature worth mentioning is the concept of *derived type*. It was supposed to resemble the *C Structure*, and first introduced in Fortran 90. One difficulty was that there was no outreach to Fortran users explaining the feature as a first step toward Object-Oriented languages style. Some old-time Fortran programmers disdained that style. It wasn't until Fortran 2003 that Fortran added procedures that resemble C++ classes, and allowed allocatable components within derived types.(A Technical Report, published in 1998, defined allocable components and allowed implementations prior to 2003.) Using pointer components instead can cause situations that resulted in memory leaks, leading to running out of memory space. The other big issue is that Fortran derived types and C++ are not compatible. To this day it seems the newer (and younger) users and members of the committee are more open to the Object-Oriented features in designing modern Fortran codes.

---

[3]As an aside, the CDC Cyber 205 already had array syntax in the early '80s, designed to support memory-to-memory architectures (already known to be ineffective). Some people ridiculed that syntax at the time.

These last two items are examples where greater interaction with the user community and earlier inclusion of language interoperability might have prevented some difficult chapters in the history of Fortran.

On Performance and Productivity

As Lionel stated, there is more emphasis on *productivity* than on *performance* when the committee considers new features for the language.[4] He also expressed the committee's philosophy of avoiding consideration of hardware architectural details, while 'doing no harm' to the compiler's ability to optimize for performance, and that users concerned with improving their code's performance can call on performance analysis tools.

Reid values safety, that is reducing the likelihood of blunders going undetected, and also the quest for high performance. He sees safety, productivity, and performance as separate of each other, and equally important.

These concepts are separate, but in practice there can be dependencies involved. For example, allowing the user to declare regions safe for vectorization or parallelization, for enhanced performance, whereas there might be possible race conditions that cause errors (safety eroded). Productivity measures through abstractions (objects) may well be to the detriment of performance. In short, the language can provide ways for productivity on code development, but to extract higher performance the developer may have to forego productivity in terms of time spent and even of features to use.

We saw that the matters of performance and productivity are complex and subjective. There is the difficulty in defining *productivity* in the context of HPC. It is generally understood to mean support for quick development of codes and easing its maintenance. However, consider the difference in approaching a research code compared to that of a *production code* that is developed once and run thousands of times and is central to an organization's mission. Optimizing for performance of the latter by utilizing complex coding techniques and tools, slows down the development phase, but increases the organization's productivity in the long run.[5]

Put coarsely, production codes, running repeatably, require high performance outcome, not fast development, for the organization's overall productivity. For one-time research code, overall productivity benefits from quick development and the resulting performance matters less. Thus, the two attributes are entangled. Both scenarios – quick development and high performance – need to be served. The challenge for the Standards committee is how to do so optimally.

---

[4]These aspects of HPC were discussed at greater length in chapters 19 and 31.

[5]It is worthwhile to repeat here the potential ambiguity of the terms *productivity* and *performance*: By higher *performance* we imply higher *productivity* of the computing system. By *high productivity* we really mean higher *performance* of the humans involved.

On Abstraction Away from Hardware Details

The principle of abstracting the language from hardware details consideration is, in my opinion, the most impactful and consequential criteria for the language. It has been a long standing principle that all the members I talked to agree should be adhered to and preserved. A legitimate principle that results in a more elegant formalism, and helps in avoiding the pitfalls of enshrining in the language features of short-lived usefulness. Other major languages are also architecture-agnostic, of course.

But for Fortran, the absolutism of not addressing important hardware features appears to be somewhat inhibiting in two areas:

- Addressing parallelism holistically

- Support for higher performance of codes

Regarding the *parallelism* item, abstracting away from the hardware architecture the presence of distributed memory was not a factor in the committee's thinking that led to the late adoption of coarrays in 2008. But that addressing distributed memory systems took so long is consistent with that philosophy. And whenever there is a hardware component, be it distributed memories or attached processors, that cannot be pointed to within the language people will, by necessity, resort to creating libraries through which these components can be accessed.

There are practical implications to relying on libraries. Library calls inhibit potential compiler optimizations for faster execution. In addition, ceding control of the language over functions such as parallelism and attached processors opens the door to multiple solutions, and away from a single standard. This has happened regarding GPUs. The Fortran standard is aligned with OpenMP and OpenACC while there is an alternative: CUDA.

It may be just an aesthetic difference, but it seems that striving for the purity of the language forced some external appendages that take away from such design purity.

A decision to not include functionality in the language means leaving it to a library. That was true at least as far as parallelism and attached processors go. With some degree of oversimplification we can say the following on the subject: Ideally (but not practical), the language should just express the processing the user wishes the computer to perform. *'This is the data. These are the equations. Apply them and present the results.'* After all, Fortran's name came from "FORmula TRANslation." Compiler front-ends transform this high-level code to an intermediate representation. At this phase architectural details can be exposed. Finally, a backend compiler and an optimizer generates machine code that is executed on the system.

Now we get to the matter of performance. The process above sounds good in theory, but the reality is more complex. There are often multiple ways to write down a numerical procedure. Multiple ways to step through the data. Multiple ways to declare data structures. And not being able to be specific about the programmer's intentions regarding aspects of the architecture can result in an intermediate representation that the backend cannot optimize well for performance.

Indeed, the reality is that, over the years, compilers and companion tools often did not live up to expectations. For users who care about performance (and what Fortran user doesn't?), providing ways to advise and assist the compiler seems like a reasonable compromise. Such features exist in some compilers, but are not universal or standard. Users who want to offer such hints and advise to the compiler also want the code to remain *portable* across compilers. For that to be true these advise capabilities have to be included in the standard. Such a capability would also allow a more consistent performance across compilers. The ability to get a performance boost through hints and directives is preferable to time-consuming tweaking of the code in a guesswork manner until the compiler does what is expected of it (recall the 'performance' discussion in Chapter 31). Fortunately, for shared memory and GPUs there is OpenMP. Its instructions appear in the Fortran code as special comment lines that the compiler can use to invoke the parallelism as requested by the programmer, and as possible on the system the code runs on. In the language itself there is the "do concurrent" statement and the "contiguous" attribute that were added in Fortran 2008.

So, we have a medley of forms and styles for expressing parallelism in Fortran. The popular tools used today are OpenMP for shared memory; OpenACC, CUDA and OpenMP for GPUs, MPI for distributed memory. An in-language model consistent in style and form that exposes the architectural features mentioned here would have been cleaner.

Adjacent to the discussion above is the difficult issue of what decisions regarding data placement to leave solely to the compiler. Should there be features that allow the performance-conscious programmer greater control over ordering and placement of data elements? Think, for example, of an ability to define an ordering of elements in an array that is different than the standard's default (and goes beyond a two-dimensional transposition). Or, how about features for controlling the mapping of the logical distribution of data to the its physical configuration, so that the program can refer to data located at the physical neighboring server (for example)? (See footnote 7, page 307)

The argument against exposing hardware details is that they are often temporary. Mostly true at the low-level details. But consider global system components and general processing modes. What I have in mind are concepts like *vector processing*, *distributed and hierarchical memory*, and *accelerators* (attached processors). These are features that have been around for a long time, even if we can't tell that it will be so forever.

Vectorization exists now for 50 years; even if missing for a short time during the transition from proprietary to commodity processors. Vectors are treated as one-dimensional arrays. An explicit and specific vector notation that implicitly recognizes the presence of vector registers would be helpful to the programmer and the compiler. The array syntax allows defining any vector within the declared arrays, be it in any dimensional direction, or as diagonal, or non-contiguous set of elements. It is a minor matter of esthetics perhaps, but a Vector type, separate from and in addition to being a one-dimensional array, would increase the code's affinity with the hardware, and indeed with the math behind the code.

Attached processors in various forms – array processors, coprocessors, accelerators, GPUs – were around since the '80s. The concept of a hosted processor continues to gain popularity. Heterogeneous computing – the mixing of different types of processors in a system, is a given in today's HPC systems. The attached processor may be a GPU, FPGA, or some specialized accelerator for machine learning. In all likelihood, quantum computing devices will show up as attached and hosted processors. They each have their own characteristics of processing and interface, but have in common that a host processor sends data to them to process, and they send results back to the host. The equivalent of the OpenMP (or CUDA) statements could have been abstracted into the language.

Multiple memory layers will stay with us, and managing data efficiently will demand awareness of the memory-storage hierarchy. Above all, parallel and distributed processing, with its piecemeal memories and interconnected nodes, will be with us as far out as we can see. The details will vary, and compilers will adjust for them, but the 'performance user' and production codes can benefit from being able to be explicit about data location on memory on-chip, directly addressable (shared), or remote (distributed) as they lay out the data and as it moves during execution of the code.

Maybe features such as vectors, accelerators, and layers of memory, that have been so fundamental to HPC systems for decades deserve to be considered non-temporary and be visible in the language.[6]

Abstracting away from the hardware architecture led to resorting to external libraries over in-language features. This affects the optimizations for performance that the compiler can perform. There is a class of users who prefer the language to reflect the problem they wish to compute and let the compiler do the best it can. There are, also, critical codes that benefit from the best performance that can be had. Often, these codes serve to select which system to acquire. It can be argued (as I do here) that if truly integrated language constructs for management of vectors (including for sparse arrays – gather and scatter and merge etc.), recognition of memory layers and proximity, and handling of attached accelerators, existed, then those performance-driven critical codes would be developed faster and perform

---

[6]Coarrays is visible and goes a long way in addressing distributed processing. Unfortunately, it is not in common use for the reasons given before.

better. The language may have a *'For Power Users'* section. The most effective aspects (memory management?, GPU access?) may well become, in time, common for all users.[7]

Perhaps I'm too optimistic and too obsessed with performance. The lesson from the coarrays story is that timing is critical. It is hard to change facts on the ground. The *facts* being the current ways of dealing with GPUs and parallelism. The Market has spoken.

Fortran is for Numerics and HPC Embraces Multi-language Programming

The HPC community is paying more attention now to AI and data analytics. That has also contributed to lessen the overall role of Fortran. Instead, users in those new areas, now integral part of HPC, tend to program in Python and C++, under frameworks such as TensorFlow, PyTorch and the like. It is not that the HPC users are strictly divided between those doing numerical simulations and those, new to the HPC world, who do AI and data analytics. There are now HPC applications that are not numerically intensive, but perhaps more significant is that 'classic' HPC applications now venture into explicitly analytics and AI domains. One such example is the use of machine learning algorithms and techniques on climate data at weather centers and climate research labs.

The Fortran guardians do not attempt to expand the applicability of Fortran to areas outside the computationally intensive one. This is not said as a criticism. Better to serve one purpose well than try to be a 'Jack of all trades'. That said, object-oriented programming features and enabling the handling of complex data types and structures were added to language. But it has taken some 20 years to get the richness of object-oriented features Fortran has today. Meanwhile, many application developers, who valued the *objects* abstraction, migrated to other languages. The community is adapting to multi-language coding practices. Especially in the area of AI.

There is, though, work in progress where Fortran is used to create neural networks for deep learning (see [158]). It is a parallel Fortran framework that makes use of the collective routines of Fortran 2018. What started as an academic exercise is becoming a useful research tool with several published research projects to its credit.

In Closing

Naturally, there is tension between the two necessities of a living programming language: The orderly and inherently consistent standard formalism of the language, on one hand, and the freedom and enabling of experimentation and innovation, on

---

[7]This is not a new idea. I am reminded that HPF – the High Performance Fortran, came out in the '90s as Fortran extension for high-performance. It did not enjoy much success and faded away.

the other. From its history so far, it is difficult not to conclude that the scales have been tipped toward the rigorous committee work and formalism. Not evolving fast enough may have been to the detriment of the popularity of Fortran among the broad HPC user community. Perhaps the repository for proposals described above is a first step in creating a more productive collaboration between users and compiler developers. And from that, innovation will spring forth.

The topic of leaving features to external libraries or including them in the language came up a number of times. The resort to libraries was presented as a failure of the language definition and standard. This was not meant as a general objection to use of libraries, but only in the cases of Fortran's expression of parallelism and of access to attached processors.

In general, the use of libraries has a place and advantages too. Gropp highlighted specific reasons for the success of MPI (page 282). Not all libraries accomplish all those attributes, but the library concept allows portability, for example, that cuts across programming languages. History shows that, compared with language features and compilers, people can experiment more easily with libraries, implement additions and changes faster, and make them accessible to the user community sooner.

There are lists of programming languages ranked by popularity. Fortran is ranked nowhere near the top. However, I am not referencing such lists because most are not confined to HPC, and their metrics are not appropriate for judging the importance of a language. They measure popularity of languages. That is, by references on search engines and chat boards. Useful information for skills needed by job seekers. For HPC the important metric is the portion of systems resources and cycles consumed by codes generated by a given language. There may be fewer Fortran programmers than for other languages, but they still account for a most significant portion of HPC cycles consumption.

There is, clearly, a much smaller universe of Fortran practitioners compared to that of some the more popular languages. Some find the need to explain and justify their loyalty to Fortran. A few years ago, Daniel C. Elton wrote a short article titled "Why physicists still use Fortran" [159]. He points to existence of legacy codes, ease of learning the language, array handling features, no need for pointers and memory allocation,[8] and a couple of other benefits. Physicists will continue to use Fortran, even as more often via calls from another language.

My attachment to Fortran goes back to my past as an applications engineer and a benchmarker. I'm happy to see a new generation of Fortran 'activists' who keep Fortran thriving.

---

[8]That is not to say that pointers and memory allocation are redundant. Memory allocation is a real necessity in modern Fortran codes. Pointers, though not compatible with C pointers, are very useful in some cases.

CHAPTER 35

Measure of HPC Impact

Quantifying the ROI on Investment in HPC

W E have seen examples of how HPC benefited society and mankind in several domains. Beyond the justifiable satisfaction and pride from the cited achievements, it is worth asking about the *measured* economic impact of the HPC enterprise.

Back in 1991 the U.S. Congress requested the General Accounting Office to report on how the Industry is using supercomputing, and to try and assess the economic impact of such use. Quantifying the impact was difficult (see [160]), but the authors give this assessment in the report's executive summary:

> *"Supercomputers contribute significantly to the oil, automobile, aerospace, and chemical and pharmaceutical industries' ability to solve complex problems. They enable companies within these industries to design new and better products in less time, and to simulate product tests that would have been impossible without spending months developing and experimenting with expensive product models. Some companies have attributed significant cost savings to the use of supercomputers. For example, although exact figures were not always available, representatives of some automobile and aerospace companies estimated that millions of dollars have been saved on specific models or vehicle parts because of reduced manufacturing or testing costs. In addition, one oil company representative estimated that over the last 10 years, supercomputer use has resulted in increased production of oil worth between $6 billion and $10 billion from two of the largest U.S. oil fields."*

The representatives from the five industries selected for the study framed the supercomputers' contributions in terms of significant cost and time savings in bringing products to market, or the ability to perform tasks we were unable to perform before and without their HPC system.

DOI: 10.1201/9781003038054-35

The oil & gas industry, only mentioned in passing in this book, applied seismic data analysis to increase the success rate of discovering locations for productive wells by some 10%, and reservoir simulation to improve the yield of extracting oil by about 25%.

Structural analysis and computational fluid dynamics applications allowed the aerospace industry to gain insights into materials and their interactions with the surroundings that could not be achieved by experiments. Simulated wind tunnel allowed quick repeats of virtual experiments, watching in slow motion, and testing with extreme conditions not possible in the physical wind tunnel (remember the 1993 supercomputer called the *Numerical Wind Tunnel* in Japan). We have mentioned before Boeing's digital design. It not only resulted in a better optimized design but also in time saving bringing it to market. Another example is the use of computational electromagnetics to reduce an aircraft's radar signature for stealth military aircraft saving millions of dollars by reducing the need for physical models and tests.

A big benefit to the automotive industry is the digital crash analysis we highlighted before. In the '90s a complete crash analysis could be done in a day on a supercomputer. Building a physical model took months and each may have cost hundreds of thousands of dollars. CFD was used to simulate airflows, but also the flow of liquids in the cooling systems. One car manufacturer reported that CFD codes helped improve the design such that manufacturing costs were reduced by 5%.

In the field of Life Sciences, that the '91 report referred to its private sector component as the Chemical and Pharmaceutical Industries, the companies used mostly applications developed in the public research domain. Molecular Modeling codes were started to get used in drug discovery. One company stated they will save millions of dollars by replacing tens of thousands of syntheses, resulting in a handful of drug candidates, with digital simulations, avoiding much time and costs of the trial-and-error process of the past. Another example describes search for a compound, done in a matter of days on a supercomputer, replacing months of ten times more costly lab experimentations.

That was the state of affairs in the early '90s, when HPC was confined to numerical simulations. The report cited above addressed how several private sector industries used and benefited from supercomputers they owned, though it recognized their extensive use of application software developed at publicly funded institutes – national labs and universities. It did not attempt to quantify the many more indirect economical benefits derived from HPC. For example, the savings to transportation costs resulting from following weather forecasts, or benefits to the agriculture sector from responding to seasonal forecasts.

Fast forward about 20 years, and we have more recent studies done by the analyst group Hyperion Research, formerly a division of International Data Corporation (IDC). They created models based on evaluation of hundreds of projects to quantify HPC's economic impact according to several metrics. Their Special Study, published in 2013 still as IDC, describes "Economic Models Showing the Relationship Between Investments in HPC and the Resulting Financial ROI and Innovation – and How It Can Impact a Nation's Competitiveness and Innovation" ([161]). Two economic models – one to quantify return on investment (in terms of revenue, profits and cost savings, and jobs), the other to identify innovations (basic and applied) supported by means to assess innovation levels, applied to a pilot study with these impressive good news about the economic value of HPC, based on over 200 entities examined (from industry, academia, and government):

- Return on investment of $356 (average) for every dollar invested in HPC.

- Profit of over $38 for every dollar invested.

- Less than 2 years before positive returns start.

- An innovation requires, on average, an investment of just over $3M, based on 160 innovations.

- Jobs were created in a sample of 40+ entities, about 30 per site, at a cost of $93K per job.

The 2013 study contains brief descriptions of a couple of hundreds successful projects done on HPC systems. For most of them the value is very clear even when quantifying it is difficult (or proprietary confidential). The Hyperion team repeated the 'benefits from HPC' study in 2017 with exascale computing on the horizon. The report, titled "Real-World Examples of Supercomputers Used for Economic and Societal Benefits: A Prelude to What the Exascale Era Can Provide" ([162]) lists case studies in the areas of:

- Health and Quality of Life

- Fuel Cost and CO2 Emissions

- Manufacturing and Competitiveness

- Disaster Mitigation and Recovery

Below are some highlights from the report cited:

Projects for improving the health of the human population show great potential and real advances, even though the impact on individuals is still a promise to be fulfilled. A detailed model of hepatitis C virus is a big step toward the development of therapies that can save $9B per year in the U.S. Supercomputers simulations used in clinical trials for long-term cancer patients helped improve success rates and saved millions in costs and months in research time. Supercomputers make possible studying how the brain synapses work and offers the promise of reducing the huge costs of treating brain related disorders. Never-before-possible simulations of the heart down to the cell level combined with digital screening of drugs improve therapies and reduce mortality of heart diseases patients.

Jet engine simulations done at national labs for the private sector improve the gas turbine fuel consumption's efficiency. One major manufacturer estimates that 1% improvement saves $2B worth of gas annually on the engines it produced. A long haul trucks manufacturer uses supercomputers to design better components that can save $5B in fuel costs. As mentioned before, seismic processing and reservoir simulations saves billions of dollars on exploration and increases yields in production.

Engine manufacturers use HPC to design engines that use new types of biofuels, estimating it will result in over $1B in savings. A university supercomputer is used to streamline and improve the manufacturing process of *continuous casting steel*, the results of which is reduction in greenhouse emissions which in turn allows keeping many jobs in the U.S. while saving some $400M per year. HPC-assisted design of CO2 compressors for future supersonic travel will save in the ten of millions dollars each year. It's worth repeating that HPC has already saved billions of dollars over the years for the aerospace and the automotive industries.

Disaster mitigation cut costs and saves life. It is supercomputers that enable about 5-days prediction of the path, intensity, and size of storms and hurricanes. Not only lives are saved, but hundreds of millions of dollars are saved in evacuation costs in an average year (and we now know these extreme events are becoming more frequent). Similarly, if not as advanced, modeling waves and storm surges will lead to more accurate and appropriate preparations, saving property and lives. Earlier alerts of earthquakes and detailed maps of their hazard locations, done on HPC systems, are also estimated to save billions of dollars annually.

The financial amounts noted above are from 2017 and expressed in that year's currency values.

We don't have a *total* for the economic benefit credited to HPC. But the examples here, and more in the Hyperion report, point to the kinds of benefits and their magnitude that are repeated and multiplied in time and all over the world.

CHAPTER 36

Looking Forward

Technology Transformation Ahead and Ongoing Expansion of Use Models and Users

T HE HPC market has always been evolving. The changes were always accompanied with the steady, if fast, march of progress in device technologies. Architecturally, we identified transitions from "big iron" monoliths with proprietary software stacks, to multi-processors, to MPPs, and to clusters with 'standard' processors, memory parts, interconnect, and a common operating system. These transitions took time and almost always were not clear cut. Some vector processor product lines survived well into the clusters period (NEC's, for example). Vector instructions found their way back into microprocessors used in clusters. Accelerators in HPC existed for a while in the '80s, then returned to HPC some 20 years later. MPPs as we knew them in the early '90s all but disappeared, but the term is now applied by the authors of the TOP500 list to systems with specialized high-performance and proprietary interprocessor networks (such as Cray XC and the Fugaku systems).

An indication of the exponential growth in complexity of HPC, and its utility in our lives is that in the '70s and the '80s supercomputers were well-defined products with only a few options of configurations. The top systems were defined by the vendors and (mostly) available to all at around \$10M (this is about \$36M in today's dollars). The top systems today, at the exascale regime, are custom-designed one-off and priced at several hundreds of millions dollars.

The New HPC

That said, the outlook for the coming decade or two is one of a much more dramatic transformative period. The reasons for saying that is the apparent end to the feature shrinking of silicon (Moore's Law), the early products using quantum computing

DOI: 10.1201/9781003038054-36

(which requires us to rethink 'programming'), and the future role of AI aspects in HPC and computing in general.

While making the statement above, I am reminded that when people contemplated exascale after we just reached petascale, the consensus was that just evolving current architectures and programming models will not be sufficient. However, as it turned out, exascale is largely an evolution of petascale systems, not a departure.

There is much more to changes of the HPC landscape than just a technological revolution. The scope of what was once considered HPC, namely numerical simulations of various phenomena, has expanded explosively with much of it driven by data and AI applications. With it the HPC user community has grown and is changing from mostly physical sciences researchers to include data analysts and data scientists, as well as AI researchers and practitioners.

With the new face of HPC – with AI and data, there is no clear line of where HPC begins. Some HPC applications use machine learning. Many apply big-data techniques. Some data analytic and AI applications unrelated to scientific computing or scientific data run on HPC-class systems. Much of AI and data centric workloads run on relatively small systems and even on desktop computers. In the past HPC was recognized both by the content of the applications (scientific, numerical simulations) and the class of systems used (high-end). Today we should include in HPC all the applications that run on HPC-class systems.

If the nature of the application is not a definitive indicator for 'membership' in the 'HPC club', then we look at the size of the system and its component features. Where HPC begins is a subjective determination. Business analysts will set a price above which systems are HPC, but exclude big datacenters where the workload does not include jobs that run on a large portion of the system. It is safe to state that computing tasks that run in parallel on hundreds of processing elements fall in this category. Somewhere on the continuum of performance and the scale of parallelism begins modern-day HPC.

Technology Trends

The transformative nature of the HPC enterprise has not escaped the hardware technologies themselves. Being outside the scope of this book, I will note only briefly on trends that have started or in progress for the last few years. The scope and varieties of the technologies and architectural developments is breathtaking, and a testament to the transformative period ahead of HPC.

AI-Targeted Chips and Lower Precision Arithmetic

Machine learning computations that have become a part of HPC, or just run on HPC-class systems, have grown so pervasive that new chips are designed to cater to its computational characteristics. There are tens of companies that offer, or shortly will offer, specialized chip for AI-type processing. They include not only the traditional chip manufacturers, but also the large internet companies. OrionX, the industry analysis and marketing company headed by Shahin Khan, identified (circa 2020) over 30 companies developing new chips, with more joining the fray. The number of companies engaged with AI-focused products numbers in the hundreds in 2023.

More than other types of processors they bring a variety of word sizes as a representation of numerical values and floating-point precision – lower than the previously desired 64-bits and even lower than the single-precision 32-bit words.

The renewed interest in machine precision and word size seems as an opportunity to John Gustafson who conceived his *universal number (unum)* and *posit arithmetic* as an answer to handling inaccuracies due to the finite word size ([163]). He says even performance will benefit:

"For machine learning what we are doing looks a lot like matrix-matrix multiply, but in really low precision. People are reexamining what we use to represent real numbers on a computer and discovering that there is a lot of efficiency to be gained. Maybe enough for two cycles of Moore's Law. We could get a factor of four by using better number representations. To me, the most exciting thing happening in high performance computing is that you can get the number sizes down to 8 bits or 16-bits, and only very selectively use higher precisions for summations and dot products, but then, immediately return to a low precision. You can get away with even 16-bit linear algebra, for example. The reasons for using double precision in the past turned out to be poorly based, if you really look at the problem carefully. And to discover we can now do the same number of operations, but with one-fourth as much data, is a big relief on the memory wall, which is the *imbalance problem.*"

Quantum Computing

Quantum computing (QC) looms large over the HPC world. Many companies – large and established, and small startups, engage in the development of QC devices, software and applications. As far as we can tell so far, QC processors will be placed, system architecture-wise, in the role of accelerators. But whereas GPUs could be absorbed within the prevailing programming models, QC applications require a whole new way of looking at, and expressing, the computational problem.

At this time, early 2020s, most HPC experts not involved with the QC development and business, are cautious and somewhat skeptics about its useful utility in the near future. Comments heard are about the devices' stability, claims that had to be walked back, not succeeding as anticipated in the decryption arena (with large integers), even that some devices are not 'real' annealing quantum devices. Mostly, the concern is about the application space that can be handled with probabilistic-in-nature computing paradigm. Rick Stevens, from Argonne Lab, summarizes QC's current status and its potential promise:

"Any useful quantum application when they eventually become so, is going to rely on a lot of classical computing as well as the quantum accelerator. The right way to think of quantum devices is as accelerators. Right now, they are physics experiments, not yet real accelerators. But when they mature, the way to access them is from a classical API. From a hardware standpoint, depending on what the latency requirement is, they'll be either embedded in a big classical machine, or they'll be right next to one. Quantum processing is also something that's going to be very important in quantum communication networks and in quantum sensing. We will have versions of processors that are actually close to sensors or close to communication devices that will not be like accelerators. They'll be an integral part of the device, whatever it is."

General-Purpose Processors

For a while, since the mid '90s and for the next 20-plus years, it was the x86 processor architecture that served as the most common processing element – excluding the accelerators. It appears that the dominant role of x86 as the HPC general-purpose processor will be challenged in the coming years by at least two alternatives.

ARM, which stands for Advanced RISC Machine, is an architecture that existed since the early '80s, but mostly for personal computers. There is no single company that produces ARM processors. It is an architecture that can be licensed. System houses who license it can create their own implementation, possibly with extensions, as long as it is compliant with the architecture definition and its instruction set. When, in 2011, a version of ARM was released that supported 64-bit arithmetic and address space it became a viable compute engine for HPC. It now enjoys an increased popularity and we saw it used in its Fujitsu implementation in the Fugaku supercomputer. It is likely ARM will gain a greater share within the HPC systems (for more see [164]).

Further behind in its development cycle is another RISC architecture, an open standard instruction set architecture named RISC-V. Some of the researchers from Berkeley University who gave us the "View from Berkeley" (Page 266) are at the core of this project. As signified by its name, this project represents the fifth generation of RISC specifications from the principals. The RISC-V Foundation became RISC-V International and moved to Switzerland due to concerns about trade regulations (more about its history in [165]). Indeed, the HPC community in Europe seem to base much of their high-end HPC efforts on RISC-V (See, or example, [166]).

Chiplets

The diversity of workloads now included in the HPC scene leads to specialization of processing components. The miniaturization of silicon features allows for enough electronic gates for multiple functions on a single chip. In past years designers chose the *System on Chip* (SoC) approach – a monolithic design that integrated the various components. This is giving way to a new trend, a modular approach where a well-defined functionality is captured on a small integrated circuit chip called *chiplet*. The chiplets methodology has two major benefits: It allows more flexibility and customization through greater freedom of choice of chiplets to include in the final chip. As a not insignificant bonus, the chiplet concept also help optimize the semiconductor manufacturing in terms of yield and cost because a manufacturing defect results in replacing a small part and not the whole chip.

Rick Stevens ties together the concepts of chiplets and fat nodes (discussed in Chapter 31). As he puts it (in 2022):

"After talking to 40 or so vendors about futures I would say it's trending toward fat nodes, but not necessarily with accelerators. With chiplets it is really easy to do integration of, say, processor, classical CPUs and GPUs, and other types of accelerators within a single package. So we will see over the next few years, packages that have CPUs and GPUs in the same package. We will see CPUs, AI accelerators, and memory accelerators in the same package. We will see FPGAs and CPUs and something else in the same package. And so fat nodes is really a packaging decision.

"It is almost always going to be shared memory within the package. The issue is how big of the shared memory domain it is going to be, and if the chiplet integration will be at wafer scale, because at wafer scale the amount of internal communication bandwidth is much higher. This is the internal surface to volume aspect. The interconnect bandwidth on the interior provides much more capability than when these components are apart and we have to go off-chip or off-module. Therefore, the trend is going to be toward bigger building blocks. With chiplets it is going to be heterogeneous, of course. What we think of as accelerators now will be thought of as chiplet building blocks, LEGO-like, that can be integrated in different ratios. Packaging will really become dependent upon your system architecture."

Innovative New Processor

Quantum computing, even when applied as an accelerator and for a narrow set of applications, forces us to re-think the programming aspect. It is also technologically challenging in terms of scale (the number of Qubits in the system) and the very low temperatures it requires. It is interesting to point to a less known but also an innovative way to design processors. Still using silicon, but making the hardware adapt to the code, not the other way around. A truly software-driven process. Such a design will be offered by a startup company from Israel, NextSilicon, now in existence for several years and as of early 2023 still in stealth mode and testing early silicon. Elad Raz, its founder and CEO, contributed two major innovations to the computing process:[1]

---

[1] I consulted with NextSilicon for a short period.

The first is an architecture featuring enough compute elements to implement an entire algorithm on the hardware, eliminating all together the need for an instruction stream.

The second is a telemetry implementation that identifies bottlenecks and dynamically optimizes the algorithms every three seconds to speed up the application.

Check [167] for progress, as the details of NextSilicon's architecture are not in the public domain yet (early 2023). Meanwhile here is a testimonial from one enthusiastic fan, John Gustafson:

"That's amazing. These two fundamental innovations represent the biggest breakthroughs I have seen in supercomputing since parallel processing. The Cray-1 was a huge leap. Going parallel was the next big leap. And it has been a long time since we have seen anything like those kinds of big leaps in HPC. It's just been cluster computing. And I think NextSilicon is going to solve the problem of arithmetic because you can drive any kind of arithmetic you want on those integer processors."

Encompassing all of the above is the certainty that the level of concurrency in HPC applications will continue to increase.

From Thought Leaders

It seems appropriate, at the end of the 50-year story of HPC, to conclude with some insights from leaders in the HPC community about its future and direction, expressed in two thoughtful articles.

Three HPC luminaries got together to present their thoughts about the future of HPC in a 2022 article titled "Reinventing High Performance Computing: Challenges and Opportunities" ([168]). They are: Daniel Reed, currently Presidential Professor and Professor of Computer Science at the University of Utah, and Chair of the National Science Board. Dennis Gannon, Emeritus Professor of Computer Science at Indiana University and retired from Microsoft Research. Jack Dongarra, Distinguished Professor at the University of Tennessee and Distinguished Research Staff at Oak Ridge National Lab. With a combined experience in HPC of well over 120 years, their many contributions to the field are too many to recite here.

The 'reinventing HPC' authors see HPC at an inflection point, and they point to the implications of the prevalence and domination of mobile devices and cloud services as a factor making satisfying the HPC needs more specialized and costly. The slowing down of the technology-driven advances that in the past were at the rate prescribed by Moore's Law and Dennard Scaling dictate a more customized processor and architectural design, shifting away from the big-volume commodity parts that power the mobile and cloud computing segments. Chiplet design is replacing monolithic chips and SoCs (System on a Chip).

Past supercomputer companies that could influence the technology and device providers are now being replaced by the much larger entities that build the small mobile instruments (smartphones, tablets, and such) and the large datacenters for internet commerce, social networks, and (some) HPC. The scale of these markets dwarfs that of scientific computing. Cloud services has changed the way compute cycles are delivered for many at the user-end of the ecosystem. Small startups can innovate without first investing in pricy HPC systems.

The trio express what they see as the implications for high performance computing in the form of six maxims. And I quote from [168]:

- Semiconductor constraints dictate new approaches

- End-to-end hardware/software co-design is essential

- Prototyping at scale is required to test new ideas

- The space of leading edge HPC applications is far broader now than in the past

- Cloud economics have changed the supply chain ecosystem

- The societal implications of technical issues really matter

Reinventing HPC, they conclude, will involve collaborative partnerships not only between users and the technology vendors but also with cloud services providers, resulting in systems more closely designed to support specific workloads.

Another influential paper about the HPC future and trends came out in the beginning of 2023. This one was led by two very well-known HPC experts: Satoshi Matsuoka, the architect of Fugaku from RIKEN (with three of his colleagues), and Torsten Hoefler from ETH Zurich. In a somewhat light-hearted, but still serious and thought provoking, they discuss "Myths and Legends in High-Performance Computing" ([169]). They propose twelve 'myths', that include active areas of investigations, discussion topics of the moment, and uncertainties. Indeed, each 'myth' discussion is followed with some questions the community does not have the answers for yet. While the paper contains loads of technical information about the state-of-the-art, each 'myth' title can be phrased as a question. The topics of the paper serves as good a list as any for what can be taken to be top-of-mind for HPC practitioners looking forward. Coming to the end of this 50-year HPC journey, the 'myths', rephrased here as factors that will impact what future HPC will look like, seem like an appropriate bookend:

- Quantum Computing

- Deep Learning

- Extreme Specialization in Supercomputers

- Accelerators

- Reconfigurable Hardware

- Getting to Zettascale

- Memory Size

- Disaggregation

- Continuation of Applications Improvements

- Fortran and/or Domain-Specific Languages

- Precision Levels

- Cloud and/or On-Prem Computing

The Last Word

This book describes the journey from the megaflops-range supercomputers of the '70s to today's exascale behemoths hundreds of millions cores, along with the accompanying transformations of programming models and the evolution of applications, punctuated by slices of events along the way.

The last 50 years of high-performance computing were exciting with an ever-evolving system architectures, programming models, and scope of applications. The multitude of transformations already in-flight foretell even faster-moving and more exciting times ahead.

Bibliography

[1] Wikipedia. *Moore's Law.* `https://en.wikipedia.org/wiki/Moore's_law`.

[2] TOP500.org. *TOP500.* `https://www.top500.org/`.

[3] William J. Kaufmann and Larry L. Smarr. *Supercomputing and the Transformation of Science.* Scientific American Library, 1993.

[4] Susan L. Graham, Marc Snir, and Cynthia A. Patterson, editors. *Getting Up to Speed: The Future Of Supercomputing.* The National Academies Press, 2005. National Research Council. `https://doi.org/10.17226/11148`.

[5] Gene H. Golub and James M. Ortega. *Scientific Computing and Differential Equations.* Academic Press, 1981.

[6] Wikipedia. *List of Numerical Analysis Topics.* `https://en.wikipedia.org/wiki/List_of_numerical_analysis_topics`.

[7] Wikipedia. *The partial differential equation.* `https://en.wikipedia.org/wiki/Partial_differential_equation`.

[8] Kenneth E. Iverson. *A Programming Language.* John Wiley & Sons, Inc., 1962.

[9] Andie Hioki. The Cray-1 Supercomputer. `http://www.openloop.com/education/classes/sjsu_engr/engr_compOrg/spring2002/studentProjects/Andie_Hioki/Cray1withAdd.htm#Intro`.

[10] Wikipedia. *Vector Processor.* `https://en.wikipedia.org/wiki/Vector_processor`.

[11] Josef T. Devreese and Piet Van Camp (editors). *Supercomputers in Theoretical and Experimental Science.* Plenum Press, 1984.

[12] Annika Reintges. *Weather and Climate @ Reading.* `https://blogs.reading.ac.uk/weather-and-climate-at-reading/2022/weather-vs-climate-prediction/`.

[13] NCAR's Research Applications Laboratory. *Climate Modeling and Downscaling.* `https://ral.ucar.edu/nsap/climate-modeling-and-downscaling`.

[14] Peter Lynch. *The origins of computer weather prediction and climate modeling*. *Journal of Computational Physics*, 227(7):3431–44, 2008. (and `https://web.archive.org/web/20100708191309/http://www.rsmas.miami.edu/personal/miskandarani/Courses/MPO662/Lynch,Peter/OriginsCompWF.JCP227.pdf`).

[15] ECMWF. *Who we are: History*. `https://www.ecmwf.int/en/about/who-we-are/history`.

[16] UK Met Office. *Who we are: Our History*. `https://www.metoffice.gov.uk/about-us/who/our-history`.

[17] Wikipedia. *Met Office*. `https://en.wikipedia.org/wiki/Met_Office`.

[18] Mike Hawkins and Isabella Weger. *Supercomputing at ECMWF*. `https://www.ecmwf.int/sites/default/files/elibrary/2015/17329-supercomputing-ecmwf.pdf`.

[19] Wikipedia. *ILLIAC*. `https://en.wikipedia.org/wiki/ILLIAC_IV`.

[20] J. E. Thornton. *Design of a Computer: The Control Data 6600*, 1970. Computer History Archive. `https://archive.computerhistory.org/resources/text/CDC/cdc.6600.thornton.design_of_a_computer_the_control_data_6600.1970.102630394.pdf`.

[21] Thomas Rosmond. *30 Years of Navy Modeling and Supercomputers: an Anecdotal History*. `http://www.ncep.noaa.gov/nwp50/Presentations/Tue_06_15_04/Session_2/jnwpu_rosmond.ppt`.

[22] Philip G. Kesel and Francis J. Winninghoff. *The Fleet Numerical Weather Central Operational Primitive-Equation Model*. *Monthly Weather Review*, 100:360–373, 1972. `https://doi.org/10.1175/1520-0493(1972)100<0360:TFNWCO>2.3.CO;2`.

[23] E. Morenoff, W. Beckett, P. G. Kesel, F. J. Winninghoff, and P. M. Wolff. *4-way parallel processor partition of an atmospheric primitive-equation prediction model*. AFIPS '71 Proceedings of the Spring Joint Computer Conference, pages 39–48, 1971. `https://doi.org/10.1145/1478786.1478793`.

[24] Enrico Clementi and Giorgina Corongiu. *Early parallelism with a loosely coupled array of processors: The lCAP experiment*. *Parallel Computing*, 25:1583–1600, 1999.

[25] N. Balram, C. Belo, and J. M. F. Moura. *Parallel Processing on Supercomputers: A Set of Computational Experiments*, 1988. In Supercomputing '88: Proceedings of the 1988 ACM/IEEE conference on Supercomputing `https://citeseerx.ist.psu.edu/viewdoc/download?doi=10.1.1.129.6837&rep=rep1&type=pdf`.

[26] R W Hockney and C R Jesshope. *Parallel Computers 2: Architecture, Programming and Algorithms*. CRC Press, 1988.

[27] John Gustafson. *Programming the FPS T Series*, 1986. `http://www.johngustafson.net/pubs/pubt1986.2/FPS.pdf`.

[28] Peter Coy. *Company Claims World's Fastest Supercomputer*, 1986. AP News. `https://apnews.com/article/c31339533c42b5ed1be5f7639bc0bbfc`.

[29] Floating Point Systems. *FPS-164 Scientific Computer*, 1982. `http://archive.computerhistory.org/resources/access/text/2010/02/102647686.05.01.acc.pdf`.

[30] Wikipedia. *Minisupercomputer*. `https://en.wikipedia.org/wiki/Minisupercomputer`.

[31] William Mahoney and J. Todd McDonald. *Enumerating x86-64 - It's Not as Easy as Counting*. `https://www.unomaha.edu/college-of-information-science-and-technology/research-labs/_files/enumerating-x86-64-instructions.pdf`.

[32] Wikipedia. *Lattice QCD*. `https://en.wikipedia.org/wiki/Lattice_QCD`.

[33] D. Barkai, K.J.M. Moriarty, and C. Rebbi. *A Modified Conjugate Gradient Solver for Very large Systems*. *Computer Physics Communications*, 36(1), 1985.

[34] D. Barkai, M. Campostrini, K.J.M. Moriarty, and C. Rebbi. *Applications Development on the ETA-10*. *Computer Physics Communications*, 46:13–33, 1987. `https://www.academia.edu/52021944/Applications_development_of_the_ETA_10`.

[35] John Markoff. *Lockout at Chen's Supercomputer Company*, 1993. The New York Times. `https://www.nytimes.com/1993/01/25/business/lockout-at-chen-s-supercomputer-company.html`.

[36] Sandia National Lab. *Sandia's ASCI Red, world's first teraflop supercomputer, is decommissioned*, 2006. `https://newsreleases.sandia.gov/releases/2006/asci-red-decom.html`.

[37] Albert M. Erisman and Kenneth W. Neves. *Advanced Computing for Manufacturing*. *Scientific American*, 1:148–155, 1988. Special Issue: Trends in Computing.

[38] NCAR's Mesoscale & Microscale Meteorology Laboratory. *The Weather Research & Forecasting Model*. `https://www.mmm.ucar.edu/weather-research-and-forecasting-model`.

[39] John Michalakes, Jimy Dudhia, D. Gill, J. B. Klemp, and W. Skaramock. *Design of a next-generation regional weather research and forecast model*, 1998. `https://www.researchgate.net/publication/240625728_Design_of_a_next-generation_regional_weather_research_and_forecast_model` and in Proceedings of the Eighth Workshop on the Use of Parallel Processors in Meteorology, European Center for Medium Range Weather Forecasting, published by World Scientific, Singapore, 1999.

[40] Jordan G. Powers et al. *The Weather Research and Forecasting Model: Overview, System Efforts, and Future Directions. Bulletin of the American Meteorological Society*, 98:1717–1737, 2017.

[41] Thomas Sterling, Paul Messina, and Paul H. Smith. *Enabling Technologies for Petaflops Computing*. The MIT Press, 1995.

[42] ExtremeTech. *The History of Supercomputers (6)*. `https://www.extremetech.com/extreme/125271-the-history-of-supercomputers/6`.

[43] Guang Gao, Konstantin K. Likharev, Paul C. Messina, and Thomas L. Sterling. *Hybrid Technology Multi-Threaded Architecture*, 1996. `https://www.hq.nasa.gov/hpcc/petaflops/paws.96/htmt/htmt.html`.

[44] Thomas L. Sterling and Larry Bergman. *A Design Analysis of a Hybrid Technology Multithreaded Architecture for Petaflops Scale Computation*, 1999. `https://cseweb.ucsd.edu/classes/sp99/cse190_C/ICSJun99Final.pdf`.

[45] Wikipedia. *Beowulf Cluster*. `https://en.wikipedia.org/wiki/Beowulf_cluster`.

[46] James R. Fischer. *The Roots of Beowulf*, 2014. In "20 Years of Beowulf: Workshop to Honor Thomas Sterling's 65th Birthday" `https://ntrs.nasa.gov/citations/20150001285`.

[47] Wikipedia. *Pentium FDIV Bug*. `https://en.wikipedia.org/wiki/Pentium_FDIV_bug`.

[48] Wikipedia. *Itanium*. `https://en.wikipedia.org/wiki/Itanium#Itanium_(Merced):_2001`.

[49] Wikipedia. *Xeon Phi*. `https://en.wikipedia.org/wiki/Xeon_Phi`.

[50] Wikipedia. *Advanced Computing Roundtable*. `https://compete.org/advanced-computing-roundtable`.

[51] Jack Dongarra, Robert Graybill, William Harrod, Robert Lucas, Ewing Lusk Piotr Luszczek, Janice McMahon, Allan Snavely, Jeffery Vetter, Katherine Yelick, Sadaf Alam, Roy Campbell, Laura Carrington, Tzu-Yi Chen, Omid Khalili, Jeremy Meredith, and Mustafa Tikir. *DARPA's HPCS*

Program: History, Models, Tools, Languages, 2008. `https://www.academia.edu/22894524/DARPAs\_HPCS\_Program\_History\_Models\_Tools\_Languages?email\_work\_card=view-paper`.

[52] David J. Kuck. *Productivity in High Performance Computing. The International Journal of High Performance Computing Applications*, 18(4):489–504, 2004. `https://doi.org/10.1177/1094342004048541`.

[53] Eugene Loh. *The Ideal HPC Programming Language*. acmqueue, 8, 2010. `https://queue.acm.org/detail.cfm?id=1820518`.

[54] NCAR's Research Applications Laboratory. *Benefits and Impact Solutions*. `https://ral.ucar.edu/solutions`.

[55] Milan Curcic. *Cloudrun is a custom weather prediction service using WRF*. `https://cloudrun.co/`.

[56] TempoQuest. *Faster Forecasts. Greater Insight. Superior Decisions*. `https://tempoquest.com/`.

[57] Jeffrey Lazo, Megan Lawson, Peter Larsen, and Donald Waldman. *U.S. Economic Sensitivity to Weather Variability*, 2011. `https://journals.ametsoc.org/bams/article/92/6/709/106985/U-S-Economic-Sensitivity-to-Weather-Variability`.

[58] NOAA Chief Economist. *NOAA's Contribution to the Economy*, 2018. `https://www.performance.noaa.gov/wp-content/uploads/NOAA-Contribution-to-the-Economy-Final.pdf`.

[59] National Human Genome Research Institute. *Biological Pathways Fact Sheet*. `https://www.genome.gov/about-genomics/fact-sheets/Biological-Pathways-Fact-Sheet`.

[60] International Union of Physiological Sciences. *Physiome Project*. `https://physiomeproject.org/about`.

[61] Riken. *Supercomputational Life Science*. `http://www.scls.riken.jp/en/research/index.html`.

[62] Cleveland Clinic. *Center for Computational Life Sciences*. `https://my.clevelandclinic.org/research/computational-life-sciences`.

[63] Yuan-Ping Pang. *Three-Dimensional Model of a Substrate-Bound SARS Chymotrypsin-Like Cysteine Proteinase Predicted by Multiple Molecular Dynamics Simulations: Catalytic Efficiency Regulated by Substrate Binding. PROTEINS: Structure, Function, and Bioinformatics*, 57:747–757, 2004. `https://onlinelibrary.wiley.com/doi/epdf/10.1002/prot.20249`.

[64] Yuan-Ping Pang. *In Silico Drug Discovery: Solving the "Target-rich and Lead-poor" Imbalance Using the Genome-to-drug-lead Paradigm*. Clinical Pharmacology & Therapeutics, 81(1):30–34, 2007. https://www.ncbi.nlm.nih.gov/pmc/articles/PMC7162381/pdf/CPT-81-30.pdf.

[65] Yuan-Ping Pang. *How fast fast-folding proteins fold in silico*. Biochemical and Biophysical Research Communications, 492:135–139, 2017. https://www.sciencedirect.com/science/article/pii/S0006291X17315462.

[66] Folding@home. *Learn more about proteins and SARS-CoV-2*. https://foldingathome.org/diseases/infectious-diseases/covid-19/?lng=en-US.

[67] National Human Genome Research Institute. *What's a Genome?* https://www.genome.gov/About-Genomics/Introduction-to-Genomics.

[68] DOE Office of Science. *Joint Genome Institute*. https://jgi.doe.gov/.

[69] DOE Office of Science. *ExaBiome Project*. https://sites.google.com/lbl.gov/exabiome/.

[70] Exascale Computing Project. *ExaBiome Brings Metagenomics into the Exascale Era*. https://www.exascaleproject.org/exabiome-brings-metagenomics-into-the-exascale-era/.

[71] Foundation Tara Ocean. *Exploring the Ocean to understand, Sharing to change*. https://fondationtaraocean.org/en/home/.

[72] Peter Kogge, Keren Bergman, Shekhar Borkar, Dan Campbell, William Carlson, William Dally, Monty Denneau, Paul Franzon, William Harrod, Kerry Hill, Sherman Karp, Stephen Keckler, Dean Klein, Robert Lucas, Mark Richards, Al Scarpelli, Steven Scott, Allan Snavely, Thomas Sterling, R. Stanley Williams, and Katherine Yelick. *ExaScale Computing Study:Technology Challenges in Achieving Exascale Systems*, 2008. https://www.academia.edu/60931242/Exascale_Computing_Study_Technology_Challenges_In_Achieving_Exascale_Systems?email_work_card=view-paper.

[73] Dongarra J., Beckman P., et al. *The International Exascale Software Roadmap*. International Journal of High Performance Computer Applications, 25(1), 2011.

[74] Jack Dongarra, Pete Beckman, Patrick Aerts, Frank Cappello, Thomas Lippert, Satoshi Matsuoka, Paul Messina, Terry Moore, Rick Stevens, Anne Trefethen, and Mateo Valero. *The International Exascale Software Project: A Call to Cooperative Action by the Global High Performance Community*. International Journal of High Performance Computer Applications, 2009. https://www.academia.edu/47026194/The_International_Exascale_

Software_Project_a_Call_To_Cooperative_Action_By_the_Global_
High_Performance_Community?email_work_card=view-paper.

[75] IESP website. *International Exascale Software Project.* https://exascale.
org/mediawiki/index.php.html.

[76] Tiffany Trader for HPCwire. *Frontier to Meet 20MW Exascale Power Target Set by DARPA in 2008.* https://www.hpcwire.com/2021/07/14/
frontier-to-meet-20mw-exascale-power-target-set-by-darpa-in-
2008/.

[77] CareerFoundry. *What is Data Analytics? A Complete Guide for Beginners.* https://careerfoundry.com/en/blog/data-analytics/what-is-
data-analytics/.

[78] Cem Dilmegani. *Graph Analytics: Types, Tools, and Top 10 Use Cases in 2022.* https://research.aimultiple.com/graph-analytics/.

[79] Yulia Gavrilova. *A Guide to Deep Learning and Neural Networks.* Serokel Labs. https://serokell.io/blog/deep-learning-and-neural-network-
guide.

[80] Sara Brown. *Machine Learning, Explained.* MIT Sloan School of Management. https://mitsloan.mit.edu/ideas-made-to-matter/machine-
learning-explained.

[81] Terence Shin. *All Machine Learning Models Explained in 6 Minutes.*
https://towardsdatascience.com/all-machine-learning-models-
explained-in-6-minutes-9fe30ff6776a.

[82] Sunil Ray. *Commonly used Machine Learning Algorithms.* Analytics Vidhya. https://www.analyticsvidhya.com/blog/2017/09/common-
machine-learning-algorithms/.

[83] Rick Stevens, Valery Taylor, Jeff Nichols, Arthur Barney Maccabe, Katherine Yelick, and David Brown. *AI for Science,* 2019. https://publications.
anl.gov/anlpubs/2020/03/158802.pdf.

[84] David Barkai. *Peer-to-Peer Computing: Technologies for Sharing and Collaborating on the Net.* Intel Press, 2002.

[85] Wikipedia. *Peer-to-peer.* https://en.wikipedia.org/wiki/Peer-to-peer.

[86] *SETI@home.* https://setiathome.berkeley.edu/.

[87] *Folding@home.* https://foldingathome.org/?lng=en.

[88] David C. Thompson and Jorg Bentzien. *Crowdsourcing and open innovation in drug discovery: recent contributions and future directions. Drug Discovery Today*, 25:2284–2293, 2020. And online: `https://www.sciencedirect.com/science/article/pii/S0006291X17315462`.

[89] *PrimeGrid*. `http://www.primegrid.com/`.

[90] *distributedcomputing.info*. `http://www.distributedcomputing.info/projects.html`.

[91] BOINC. *Compute for Science*. `https://boinc.berkeley.edu/`.

[92] distributedcomputing.info. *BOINC*. ([91]).

[93] Globus. *Introducing Flows*. `https://www.globus.org/`.

[94] Ian Foster and Carl Kesselman. *The History of the Grid. Advances in Parallel Computing*, 20:3–30, 2011. Also via `https://www.sciencedirect.com/science/article/pii/S0006291X17315462`, or `http://www.ianfoster.org/wordpress/wp-content/uploads/2014/01/History-of-the-Grid-numbered.pdf`.

[95] NCAR's Computational & Information Systems Lab. *NCAR supercomputing history*. `https://www2.cisl.ucar.edu/ncar-supercomputing-history`.

[96] NCAR's Computational & Information Systems Lab. *Computational systems*. `https://www2.cisl.ucar.edu/computing-data/computing`.

[97] NCAR. *Simulating a Complex World*. `https://ncar.ucar.edu/what-we-offer/models`.

[98] Met Office. *The Met Office ensemble system*. `https://www.metoffice.gov.uk/research/weather/ensemble-forecasting/mogreps`.

[99] ECMWF. *Forecast upgrade innovates on single precision and ensemble resolution*. `https://www.ecmwf.int/en/about/media-centre/news/2021/forecast-upgrade-innovates-single-precision-and-ensemble-resolution`.

[100] John Michalakes. HPC for Weather Forecasting. In A. Grama and A. Sameh, editors, *Parallel Algorithms in Computational Science and Engineering*. Birkhauser-Science, Basel, 2020.

[101] Y. Miyamoto, Y. Kajikawa, R. Yoshida, T. Yamaura, H. Yashiro, and H. Tomita. *Deep moist atmospheric convection in a subkilometer global simulation. Geophysical Research Letters*, 40:4922, 2013. `http://doi.org/10.1002/grl.50944`.

[102] David Barkai and Achi Brandt. *Vectorized multigrid poisson solver for the CDC cyber 205*. Applied Mathematics and Computation, 13:215–227, 1983. Published by Elsevier. `https://www.sciencedirect.com/science/article/abs/pii/0096300383900139`, or `https://ntrs.nasa.gov/api/citations/19840012162/downloads/19840012162.pdf`.

[103] University of Bath Institute for Mathematical Innovation. *Multi-grid methods for speeding up weather forecasts*. `https://imibath.ac.uk/projects/multi-grid-methods-for-speeding-up-weather-forecasts/`.

[104] Anton Afanasyev, Mauro Bianco, Lukas Mosimann, Carlos Osuna, Felix Thaler, Hannes Vogt, Oliver Fuhrer, Joost VandeVondele, and Thomas C. Schulthess. *GridTools: A framework for portable weather and climate applications*. SoftwareX, 15:100707, 2021. Published by Elsevier. `https://www.sciencedirect.com/science/article/pii/S2352711021000522`.

[105] Thomas C. Schulthess. *Reflecting on the Goal and Baseline for Exascale Computing*. `https://bluewaters.ncsa.illinois.edu/liferay-content/document-library/18symposium-slides/talk-schulthess.pdf`.

[106] ECMWF. *About our forecasts*. `https://www.ecmwf.int/en/forecasts/documentation-and-support#ERA`.

[107] ECMWF. *Destination Earth*. `https://www.ecmwf.int/en/about/what-we-do/environmental-services-and-future-vision/destination-earth`.

[108] Xiaoxiang Zhu. *Artificial Intelligence and Data Science in Earth Observation*. `https://az659834.vo.msecnd.net/eventsairwesteuprod/production-nikal-public/940a052ebdba4573855d738faa2ec946`.

[109] Oliver Peckham. *Supercomputing Experts React to Dire Climate Report*. HPCwire. `https://www.hpcwire.com/2021/08/26/supercomputing-experts-react-to-dire-climate-report/`.

[110] M. G. Schultz, C. Betancourt, B. Gong, F. Kleinert, M. Langguth, L. H. Leufen, A. Mozaffari, and S. Stadtler. *Can deep learning beat numerical weather prediction?* Philosophical Transactions of the Royal Society A: Mathematical, Physical and Engineering Sciences, 379, 2021. `https://royalsocietypublishing.org/doi/abs/10.1098/rsta.2020.0097`.

[111] Matthew Chantry and Peter Dueben. *Machine learning to emulate components of ECMWF's Integrated Forecasting System*. ECMWF, Science Blog. `https://www.ecmwf.int/en/about/media-centre/science-blog/2021/machine-learning-emulate-components-ecmwfs-integrated`.

[112] IPCC. *Sixth Assessment Report*. `https://www.ipcc.ch/assessment-report/ar6/`.

[113] Air Force Magazine. *Rolls-Royce Digitally Modeled Wing and Pylon With Engine to Win B-52 Contract.* https://www.airforcemag.com/rolls-royce-digitally-modeled-entire-wing-pylon-to-win-b-52-engine-contract/.

[114] Jeffrey Slotnick, Abdollah Khodadoust, Juan Alonso, David Darmofal, William Gropp, Elizabeth Lurie, and Dimitri Mavriplis. *CFD Vision 2030 Study: A Path to Revolutionary Computational Aerosciences*, 2014. NASA/CR-2014-218178. https://ntrs.nasa.gov/api/citations/20140003093/downloads/20140003093.pdf.

[115] Andrew Carey, John Chawner, Earl Duque, William Gropp, Bil Kleb, Ray Kolonay, Eric Nielsen, and Brian Smith. *The CFD Vision 2030 Roadmap: 2020 Status, Progress and Challenges*, 2021. https://cfd2030.com/report/CFD-Vision-2030-Roadmap-2020-Report.pdf.

[116] Xuan Liu, David Furrer, Jared Kosters, and Jack Holmes. *Vision 2040: A Roadmap for Integrated, Multiscale Modeling and Simulation of Materials and Systems*, 2018. https://ntrs.nasa.gov/api/citations/20180002010/downloads/20180002010.pdf.

[117] Rolls-Royce. *How Digital Twin technology can enhance Aviation.* https://www.rolls-royce.com/media/our-stories/discover/2019/how-digital-twin-technology-can-enhance-aviation.aspx.

[118] Ansys. *Why High-Performance Computing (HPC) Is Critical to Autonomous Vehicle Development.* https://www.ansys.com/blog/why-hpc-is-critical.

[119] LLNL HPC for Energy Innovation. *Success Stories.* https://hpc4energyinnovation.llnl.gov/success-stories.

[120] Dennis Overbye. Gravitational Waves Detected, Confirming Einstein's Theory. The New York Times. https://www.nytimes.com/2016/02/12/science/ligo-gravitational-waves-black-holes-einstein.html.

[121] Department of Energy. *DOE National Laboratory Makes History by Achieving Fusion Ignition.* https://www.energy.gov/articles/doe-national-laboratory-makes-history-achieving-fusion-ignition.

[122] Doug Black. *Due Credit: Sierra, JADE and HPC's Role in Livermore's Fusion Ignition Breakthrough.* InsideHPC. https://insidehpc.com/2022/12/due-credit-sierra-jade-and-hpcs-role-in-livermores-fusion-ignition-breakthrough/.

[123] James Brase, Nancy Campbell, Barbara Helland, Thuc Hoang, Manish Parashar, Michael Rosenfield, and John Towns. *The COVID-19 High*

Performance Computing Consortium - Overview Paper. https://s3.us-south.cloud-object-storage.appdomain.cloud/covid-19-hpc-object-storage-production/Consortium_Overview_Paper_03_2022_1f72939a70.

[124] COVID-19 HPC Consortium. *The COVID-19 High Performance Computing Consortium.* ([123]).

[125] John L. Gustafson. *The Consequences of Fixed Time Performance Measurement.* Proceedings of the Twenty-Fifth Hawaii International Conference on System Sciences, 1992. https://ieeexplore.ieee.org/document/183285 and http://www.johngustafson.net/pubs/pub35/FixedTime.pdf.

[126] John L. Gustafson. *Reevaluating Amdahl's Law,* 1988. http://www.johngustafson.net/pubs/pub13/amdahl.htm.

[127] David H. Bailey. *Twelve Ways to Fool the Masses When Giving Performance Results on Parallel Computers,* 1991. Supercomputing Review, Aug. 1991, pg. 54–55, and https://www.davidhbailey.com/dhbpapers/twelve-ways.pdf.

[128] Wikipedia. *Benchmark (computing).* https://en.wikipedia.org/wiki/Benchmark_(computing).

[129] J.L. Gustafson and R. Todi. *Conventional benchmarks as a sample of the performance spectrum.* In *Proceedings of the Thirty-First Hawaii International Conference on System Sciences,* volume 7, pages 514–523, 1998.

[130] J. D. McCalpin. *Memory Bandwidth and Machine Balance in Current High Performance Computers.* IEEE Technical Committee on Computer Architecture (TCCA) Newsletter, 1995. Also at https://www.researchgate.net/publication/51992086_Memory_bandwidth_and_machine_balance_in_high_performance_computers.

[131] T. Chen, M. Gunn, B. Simon, L. Carrington, and A. Snavely. *Metrics for Ranking the Performance of Supercomputers.* In *Cyberinfrastructure Technology Watch Journal: Special Issue on High Productivity Computer Systems,* volume 2, 2007. Also at https://users.sdsc.edu/~lcarring/Papers/2007_CTWJ.pdf.

[132] hpcchallenge.org. *HPC Challenge Benchmark.* https://hpcchallenge.org/hpcc/index.html.

[133] NASA Advanced Supercomputing (NAS) Division. *NAS Parallel Benchmarks.* https://www.nas.nasa.gov/software/npb.html.

[134] David Bailey. *The NAS Parallel Benchmarks.* https://www.osti.gov/servlets/purl/983318.

[135] Krste Asanović, Ras Bodik, Bryan Christopher Catanzaro, Joseph James Gebis, Parry Husbands, Kurt Keutzer, David A. Patterson, William Lester Plishker, John Shalf, Samuel Webb Williams, and Katherine A. Yelick. *The Landscape of Parallel Computing Research: A View from Berkeley*. Technical Report UCB/EECS-2006-183, EECS Department, University of California, Berkeley, Dec 2006.

[136] David Barkai. *The Application Perspective: Seeking Productivity and Performance*. International Exascale Software Project. `https://exascale.org/mediawiki/images/2/2d/BarkaiIESP.pdf`.

[137] David Nelson et al. *Grand Challenges: High Performance Computing and Communications*, 1992. The FY 1992 U.S. Research and Development Program. Committee on Physical, Mathematical, and Engineering Sciences. Office of Science and Technology Policy. `https://www.nitrd.gov/pubs/bluebooks/1992/pdf/bluebook92.pdf`.

[138] John Reid, Bill Long, and Jon Steidel. *History of Coarrays and SPMD Parallelism in Fortran*. Proc. ACM Program. Lang., 4:72:1, 2020. `https://dl.acm.org/doi/pdf/10.1145/3386322`.

[139] Robert W. Numrich. *F⁻⁻: A Parallel Extension to Cray Fortran*. Scientific Programming, 6(3):275–284, 1997.

[140] Robert W. Numrich, John Reid, and Kieun Kim. *Writing a Multigrid Solver Using Co-Array Fortran*. In Bo Kågström, Jack Dongarra, Erik Elmroth, and Jerzy Waśniewski, editors, *Applied Parallel Computing: Large Scale Scientific and Industrial Problems*, pages 390–399. 4th International Workshop, PARA98, Umeå, Sweden, June 1998, Springer, 1998. Lecture Notes in Computer Science 1541.

[141] Robert W. Numrich and John Reid. *Co-array Fortran for parallel programming*. ACM SIGPLAN Fortran Forum, 17(2), 1998.

[142] Robert Numrich. *Parallel Programming with Co-Arrays*. CRC Press, 2018.

[143] Wikipedia. *Message Passing Interface*. `https://en.wikipedia.org/wiki/Message_Passing_Interface`.

[144] Wikipedia. *Parallel Virtual Machine*. `https://en.wikipedia.org/wiki/Parallel_Virtual_Machine`.

[145] Marc Snir. *MPI is too High-Level; MPI is too Low-Level*. `https://www.mcs.anl.gov/mpi-symposium/slides/marc_snir_25yrsmpi.pdf`.

[146] William Gropp. *Learning from the Success of MPI*. `https://wgropp.cs.illinois.edu/bib/papers/pdata/2001/mpi-lessons.pdf`.

[147] OpenCoarrays.org. *OpenCoarrays*. http://www.opencoarrays.org/.

[148] Robert W. Numrich. *A Team Object for CoArray Fortran*. In R. Wyrzykowski, J. Dongarra, K. Karczewsk, and J. Wasniewski, editors, *Proceedings of the 8th International Conference on Parallel Processing and Applied Mathematics (PPAM 2009)*, Lecture Notes in Computer Science, Vol. 6068, Part 2, pages 68–73. Springer-Verlag, 2010.

[149] The Organizing Committee of FortranCon 2020. University of Zurich. *International Fortran Conference 2020*. https://www.youtube.com/playlist?list=PLeKbr7eYHjt77h90hDVC-vGzrWmvDhCAf.

[150] fortran-lang.org. *High-performance parallel programming language*. https://fortran-lang.org/.

[151] Ondřej Čertík. *Resurrecting Fortran*. https://ondrejcertik.com/blog/2021/03/resurrecting-fortran/.

[152] LLVM. *The LLVM Compiler Infrastructure*. https://llvm.org/.

[153] Wikipedia. *LLVM*. https://en.wikipedia.org/wiki/LLVM.

[154] LFortran. *Modern interactive LLVM-based Fortran compiler*. https://lfortran.org/.

[155] Rob Farber. *LLVM Holds the Keys to Exascale Supercomputing*, 2021. Exascale Computing Project by insideHPC https://insidehpc.com/2021/07/llvm-holds-the-keys-to-exascale-supercomputing/.

[156] Alessandro Fanfarillo, Tobias Burnus, Valeria Cardellini, Salvatore Filippone, Dan Nagle, and Damian W I Rouson. *OpenCoarrays: Open-source Transport Layers Supporting Coarray Fortran Compilers*. *PGAS '14: Proceedings of the 8th International Conference on Partitioned Global Address Space Programming Models*, 2014. https://dl.acm.org/doi/10.1145/2676870.2676876.

[157] Steve Lionel. *Doctor Fortran*. https://stevelionel.com/drfortran/.

[158] Milan Curcic. *A parallel Fortran framework for neural networks and deep learning*. *ACM SIGPLAN Fortran Forum*, 38(1), 2019. https://dl.acm.org/doi/abs/10.1145/3323057.3323059.

[159] Daniel C. Elton. *Why Physicists Still Use Fortran*, 2015. http://www.moreisdifferent.com/2015/07/16/why-physicsts-still-use-fortran/.

[160] United States General Accounting Office. *High Performance Computing. Industry Uses of Supercomputers and High-Speed Networks*, 1991. https://www.gao.gov/assets/imtec-91-58.pdf.

[161] Earl C. Joseph, Steve Conway, and Chirag Dekate. *Creating Economic Models Showing the Relationship Between Investments in HPC and the Resulting Financial ROI and Innovation - and How It Can Impact a Nation's Competitiveness and Innovation.* IDC, 2013. `https://www.osti.gov/servlets/purl/1156830`.

[162] Earl C. Joseph, Steve Conway, and Bob Sorensen. *Real-World Examples of Supercomputers Used For Economic and Societal Benefits: A Prelude to What the Exascale Era Can Provide*, 2017. `https://www.hpcuserforum.com/wp-content/uploads/2022/02/Hyperion-Research-Benefits-of-Supercomputers_2017.pdf`.

[163] John L. Gustafson and Isaac Yonemoto. *Beating Floating Point at its Own Game: Posit Arithmetic.* `http://www.johngustafson.net/pdfs/BeatingFloatingPoint.pdf`.

[164] ARM. *The Future is Built on Arm.* `https://www.arm.com/company`.

[165] RISC-V International. *History of RISC-V.* `https://riscv.org/about/history/`.

[166] John D. Davis. *RISC-V in Europe: The Road to an Open Source HPC Stack.* `https://www.european-processor-initiative.eu/wp-content/uploads/2022/03/EPI-@-HPC-User-Forum.pdf`.

[167] NextSilicon. *We Are getting Our Chip Together.* `https://www.nextsilicon.com/`.

[168] Daniel Reed, Dennis Gannon, and Jack Dongarra. *Reinventing High Performance Computing: Challenges and Opportunities*, 2022. `https://arxiv.org/pdf/2203.02544.pdf`.

[169] Satoshi Matsuoka, Jens Domke, Mohamed Wahib, Aleksandr Drozd, and Torsten Hoefler. *Myths and Legends in High-Performance Computing*, 2023. `https://arxiv.org/pdf/2301.02432v1.pdf`.

Index